# Nanofibrous Membrane for Biomedical Application

# Nanofibrous Membrane for Biomedical Application

Editor

**Andrea Ehrmann**

MDPI • Basel • Beijing • Wuhan • Barcelona • Belgrade • Manchester • Tokyo • Cluj • Tianjin

*Editor*
Andrea Ehrmann
Faculty of Engineering and
Mathematics
Bielefeld University of
Applied Sciences and Arts
Bielefeld
Germany

*Editorial Office*
MDPI
St. Alban-Anlage 66
4052 Basel, Switzerland

This is a reprint of articles from the Special Issue published online in the open access journal *Membranes* (ISSN 2077-0375) (available at: www.mdpi.com/journal/membranes/special_issues/electrospun_membrane_bio).

For citation purposes, cite each article independently as indicated on the article page online and as indicated below:

LastName, A.A.; LastName, B.B.; LastName, C.C. Article Title. *Journal Name* **Year**, *Volume Number*, Page Range.

**ISBN 978-3-0365-7787-6 (Hbk)**
**ISBN 978-3-0365-7786-9 (PDF)**

# Contents

**Sarah Vanessa Langwald, Andrea Ehrmann and Lilia Sabantina**
Measuring Physical Properties of Electrospun Nanofiber Mats for Different Biomedical Applications
Reprinted from: *Membranes* **2023**, *13*, 488, doi:10.3390/membranes13050488 . . . . . . . . . . . . 1

**Nikoleta Stoyanova, Mariya Spasova, Nevena Manolova, Iliya Rashkov, Sabina Taneva and Svetlana Momchilova et al.**
Physico-Chemical, Mechanical, and Biological Properties of Polylactide/*Portulaca oleracea* Extract Electrospun Fibers
Reprinted from: *Membranes* **2023**, *13*, 298, doi:10.3390/membranes13030298 . . . . . . . . . . . . 25

**Wilson Agudelo, Yuliet Montoya, Alejandra Garcia-Garcia, Adriana Restrepo-Osorio and John Bustamante**
Electrochemical and Electroconductive Behavior of Silk Fibroin Electrospun Membrane Coated with Gold or Silver Nanoparticles
Reprinted from: *Membranes* **2022**, *12*, 1154, doi:10.3390/membranes12111154 . . . . . . . . . . . . 39

**Suzan Ozdemir, Ipek Yalcin-Enis, Baturalp Yalcinkaya and Fatma Yalcinkaya**
An Investigation of the Constructional Design Components Affecting the Mechanical Response and Cellular Activity of Electrospun Vascular Grafts
Reprinted from: *Membranes* **2022**, *12*, 929, doi:10.3390/membranes12100929 . . . . . . . . . . . . 59

**Shu-Ting Chen, Sumith Ranil Wickramasinghe and Xianghong Qian**
Electrospun Hydrophobic Interaction Chromatography (HIC) Membranes for Protein Purification
Reprinted from: *Membranes* **2022**, *12*, 714, doi:10.3390/membranes12070714 . . . . . . . . . . . . 87

**Dianney Clavijo-Grimaldo, Ciro Alfonso Casadiego-Torrado, Juan Villalobos-Elías, Adolfo Ocampo-Páramo and Magreth Torres-Parada**
Characterization of Electrospun Poly($\varepsilon$-caprolactone) Nano/Micro Fibrous Membrane as Scaffolds in Tissue Engineering: Effects of the Type of Collector Used
Reprinted from: *Membranes* **2022**, *12*, 563, doi:10.3390/membranes12060563 . . . . . . . . . . . . 103

**Anton M. Manakhov, Natalya A. Sitnikova, Alphiya R. Tsygankova, Alexander Yu. Alekseev, Lyubov S. Adamenko and Elizaveta Permyakova et al.**
Electrospun Biodegradable Nanofibers Coated Homogenously by Cu Magnetron Sputtering Exhibit Fast Ion Release. Computational and Experimental Study
Reprinted from: *Membranes* **2021**, *11*, 965, doi:10.3390/membranes11120965 . . . . . . . . . . . . 121

**Aysegul Gul, Izabela Gallus, Akshat Tegginamath, Jiri Maryska and Fatma Yalcinkaya**
Electrospun Antibacterial Nanomaterials for Wound Dressings Applications
Reprinted from: *Membranes* **2021**, *11*, 908, doi:10.3390/membranes11120908 . . . . . . . . . . . . 141

**Pattawika Sinsup, Veerawat Teeranachaideekul, Arthit Makarasen, Laemthong Chuenchom, Pongthep Prajongtat and Supanna Techasakul et al.**
*Zingiber cassumunar* Roxb. Essential Oil-Loaded Electrospun Poly(lactic acid)/Poly(ethylene oxide) Fiber Blend Membrane for Antibacterial Wound Dressing Application
Reprinted from: *Membranes* **2021**, *11*, 648, doi:10.3390/membranes11090648 . . . . . . . . . . . . 163

**Seung-Jun Lee, Perry Ayn Mayson A Maza, Gyu-Min Sun, Petr Slama, In-Jeong Lee and Jong-Young Kwak**
Bacterial Infection-Mimicking Three-Dimensional Phagocytosis and Chemotaxis in Electrospun Poly($\varepsilon$-caprolactone) Nanofibrous Membrane
Reprinted from: *Membranes* **2021**, *11*, 569, doi:10.3390/membranes11080569 . . . . . . . . . . . . **177**

membranes

Review

# Measuring Physical Properties of Electrospun Nanofiber Mats for Different Biomedical Applications

Sarah Vanessa Langwald [1], Andrea Ehrmann [1,*] and Lilia Sabantina [2]

[1] Faculty of Engineering and Mathematics, Bielefeld University of Applied Sciences and Arts, 33619 Bielefeld, Germany; sarah_vanessa.homburg@hsbi.de
[2] Faculty of Clothing Technology and Garment Engineering, School of Culture + Design, HTW Berlin—University of Applied Sciences, 12459 Berlin, Germany
* Correspondence: andrea.ehrmann@hsbi.de

**Abstract:** Electrospun nanofiber mats are nowadays often used for biotechnological and biomedical applications, such as wound healing or tissue engineering. While most studies concentrate on their chemical and biochemical properties, the physical properties are often measured without long explanations regarding the chosen methods. Here, we give an overview of typical measurements of topological features such as porosity, pore size, fiber diameter and orientation, hydrophobic/hydrophilic properties and water uptake, mechanical and electrical properties as well as water vapor and air permeability. Besides describing typically used methods with potential modifications, we suggest some low-cost methods as alternatives in cases where special equipment is not available.

**Keywords:** ImageJ; apparent density; porometer; scanning electron microscopy (SEM); specific surface area; fast Fourier transform (FFT); water contact angle; surface roughness; tensile test; conductivity

**Citation:** Langwald, S.V.; Ehrmann, A.; Sabantina, L. Measuring Physical Properties of Electrospun Nanofiber Mats for Different Biomedical Applications. *Membranes* **2023**, *13*, 488. https://doi.org/10.3390/membranes13050488

Academic Editor: Benjamin S. Hsiao

Received: 31 March 2023
Revised: 26 April 2023
Accepted: 27 April 2023
Published: 30 April 2023



## 1. Introduction

Electrospinning allows for producing nanofiber mats from diverse polymers or polymer blends, including various nanoparticles, and in this way tailoring the nanofiber materials in a broad range [1–3]. Their large specific surfaces as well as other physical and chemical properties make such nanofiber mats highly suitable for biotechnological and biomedical applications, such as wound healing or tissue engineering [4–6].

Naturally, nanofiber mats for biomedical applications need special properties, especially being not cytotoxic, but depending on the exact application, they can be desired to be biodegradable or waterproof, have antibacterial of fungicide properties, etc. [7–10]. However, their morphological, mechanical and other physical properties may also be important for the planned application, although these values are often less intensively investigated than chemical and biochemical properties and often only briefly described in the methodic sections. Nevertheless, the mechanical properties are decisive for the lifetime of a nanofibrous product and the limits of its potential application, while cell adhesion depends on morphological parameters, hydrophobicity and water uptake are among the parameters controlling liquid transport, which is important for wound dressing, and porosity and water vapor/air permeability are physical parameters influencing the filtration of liquids or gases, respectively. The porosity is often mentioned as an important parameter for wound exudate transport and cell adhesion [11–17]. While the porosity describes the amount of porous volume inside the nanofiber mat volume, the pore size distribution is also often taken into account [14–16,18,19]. Other morphological parameters are the nanofiber diameters [16,20–24] and their orientation [25–27] as well as the surface roughness and nanofiber mat thickness [28]. Besides such structural features, the hydrophobic/hydrophilic properties of nanofiber mats [11,16] and their water uptake [17,24,29] are often reported. Other often-mentioned parameters are mechanical [11,15,17–21,23,24] and electrical properties [18,30] as well as water vapor and air permeability [31,32]. This review gives an overview of the different measurement methods for these parameters, discusses differences in the gained results and suggests some less well-known inexpensive alternatives to the typically used instruments that are not always available for each study.

## 2. Porosity

The porosity describes the volume of voids inside a given volume of a nanofiber mat. Firstly, it must be mentioned that there can be open as well as closed pores, the latter of which are not accessible for all methods described below [33–35]. However, for typical nanofiber mats, pores can be expected to be openly accessible to any test fluid, so that for most nanofibrous membranes, no differences between the measurement principles are expected, whether they take into account closed pores or not.

One of the methods that would also measure closed pores is the Archimedean principle [36]. Pati et al. used a specific gravity bottle filled with ethanol in which the nanofibrous scaffold was dipped and afterwards removed again [37]. The porosity was then calculated according to

$$Porosity = \frac{(m_2 - m_3 - m_S)}{m_1 - m_3} \times 100\% \quad (1)$$

with the mass $m_1$ of the specific gravity bottle filled with ethanol, the mass $m_2$ of the bottle with ethanol and the scaffold, the mass $m_3$ of the bottle after taking out the scaffold again, and $m_S$ the mass of the scaffold. Dividing both numerator and denominator by the density of ethanol, it is observable that the porosity is determined as the volume of the ethanol taken out of the bottle with the scaffold, i.e., of the ethanol that was sticking in its pores, divided by the volume of the scaffold with ethanol. Safari et al. used the same principle based on deionized water in which their nanofiber mats were immersed for 15 min, taken out, quickly dried at the sample surface and weighed, so that the porosity could be calculated as the mass of the uptaken water, divided by the sample mass [38].

Without directly using the Archimedean principle, Kahdim et al. calculated the porosity by soaking their nanofiber mats in phosphate-buffered saline (PBS) solution for 24 h, measuring the sample mass before and after PBS uptake and calculating the porosity according to this fluid uptake and the PBS density [39]. Here, it is not mentioned whether the samples were also dried on both surfaces before weighing them. Using immersion of the dried nanofiber mat in n-butanol for 2 h, Wang et al. as well as Chen et al. calculated the porosity from the densities of membrane and n-butanol as well as the measured dry and wet mass of the nanofiber mat [40,41].

Other research groups used a similar technique of wetting a sample in a fluid, but measured the volumes of the fluid instead of the immersed nanofibrous mat. Salehi et al. calculated the porosity of poly(ε-caprolactone)(PCL)/gelatin nanofiber mats by immersion in ethanol and calculated the porosity as

$$Porosity = \frac{V1 - V3}{V2 - V3} \times 100\% \quad (2)$$

with the initial volume $V1$ of ethanol, the volume $V2$ after immersion of the nanofiber mat and the volume $V3$ of the ethanol without the soaked mat, taken out after 10 min [42]. Ghaee et al. also used ethanol to investigate the porosity of their PCL nanofiber mats by this liquid displacement technique [43], similarly to Esmaeili et al. for cellulose acetate/polyurethane nanofiber mats [44], while Chen et al. used the same method for their poly(lactic acid) (PLA)/regenerative cellulose composite scaffolds with hexane instead of ethanol [45].

Several papers mention a gas pycnometer as a possibility to measure the volume of a porous sample which enables calculating the theoretical density of a sample and correspondingly the porosity by the formula

$$Porosity = \left(1 - \frac{\rho_{exp}}{\rho_{theo}}\right) \times 100\% \quad (3)$$

with the measured density $\rho_{exp}$ and the theoretical density $\rho_{theo}$ of the material under investigation [46]. The easiest form of a gas pycnometer contains two chambers, one of them with well-known reference volume, while the sample is introduced into the other one. A measuring gas is introduced into one of the chambers and allowed to expand into the second chamber through a valve. The sample volume can then be calculated from the previously known volumes of the empty sample chamber and the reference chamber as well as the pressure of the firstly filled chamber and the equilibrium pressure after gas expansion [47]. While this method thus necessitates more sophisticated equipment than the previously described methods based on fluids filling the pores of the nanofiber mat, the latter take more time and are more error prone, especially when the experimental procedure is not perfectly described, e.g., regarding drying the sample surfaces after dipping or not.

Equation (3) can also be used for other ways to determine the apparent density of a sample, in the easiest way by measuring its mass as well as its thickness and lateral dimensions, where the error range is mostly influenced by the thickness measurement, which will be discussed in Section 8. Nevertheless, this relatively simple method can be used to give an estimate of the porosity, keeping in mind that irregularities of the sample thickness and its compressibility will potentially cause deviations from the real value. Porosity calculations by the apparent density, calculated from the sample volume, were reported by several research groups for different nanofiber mat materials, such as polyamide-6/polyvinylpyrrolidone [48], polyurethane [49] or collagen-coated poly(l-lactic acid)-*co*-poly(ε-caprolactone) [50].

Besides these methods, which are used to determine 3D pore structures, some papers also mention calculating the surface porosity, typically based on scanning electron microscope (SEM) images and their evaluation by ImageJ (National Institute of Health, Bethesda, MD, USA) or partly automated by the plugin DiameterJ [49,51–53].

Finally, a direct measurement of the porosity is enabled by laser metrology, measuring the surface of an electrospun nanofiber mat on the collector, followed by completely densifying via heat treatment and afterwards measuring the surface profile again, so that the porosity can be calculated from the vertical shrinkage [54,55].

## 3. Pore Size Distribution

While the porosity describes the overall volume of the pores in a given sample, the pore size distribution is sometimes even more important in biomedical scaffolds since it defines which pores are available for cells or can release a drug. In the easiest way, pore sizes are measured on the surface or along cross-sections of samples, typically from SEM images. Agueda et al. describe that they used ImageJ to investigate pore sizes from 3 areas per sample from SEM images taken with magnification of 2000× and 5000×, measuring 30 pores per sample [56]. Liu et al. similarly examined pore sizes from SEM images of their nanofiber mats, taken with magnifications of 5000× and 20,000×, averaging over 100 pore areas [57]. Tahami et al. also used ImageJ to measure pore sizes in SEM images, while not exactly describing the number of measurements [58], while Stella et al. showed histograms of the pore size distributions, which in principle allow for counting the number of measurements per sample, again taken with ImageJ from SEM images [59].

Only a few groups describe how they defined the pore size that they measured. Zhang et al. described measuring a reversible change in pore size by analyzing 30 pores per sample with ImageJ in their SEM images by measuring the longest diameter, as shown in Figure 1a [60]. Havlícek et al., on the other hand, used Matlab to determine the pore sizes as equivalent circle diameters, i.e., they measured the pore areas and calculated the diameter of a circle with identical area [61]. Krysiak et al. similarly fitted ellipses to the pores in the SEM images by ImageJ and calculated their areas [62]. Nejad et al. also worked with ellipses fitted into the pores (Figure 1b), but gave the larger diameter as the pore size [53]. In some papers, the average pore size could be estimated from SEM images, without a detailed explanation of how this value was determined [63].

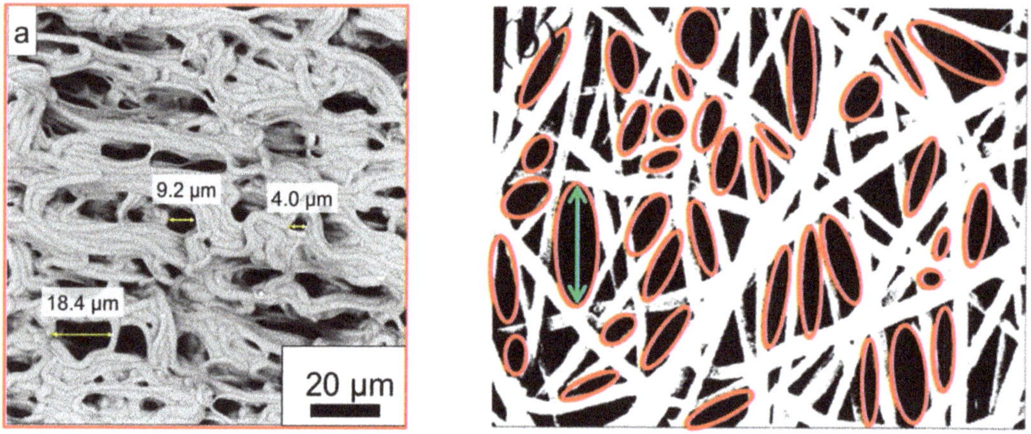


**Figure 1.** Scanning electron microscope (SEM) images of nanofiber mats for pore size determination: (**a**) covalently crosslinked poly(ε-caprolactone) at a temperature of 60 °C, from [58], originally published under a CC-BY license; (**b**) polyethylene terephthalate/polycaprolactone blends. From [53], copyright (2020), with permission from Elsevier.

Due to the broad range of possibilities to define the pore size, it is strongly suggested to always clearly mention the chosen definition as well as the number of investigated pores in a paper.

Besides these 2D methods, some groups chose 3D pore measurement methods. One of them is the Barrett–Joyner–Halenda (BJH) technique, allowing analysis of pores between 1.7 nm and 300 nm [64]. This method is based on $N_2$ adsorption–desorption isotherms, taken at liquid nitrogen temperature [65–68], i.e., similarly to the Brunauer–Emmett–Teller (BET) surface area measurements described in the next section. Generally, the BJH method as well as further developments are based on measuring the film formation on the mesopore walls in dependence of the condensation pressure, taking into account the so-called Kelvin-type relation describing capillary condensation, meaning that mesopores covered with an absorbed fill will instantaneously be filled [68]. In particular, the extended BJH-KJS (Kruk-Jaroniec and Sayari) method was found to allow for accurately calculating mesopore volumes [69].

A capillary flow porometer can also be used to investigate the pore sizes of nanofiber mats [41,70]. In this method, the sample pores are filled with a wetting liquid that is afterwards blown out of the pores by a pressurized gas or liquid [71], where smaller pores need a higher pressure to be emptied, i.e., the measured flow rate depends on the proportion of filled pores that block the flow so that there is zero flow at low pressure, while at a certain high pressure, all pores are emptied, and the flow rate becomes identical to the value measured for the dry sample at the same pressure [72]. It should be mentioned that this method may depend on the wetting fluid [73,74] and the used instrument [75], and thus, the results should be compared with other methods to evaluate their reliability.

Generally, some other methods are available, although less often reported in the recent literature, to evaluate the pore size distribution of nanofiber mats, such as mercury intrusion porosimetry [76]. Another method that is less well-known but often more readily available than a porometer or nitrogen absorption techniques, is thermoporometry, also known as thermoporosimetry or cryoporometry [77–80]. This calorimetric method is based on the melting or freezing point depression of the pore liquid, which can be measured with a laboratory differential scanning calorimetry (DSC) instrument by fast cooling the sample wetted with deionized water to −30 °C or lower and then slowly (e.g., with a heating rate of 0.1–1 K/min) heating it up to a temperature slightly above 0 °C [81,82]. A summary of the theory behind the technique can be found elsewhere [83]. Although the DSC results are less straightforward to interpret than the results of other techniques [84], some authors report thermoporometry measurements of nanofiber mats.

Abolhasani et al. used thermoporometry to measure the porosity of poly(vinylidene fluoride-trifluoroethylene) (P(VDF-TrFE)) nanofiber mats [85]. Gustafsson et al. compared dry- and wet-state porometry methods for the analysis of virus removal filter paper [86]. They found that thermoporometry by DSC is particularly useful to characterize the pore-size distribution in the wet state. Fashandi et al. mentioned that their originally hydrophobic polystyrene nanofiber mat was hydrophilized by oxygen plasma treatment to enable thermoporometric measurements, showing a pore size radius distribution around 20–50 nm [87]. As these few examples show, characterization of the pore size distribution of nanofiber mats is also possible by using a DSC instrument, which is more often available in laboratories than more-specialized porometers, etc.; however, care should always be taken when comparing the results gained with different methods, as the simple comparison of different pore size definitions in 2D optical methods already showed.

## 4. Specific Surface Area

Among the very special properties of nanofiber mats is their large specific surface area. While this value is often mentioned as a reason why nanofiber mats are especially useful for a certain application, its value is scarcely measured. The most common measurement technique is based on the aforementioned BET adsorption–desorption isotherms of $N_2$ gas on the sample surface [68,88–94]. Generally, a wider hysteresis loop in the adsorption–desorption curve indicates a more mesoporous structure of the nanofiber mat [68].

To evaluate these curves, it is necessary to differentiate between the different adsorption isotherms. Many nanofiber mats reported in the literature belong to type IV [95], while type I or a pressure-dependent change between these types are also found [96,97]. Differentiation between the different types of adsorption isotherms is possible by fitting a range of possible equations to the measured curves, while a first idea of the type can already be gained by looking at the slopes of the measured adsorption–desorption curves (Figure 2) [98]. For a comprehensive overview of mono- and multi-parametric isotherm models with the corresponding regression equations, the reader is referred to the review paper of Al-Ghouti and Da'ana [98].

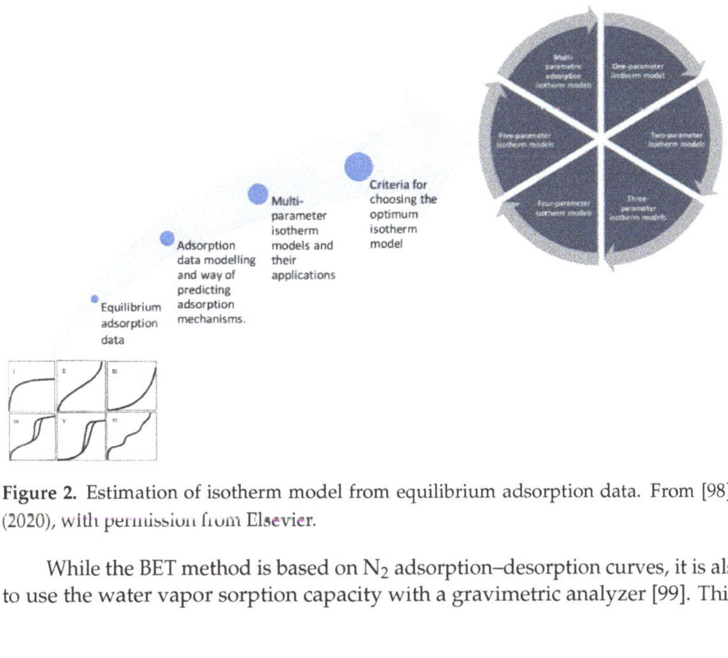


**Figure 2.** Estimation of isotherm model from equilibrium adsorption data. From [98], copyright (2020), with permission from Elsevier.

While the BET method is based on $N_2$ adsorption–desorption curves, it is also possible to use the water vapor sorption capacity with a gravimetric analyzer [99]. This dynamic measurement technique by an automated gravimetric analyzer is based on an ultrasensitive micro-balance, measuring the mass change of a sample while the humidity in the sample chamber is increased from less than 1 to 90% in steps of 10%, where the sample is allowed to reach equilibrium for 10–20 min per step [100]. Similarly, desorption curves were measured during decreasing relative humidity. The results, depicted in Figure 3, show hysteresis curves, but are partly not closed, as opposed to the $N_2$ isotherm curves visible in Figure 2. These curves were interpreted as type IV, and average pore sizes around 1 nm were calculated from them [100].

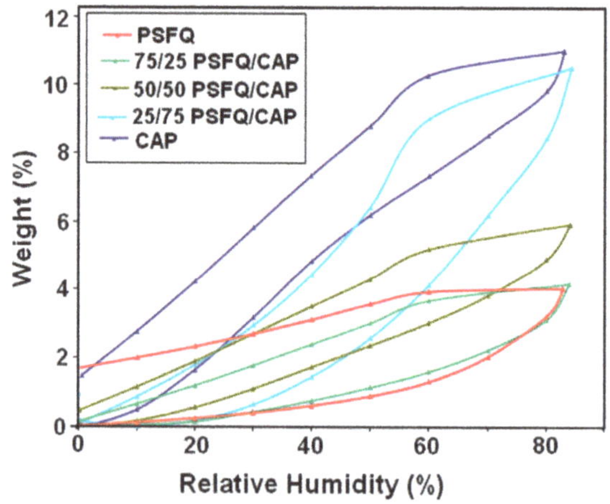


**Figure 3.** Moisture sorption isotherms for different quaternized polysulfone-based blends with tunable cellulose acetate phthalate content. From [100], copyright (2018), with permission from Elsevier.

While the specific surface area is one of the parameters that need special equipment to being measured, the diameters and orientations of nanofibers in an electrospun membrane are usually measured from SEM images, as discussed in the next sections.

## 5. Nanofiber Diameter

The diameters of nanofibers in an electrospun membrane are usually obtained from SEM images and either given as average with standard deviation or as distribution, sometimes as distribution boxplots [101], but mostly as a histogram. In the latter case, typically 100 or more fiber diameters per sample are measured to prepare a histogram [39,41,48,49,90], as shown in Figure 4 [53]. In most cases, the diameters are measured manually by ImageJ [14,43,56,57,68], while a few groups mention other software [41,44,90] or do not mention the software used [70]. Only a few papers mention the use of the ImageJ plugin DiameterJ or Super Pixel, which can in principle be used to automatically measure fiber diameters from SEM images [51,102–104], possibly because of problems with this automatic fiber detection caused by partial fibers or fiber intersections with dark spots [105]. On the other hand, a few groups suggested their own image analysis tools for this purpose [106,107].

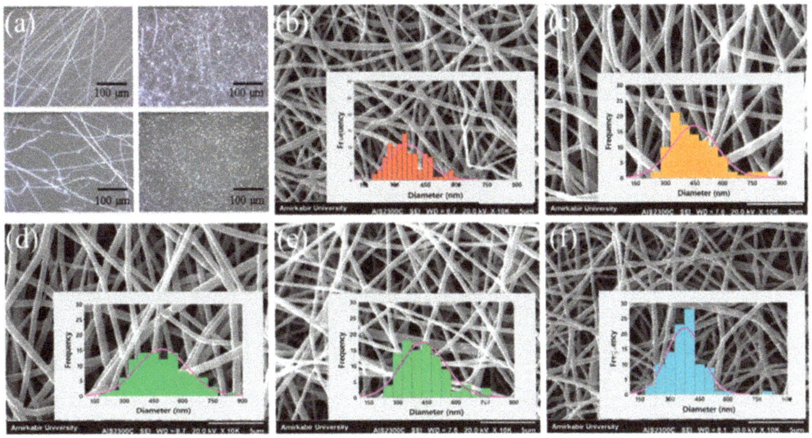


**Figure 4.** (**a**) Digital microscope images for pre-assessing the electrospinning conditions (scale bar: 100 µm, 370×), SEM images (scale bar: 5 µm) and nanofiber diameter histograms of (**b**) polyethylene terephthalate (PET), (**c**) PET/polycaprolactone (PCL) (3:1), (**d**) PET/PCL (1:1), (**e**) PET/PCL (1:3), (**f**) PCL. From [53], copyright (2020), with permission from Elsevier.

## 6. Nanofiber Orientation

Oriented nanofibers can be produced, e.g., by a fast rotating collector cylinder. Similarly to the nanofiber diameter distribution, the orientation of the nanofibers in an electrospun membrane is also usually determined from SEM images. The fiber orientations can be measured manually in ImageJ [108] or other software [109]. An interesting possibility to automatically detect fiber orientations is given by the ImageJ plugin OrientationJ, as depicted in Figure 5 [105]. While the color-coded fiber images (Figure 5B,C) enable checking the correctness of the detected orientation, the orientation graphs show the quantitative evaluation of the fiber orientation.

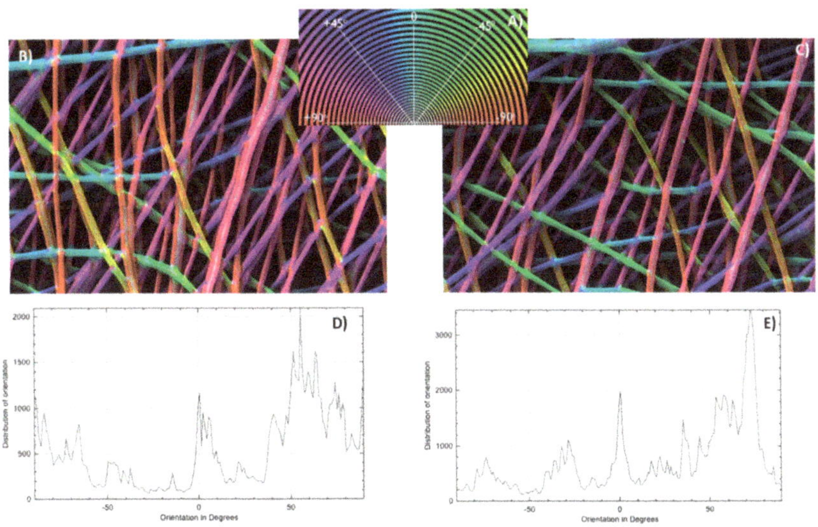


**Figure 5.** OrientationJ results: (**A**) circular color map coding; (**B**) color map image of batch 1, electrospun with 262 rad/s mandrel rotation speed; (**C**) color map image of batch 8, electrospun with 183 rad/s; (**D**) trace of orientation of batch 1; (**E**) trace of orientation of batch 8. From [105], copyright (2020), with permission from Elsevier.

Another possibility to evaluate fiber orientation automatically in ImageJ is given by the inbuilt fast Fourier transform (FFT) function as well as the Oval Profile plugin to receive a radial direction intensity plot [110–112], as depicted in Figure 6 [113]. The latter can also be given as a polar plot [113], which is often more intuitively understandable.

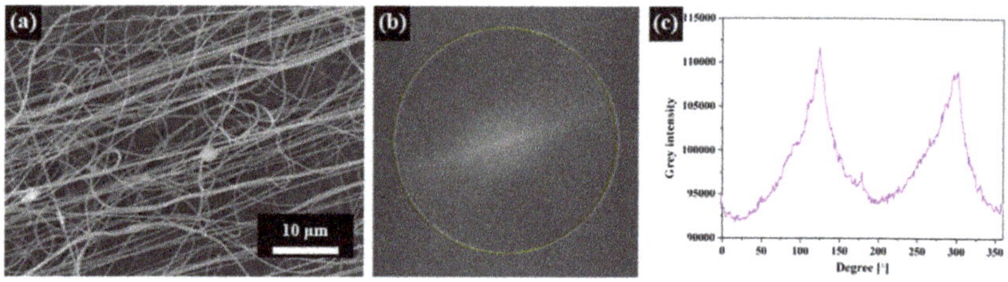


**Figure 6.** Fast Fourier transform (FFT) conversion from an SEM image to the intensity spectrum: (**a**) SEM image of nanofiber mat; (**b**) FFT frequency spectrogram; (**c**) grey intensity spectrum. From [113], copyright (2020), originally published under a CC-BY license.

These automatic orientation examinations have the advantage of taking into account all fiber parts, while manual calculations naturally have to be limited to certain parts of the fibers and are thus susceptible to subjective decisions of the evaluator. On the other hand, automatic calculations of the fiber orientations are highly error-prone if the fibers are too thin, i.e., only a few pixels per diameter, which will lead to favoring 0°, ±45° and ±90° orientations [110]. Thus, the choice of the images will potentially influence the results and has to be done with care.

## 7. Surface Roughness

The surface roughness of electrospun nanofiber mats can influence their hydrophobicity to a certain extent. When researchers mention measuring the roughness related to electrospun nanofiber mats, sometimes the roughness of the whole membrane is meant, while in other cases the roughness of single nanofibers is addressed. Correspondingly, different measurement methods are necessary to detect these different orders of magnitude of roughness.

Havlícek et al., e.g., show roughness measurements based on confocal laser scanning microscope (CLSM) images [61]. As preparation, they coated the investigated samples with a thin gold layer to enable better visibility of the relatively transparent nanofibers. In this way, 3D maps of the nanofibrous surfaces were prepared, as depicted in Figure 7, from which different roughness parameters could be calculated [61]. As the images show, the resolution of these images is much lower than in SEM images, so that in the lateral direction, only thicker fibers with diameters of some hundred nanometers are visible. This technique is thus only suitable to detect the roughness of a whole nanofiber mat, not of a single nanofiber surface.

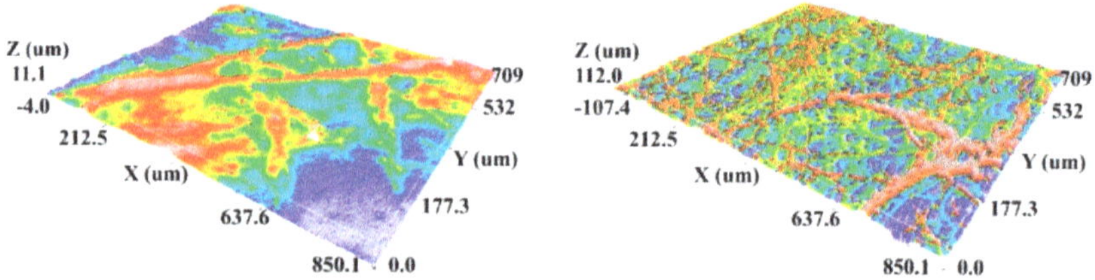


**Figure 7.** Confocal laser scanning microscopy (CLSM) images of needle-electrospun nanofiber mats from polyurethane (PUR, **left image**) and polyvinyl butyral (PVB, **right image**). From [61], copyright (2020), originally published under a CC-BY license.

If the latter is required, usually SEM or atomic force microscopy (AFM) images are investigated. However, some research groups also investigated the surface roughness of whole nanofiber mats by SEM or AFM.

Field emission SEM (FE-SEM) images were used by Shahverdi et al., who investigated nanofiber mat surfaces with Fiji software (National Institute of Health, 9000 Rockville Pike, Bethesda, MD 20892, USA), leading to relatively noisy 3D images for most samples on which a qualitative comparison of the fiber roughness was performed [114]. El-Morsy et al. used Gwyddion (http://gwyddion.net/ (accessed on 25 March 2023)) to evaluate FE-SEM images, showing average roughness $R_a$ of around 100 nm for different fiber material compositions, again with high noise [115]. Other studies show similarly noisy 3D maps, created by Gwyddion or other software from SEM images [116], although the noise could be reduced by using SEM images with higher magnification [117]. Nevertheless, this problem generally occurs during the transfer of SEM images into 3D maps according to the SEM grey scales; more realistic 3D maps need more sophisticated model creation [118].

This is why many groups use AFM measurements instead, which directly measure the fiber heights and where the color code thus directly gives a 3D map of the nanofiber surfaces [119]. Beigmoradi et al. measured the roughness along the fiber axis, as indicated in Figure 8, and found an average roughness $R_a$ in the range of 0.5–8 nm for different fibers [120]. As Figure 8 shows, there is no problematic noise that would make the evaluation unreliable. Nevertheless, it is necessary to obtain AFM images with sufficient resolution [121]; otherwise, nanofiber surface evaluation is not possible with AFM, either [122].

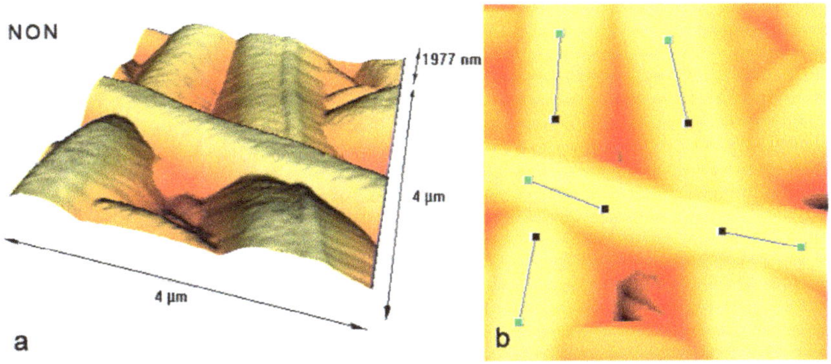


**Figure 8.** AFM mages of free-surfactant ("non") electrospun nanofiber mats. (**a**) 3D image and (**b**) 2D image. From [120], copyright (2021), originally published under a CC-BY license.

Besides the aforementioned methods based on surface images, taken by SEM or AFM, it is also possible to use a laser surface profilometer. In this way, Kichi et al. reported roughness $R_a$ in the range of 3–6 µm, i.e., apparently taking into account a larger area of the nanofiber mat, as could be expected due to the optical measurement and the correspondingly limited resolution [123]. Even a mechanical stylus-based profilometer was used to measure the roughness of nanofiber mats, finding $R_a$ values around 160–260 nm, however, for a lateral resolution of approx. 60 µm [124].

As this short overview shows, roughness values can be detected with a broad range of different techniques, although it is important to mention which resolution can be expected and whether the measurement was performed over a whole nanofiber mat area or along single nanofibers.

## 8. Nanofiber Mat Thickness

While the macroscopic thickness of a nanofiber mat seems to be simply measurable at first glance, there are, nevertheless, diverse methods with their advantages and disadvantages, sometimes influencing the result by the measurement. One of the problematic methods is using a micrometer caliper since its pressure limitation is usually not sufficient to avoid compression of fine nanofiber mats, similarly to microscopic textile fabrics. Liu et al. tried to compensate for this effect by folding the membranes twice before measuring, i.e., by measuring four layers instead of only one [57], while other groups did not comment on this problem [125]. A typical textile thickness measurement instrument, which has a larger measurement area and causes less pressure on the investigated sample, was applied by Pakolpakcil et al., who used a digital thickness gauge for nonwovens and measured at 10 points on the nanofiber mat [126].

To fully avoid this influence of the measurement on the measured value, some studies used optical methods to investigate the thickness of an electrospun membrane. Ryu et al. applied light transmittance measurements to investigate the sample thickness [127]. For this, they prepared nanofiber mats with different electrospinning times between 15 min and 75 min, measured their light transmittance and the thickness, the latter by cross-sectional microscopic images, and used the Beer–Lambert law correlating both values. This enabled a real-time thickness measurement during electrospinning.

Similarly to the aforementioned roughness measurements, but on larger length scales, Adhikari used a confocal microscope to measure the height of samples from the z-stacks along the sample edges [128]. A profilometer was also used to estimate the depth of a cross-sectional cut by scanning the surface perpendicular to the cut [129]. Naturally, it is also possible to directly investigate the cross-section along a cut through the nanofiber mats by SEM [130,131].

## 9. Hydrophobicity/Hydrophilicity

Hydrophobic or hydrophilic properties of nanofiber mats can be significantly influenced by surface functionalization, e.g., by plasma treatment [132]. The hydrophobic or hydrophilic properties of nanofiber mats are mostly determined by contact angle measurements, mostly applying the sessile-drop method in which a small droplet (e.g., with 5 μL volume [49,133], sometimes less [43]) is placed on the sample, and a microscope with camera is used to take photographs from the side, often at defined times, enabling fitting the contact angles on the photographs. There are also contact measurement instruments that are commercially available [38,42,44,46,48,108,134]. For custom-made setups, evaluation of the contact angles is possible with ImageJ, either manually or with a plugin such as DropSnake [57].

As an example of the time-dependent change in the hydrophobicity of a nanofiber mat surface, Figure 9 shows measurements taken 5 s and 60 s after the droplet was placed on the membrane, respectively. While most groups used (deionized) water for their contact angle measurements [45,53,70,93,133], others chose PBS solution as the right medium for these examinations [101].

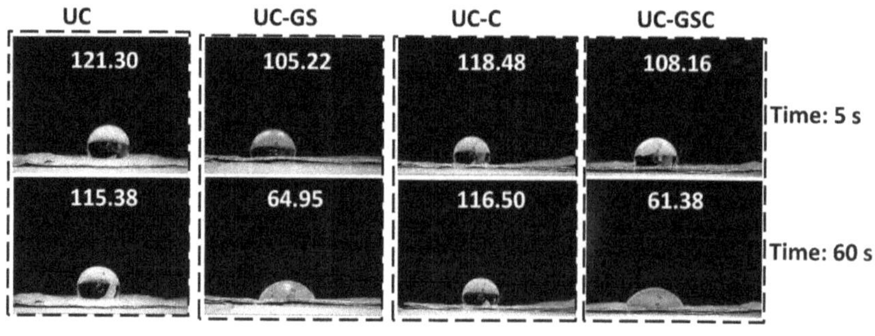


**Figure 9.** Contact angle measurements of different electrospun nanofiber mats prepared from polyurethane/cellulose acetate, 5 s and 60 s after contact with water droplet. From [44], copyright (2020), with permission from Elsevier.

An interesting alternative to the contact angle measurements is given by the Wilhelmy plate method, as depicted in Figure 10, for advancing and receding motion [135]. In this method, a plate is immersed into a fluid or retracted from it, allowing measuring the dynamic contact angle. This method was used by Kahdim for contact angle measurements on their nanofibrous scaffolds and showed relatively small standard deviations, i.e., gave reproducible results [39].

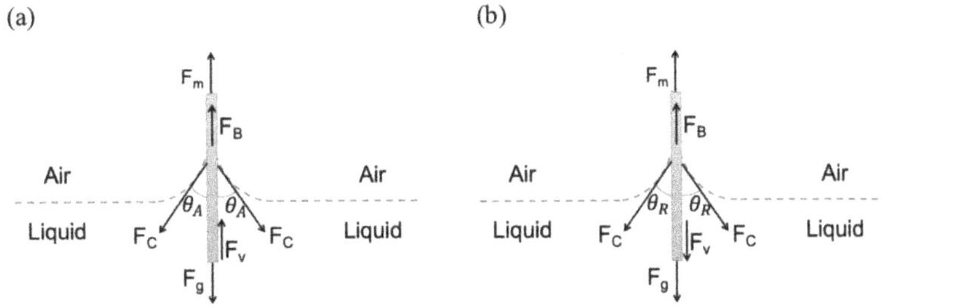


**Figure 10.** (**a**) Schematic of the advancing motion of the solid plate into the pool of the liquid. (**b**) Schematic of the receding motion of the solid plate out of the pool of the liquid. From [135], copyright (2018), with permission from Elsevier.

Interestingly, other typical textile tests regarding the water repellence of macroscopic textiles, such as the spray test according to AATCC standard test 22 and the water/alcohol solution resistance test according to AATCC 193 and 118 [136–138], were not found for electrospun nanofiber mats.

## 10. Water Uptake

The water uptake of an electrospun nanofiber mat can be defined in different ways—by the uptake in the pores around the fibers (cf. Section 2), or by the uptake inside the fibers, causing swelling, which is especially the case for electrospun hydrogels [139]. The water uptake of the material itself can be tested in the bulk form, e.g., by measuring the water uptake of a film of the examined material [38]. It is calculated by

$$Water\ uptake = \frac{m_1 - m_0}{m_0} \times 100\% \quad (4)$$

with the masses $m_0$ of the dry sample and $m_1$ of the sample after immersion in water for a defined time. Often, distilled or deionized water is used, and immersion times are usually around 1–2 days [38,42,140]. The water uptake may vary upon adding fillers, such as nanoclays [138]. While typical values of water uptake are around several percent for many materials, it can also be in the range of 200–600% for very hydrophilic, porous scaffolds [38,43,49,134,141]. For hydrogels, even increasing values of several thousand percent during a few minutes were measured [45,142,143]. Besides water, several papers used PBS solution for fluid uptake tests since cell cultivation often occurs in this medium [38,49,101,134].

## 11. Mechanical Properties

The mechanical properties of electrospun nanofiber mats depend on the fiber material and orientation, but also on the crystallinity of the fibers. They are mostly investigated by tensile tests [56,57], often with test speeds of 1–10 mm/min [41,42,90], sometimes even 20–30 mm/min [46,101], depending on the sample size and elongation at break. A few papers report measuring stress–strain curves with a constant force ramp rate, e.g., 0.3 N/min [88].

Safari et al. examined the difference between dry and wet state and found a significantly higher elongation at break and lower tensile strength in the wet state of Poly(*N*-vinylcaprolactam)/poly(vinyl acetate) copolymer nanofiber mats [38]. Zadeh et al. found that the percentage of carbon nanotubes in the polyurethane nanofiber mats influenced Young's modulus of the investigated samples [49]. Zhang et al. reported high tensile strength and modulus for PLA, high elongation at break for PCL, and averaged values for PLA/PCL blends [70]. Bazzi et al. found a significant increase in Young's modulus and toughness by adding a small amount of graphene nano-platelets to chitosan/polyvinyl alcohol electrospun nanofiber mats [68].

A very special tensile test, based on single electrospun fibers, was reported by Munawar and Schubert [144]. Working with well-aligned fibers, they rolled a fiber bundle for the tensile testing, clamped it in a single-fiber tensile tester, and afterwards cut the tested area and weighed the tested part of the fiber bundle to enable calculation of Young's modulus. In this way, they could measure along the fiber axes instead of taking the mechanical properties of a whole nanofiber mat, averaging over arbitrary fiber orientation.

Besides these typical textile tests, some authors also report compressive tests of their samples. This was the case for 3D specimens, e.g., prepared by combining electrospinning and freeze-drying [43,45,133]. Chen et al. investigated cyclic compressive stress–strain curves on their 3D scaffolds, showing the usual hysteresis loops, as depicted in Figure 11 [45]. Chen, on the other hand, found super-elastic and shape-recovery properties and reported enhanced elastic modulus and reduced energy loss during cyclic testing with a gelatin coating [133].

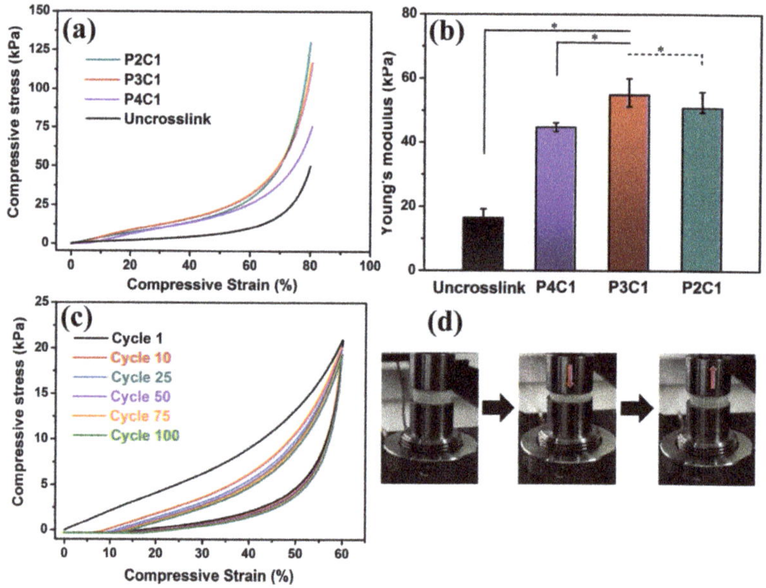


**Figure 11.** Mechanical tests on scaffolds from poly(lactic acid)/regenerated cellulose/citric acid in a ratio of x:1:1, named PxC1, and un-crosslinked without citric acid. (**a**) Compressive stress–strain curves and (**b**) Young's modulus of the different scaffolds, * indicating significant differences; (**c**) 100 cyclic compressive fatigue tests of P3C1 scaffolds under a compressive strain of 60%; (**d**) photographs of the P3C1 scaffolds under a compressing and releasing cycle. From [45], copyright (2020), with permission from Elsevier.

Besides tensile and compressive tests, some groups also reported bursting tests of nanofiber mats. Jalalah et al. applied the standard bursting strength test according to ISO 13938-2:1999 and found a linear correlation between nanofiber mat thickness and bursting strength [89]. Nejad et al. found a significant increase in the bursting strength of their nanofiber mats by adding PCL to poly(ethylene terephthalate) (PET). It was, however, calculated from the tensile strength tests [53]. The latter also tested suture retention according to ISO 7198:2016 by tensile tests with a well-defined suture thread that was inserted 2 mm from the top edge of the electrospun strip, so that the tensile test led to pulling the suture through the graft [53].

As these examples show, tensile tests are not the only tests possible to be performed on electrospun nanofiber mats; however, the most meaningful tests should be chosen for the planned application.

## 12. Electrical Conductivity

The conductivity of nanofibers depends on their material, thickness, crystallinity, etc. Conductive nanofiber mats can stimulate cell attachment, proliferation and differentiation [145]. This is why the conductivity of electrospun membranes is often measured [9]. On the other hand, soft and compressible textile fabrics generally pose a challenge to measurements of their conductivity since the contact between the measuring instrument and conductive parts of the sample may be prohibited by non-conductive fibers, and the fibrous structure reduces the contact area if contact pins are used, as is usual for multimeters [146]. Generally, samples can be measured with the two-electrode methods (as in common multimeters), the four-electrode method, which is capable of eliminating the contact resistance, and methods with even more electrodes [147].

The four-wire measurement (also known as four-terminal sensing or four-point probe) uses two outer current-introducing and two inner voltage-sensing electrodes, in this way becoming independent from the contact resistances. The van der Pauw method works similarly: while the four contacts are not aligned but positioned along the sample perimeter, the van der Pauw method measures the average sample resistivity, whereas the linear four-point probe method measures the resistivity along the electrode orientation [148]. Due to the expected high contact resistance, textile fabrics should normally be measured with the linear four-point probe or the van der Pauw method, depending on the desired information and the sample geometry.

Nevertheless, multimeters are often used to investigate the resistance of electrospun nanofiber mats, either with constant voltage [133] or by measuring the voltage-dependent current [68]. Zadeh et al. used a special four-probe cell to measure the sample impedance in a frequency range from 1 Hz to 100 kHz and calculated the sample resistance from this curve [49], while Zarei et al. [130] as well as Simsek et al. [149] directly measured the resistance using a four-probe method.

Munawar et al. decided to use another solution for the potentially high contact resistance—they coated the ends of the measured nanofiber bundles with silver ink to increase the conductivity of the contacts, so that they could perform reliable two-point resistance measurements [144].

Finally, it should be mentioned that in spite of the importance of the conductivity of electrospun scaffolds, most papers only report on conductivity measurements performed on the spinning solutions, since this parameter directly influences the electrospinning process.

## 13. Water Vapor Permeability

The water vapor permeability of an electrospun membrane is correlated with its porosity and is especially important for wound dressing applications, where too high water vapor permeability results in fast hydration and thus scars, while too low values let exudates accumulate and thus increase the risk of infection [150,151]. Quantitatively, the water vapor transmission should be between 76 and 9360 g/($m^2$ day) to improve wound healing [152,153]. Gu et al. reduced this optimum window to 2000–2500 g/($m^2$ day) [154].

Mostly, the water vapor transmittance is measured by gravimetry, where the sample is fixed on the opening, with defined diameter (e.g., 1.23 cm), of a round bottle that is filled with a defined volume of distilled water (e.g., 5 mL) and placed in an oven at typically 37 °C for 24 h [155]. The water vapor transmittance (WVTR) is calculated as

$$\text{WVTR} = \frac{-\Delta W}{AT} \quad (5)$$

with the mass change $-\Delta W$ of the water in the container, the exposed area $A$ and the measurement time $T$ [42]. Usually, the same measurement is performed with an open container as the reference [155].

Chaiarwut et al. found a WVTR of 9335 g/(m$^2$ day) for the positive control (open bottle), i.e., the upper border of the desired water vapor transmission to improve wound healing, and values around 2500 g/(m$^2$ day) for different nanofiber mats from PCL [101]. Samadian et al. reported a similar value for cellulose acetate/gelatin nanofibers [156], as did Zheng et al. for crosslinked pectin nanofiber mats [157]. Slightly higher values of around 3500 g/(m$^2$ day) were found for polyamide-6/polyvinylpyrrolidone nanofibers [48], while Esmaeili reported smaller values of around 1300 g/(m$^2$ day) [44]. Salemi et al. used a test temperature of only 33 °C, resulting in lower reference values of around 7440 g/(m$^2$ day) and around 3600 g/(m$^2$ day) for their poly (caprolactone)/poly (vinyl alcohol)/collagen nanofiber mats [158].

While this test method, also known as the cup test [159], is most often used to investigate the water vapor permeability of electrospun nanofiber mats, it is nevertheless not free of potential errors. As Mustapha et al. discussed, the measured water vapor permeability actually consists of three different resistances: those of the air cavity over the water, then the actual intrinsic membrane resistance, and finally the boundary air layer resistance, the latter of which the authors suggested to reduce by introducing a fan blowing the air above the cups away, while they used an open control cup to measure the air resistance by comparing the evaporation from this cup with an analytical model [160]. Slightly different methods are described in ASTM E96-95, where the cup is placed in a desiccator containing saturated $MgCl_2$ or $Mg(NO_3)_2 \cdot 6H_2O$ solution to provide a constant relative humidity, and the water transferred through the film is measured by an analytical balance [161,162]. Other tests, such as the sweating guarded hot plate test according to ISO 11092, the inverted cup method, the dynamic moisture permeation cell test method, the desiccant inverted cup test method, etc., are only scarcely reported for nanofiber mats in the literature [163]. Nevertheless, it is important to mention the method used, since even the methods giving results in the same units (typically g/(m$^2$ day)) deliver quite different results for identical samples [164].

## 14. Air Permeability

Air permeability is one of the parameters often measured for macroscopic textiles, but less often for electrospun nanofiber mats, potentially because it can partly be estimated from water vapor transmission tests [153]. Nevertheless, some studies report measuring the air permeability of nanofibrous scaffolds directly, usually giving the transmitted air volume per area and time, i.e., in the unit cm$^3$/(cm$^2$ s) or cm/s. Pakolpakcil et al. used a commercial air permeability tester at a fixed pressure of 100 Pa and a test area of 20 cm$^2$, resulting in values of around 10–12 cm/s [126]. Using the same parameters, Sun et al. found values of around 1–3 cm/s for their nanofiber mats electrospun from polyamide and multi-wall carbon nanotubes [165]. Yardimci et al. combined the same pressure with a test area of 50 cm$^2$ and found values of around 2.7–2.9 cm/s [166]. Slightly different parameters of 125 Pa and 38.3 cm$^2$ were used by Kim et al., who measured air permeability values in the range of 3.6–5.3 cm/s for their polyurethane-coated nanofiber mats [167]. Sarwar et al., on the other hand, applied a constant air flow of 2 cm/s and measured the air permeation in liters per minute [168].

Other methods, such as methods based on falling pistons in a chamber closed by the investigated textile fabric [169–173], are usually not reported for nanofiber mats.

## 15. Thermal Properties

The thermal conductivity of an electrospun nanofiber mat is often correlated with its electrical conductivity; however, in many cases, only the thermal conductivity is measured. Depending on the planned application, sometimes a high thermal conductivity is sought, while often the high porosity of a nanofiber mat, combined with the low thermal conductivity of most polymers, is used to prepare heat-blocking nanofibrous membranes instead.

Thermal conductivity can be evaluated, e.g., by a diffusivity measurement instrument at a defined temperature, often not far above the room temperature [174]. Measurements at high temperatures, however, are also possible using the hot disk method, e.g., using infrared thermography on the upper side of the sample, which is placed on the hot disk [175–177]. Besides a hot plate to heat up one sample side, a light flash, e.g., from a pulsed xenon lamp, can also be applied to heat one side of the sample [178–180].

## 16. Conclusions

The physical parameters that are typically measured on electrospun nanofiber mats for biotechnological applications can be described as morphology-related ones (porosity, pore size and specific surface area, fiber diameter and orientation, roughness and thickness of the nanofiber mat), hydrophobic/hydrophilic properties and water uptake, mechanical properties, and electric conductivity as well as water vapor and air permeability. Other potentially interesting properties, such as solute transport, which could be measured in a side-by-side diffusion chamber [57], are only scarcely reported. Table 1 gives a brief overview of these properties and typical measurement procedures as well as sample dimensions and test standards, if mentioned in the papers.

**Table 1.** Physical properties, typical measurements, sample dimensions and standards.

| Physical Property | Test Procedures | Dimensions, Standards | References |
|---|---|---|---|
| Porosity | Fluid uptake | | [39–45] |
| | Gas pycnometer | ASTM D2000 | [47] |
| | Apparent density | | [57] |
| Pore size distribution | SEM images, ImageJ | | [52,61] |
| | Thermoporometry | | [85–87] |
| Specific surface area | BET isotherms | | [88–90] |
| Nanofiber diameter | SEM images, DiameterJ | | [51,60] |
| Nanofiber orientation | SEM images, ImageJ | | [54] |
| Surface roughness | CLSM | | [61] |
| | SEM and Fiji software | | [114] |
| | SEM and Gwyddion software | | [115,116] |
| | Atomic force microscopy | | [119–121] |
| Nanofiber mat thickness | Textile thickness tester | | [52] |
| | Laser profilometer | | [54] |
| | Micrometer caliper | | [57] |
| Hydrophobicity | Sessile drop | | [43,44,50,53,88] |
| Water uptake | Mass difference dry/wet | | [140–143] |
| Mechanical properties | Tensile tests | 10 mm × 10 mm | [131] |
| | | 15 mm × 20 mm | [45] |
| | | Length 100 mm, ASTM D882 | [46] |
| | | 10 mm × 30 mm, ASTM D882 | [48] |
| | | 10″ × 3″, EN ISO 13934:1:1999 | [89] |
| Electrical conductivity | Impedance measurement | | [88] |
| | Conductivity meter | | [114] |
| Water vapor permeability | Bottle permeation test | 1.18 $cm^2$ | [42] |
| | | 1.77 $cm^2$ | [44] |
| Air permeability | Air permeability tester | 20 $cm^2$ | [126] |
| | | 50 $cm^2$ | [166] |
| | | 38.3 $cm^3$ | [167] |
| Thermal conductivity | Hot plate | | [175–177] |
| | Light flash | | [178–180] |

Since most of these parameters can be measured in different ways, it is generally highly recommended to precisely define the measurement technique, regarding the physical principle, environmental conditions and all parameters that can be modified. This should also become a standard when commercial instruments are used but whose exact function is often not completely known even to the user and cannot be reproduced by other researchers who do not own the same instrument, or for country-based standards that are not necessarily available worldwide. Generally, in many cases, more exact descriptions or definitions of the measured parameters are necessary, e.g., regarding the term "roughness", which can mean the surface roughness of a single nanofiber, but also the areal roughness of a whole nanofiber mat.

Moreover, some of the typical textile test methods, such as the simple spray test or water/alcohol test for the determination of the hydrophobicity of a fabric, or tests well-known from other research areas, such as thermoporometry, should be investigated regarding their usability for electrospun nanofiber mats, and their limits should be discussed as well as the possibilities they offer by enabling more tests if highly specialized equipment to measure a certain parameter is not available.

We hope that this review will encourage colleagues to test some new measurement techniques and extend their experimental descriptions so that all experiments can be reproduced by other research groups.

**Author Contributions:** Conceptualization, L.S. and A.E.; methodology, all authors; formal analysis, all authors; investigation, all authors; writing—original draft preparation, S.V.L. and A.E.; writing—review and editing, L.S.; visualization, A.E. All authors have read and agreed to the published version of the manuscript.

**Funding:** This research was partly funded by the German Federal Ministry for Economic Affairs and Energy as part of the Central Innovation Program for SMEs (ZIM) via the AiF, based on a resolution of the German Bundestag, grant number KK5129703CR0.

**Institutional Review Board Statement:** Not applicable.

**Data Availability Statement:** No data were produced in this review paper.

**Conflicts of Interest:** The authors declare no conflict of interest.

## References

1. Teo, W.E.; Ramakrishna, S. A review on electrospinning design and nanofibre assemblies. *Nanotechnology* **2006**, *17*, R89. [CrossRef] [PubMed]
2. Ramakrishna, S.; Fujihara, K.; Teo, W.E.; Yong, T.; Ma, Z.; Ramaseshan, R. Electrospun nanofibers: Solving global issues. *Mater. Today* **2006**, *9*, 40–50. [CrossRef]
3. Grothe, T.; Wehlage, D.; Böhm, T.; Remche, A.; Ehrmann, A. Needleless electrospinning of PAN nanofibre mats. *Tekstilec* **2017**, *60*, 290–295. [CrossRef]
4. Klinkhammer, K.; Seiler, N.; Grafahrend, D.; Gerardo-Nava, J.; Mey, J.; Brook, G.A.; Möller, M.; Dalton, P.D.; Klee, D. Deposition of electrospun fibers on reactive substrates for In Vitro investigations. *Tissue Eng. Part C* **2009**, *15*, 77–85. [CrossRef] [PubMed]
5. Mamun, A. Review of possible applications of nanofibrous mats for wound dressings. *Tekstilec* **2019**, *62*, 89–100. [CrossRef]
6. Gao, S.T.; Tang, G.S.; Hua, D.W.; Xiong, R.H.; Han, J.Q.; Jiang, S.H.; Zhang, Q.L.; Huang, C.B. Stimuli-responsive bio-based polymeric systems and their applications. *J. Mater. Chem. B* **2019**, *7*, 709–729. [CrossRef]
7. Rasouli, R.; Barhoum, A.; Bechelany, M.; Dufresne, A. Nanofibers for biomedical and healthcare applications. *Macromol. Biosci.* **2019**, *19*, 1800256. [CrossRef]
8. Wehlage, D.; Blattner, H.; Mamun, A.; Kutzli, I.; Diestelhorst, E.; Rattenholl, A.; Gudermann, F.; Lütkemeyer, D.; Ehrmann, A. Cell growth on electrospun nanofiber mats from polyacrylonitrile (PAN) blends. *AIMS Bioeng.* **2020**, *7*, 43–54. [CrossRef]
9. Tanzli, E.; Ehrmann, A. Electrospun nanofibrous membranes for tissue engineering and cell growth. *Appl. Sci.* **2021**, *11*, 6929. [CrossRef]
10. Zahedi, P.; Khatibi, A.; Fallah-Darrehchi, M. ANtimicrobial electrospun membranes. In *Electrospun and Nanofibrous Membranes*; Elsevier: Amsterdam, The Netherlands, 2023; pp. 501–519.
11. Jafari, S.; Hosseini Salekdeh, S.S.; Solouk, A.; Yousefzadeh, M. Electrospun polyethylene terephthalate (PET) nanofibrous conduit for biomedical application. *Polym. Adv. Technol.* **2020**, *31*, 284–296. [CrossRef]
12. More, N.; Ranglani, D.; Kharche, S.; Kapusetti, G. Electrospun mat of thermal-treatment-induced nanocomposite hydrogel of polyvinyl alcohol and cerium oxide for biomedical applications. *J. Appl. Polym. Sci.* **2020**, *137*, 49426. [CrossRef]
13. Ramos, C.; Lanno, G.-M.; Laidmäe, I.; Meos, A.; Härmas, R.; Kogermann, K. High humidity electrospinning of porous fibers for tuning the release of drug delivery systems. *Int. J. Polym. Mater. Polym. Biomater.* **2021**, *70*, 880–892. [CrossRef]

14. Chen, S.X.; John, J.V.; McCarthy, A.; Xie, J.W. New forms of electrospun nanofiber materials for biomedical applications. *J. Mater. Chem. B* **2020**, *8*, 3733–3746. [CrossRef]
15. Zou, S.Z.; Wang, X.R.; Fan, S.N.; Yao, X.; Zhang, Y.P.; Shao, H.L. Electrospun regenerated *Antheraea pernyi* silk fibroin scaffolds with improved pore size, mechanical properties and cytocompatibility using mesh collectors. *J. Mater. Chem. B* **2021**, *9*, 5514–5527. [CrossRef]
16. Selvaras, T.; Alshamrani, S.A.; Gopal, R.; Jaganathan, S.K.; Sivalingam, S.; Kadiman, S.; Saidin, S. Biodegradable and antithrombogenic chitosan/elastin blended polyurethane electrospun membrane for vascular tissue integration. *J. Biomed. Mater. Res. B* **2023**, *111*, 1171–1181. [CrossRef]
17. Entekhabi, E.; Nazarpak, M.H.; Shafieian, M.; Mohammadi, H.; Firouzi, M.; Hassannejad, Z. Fabrication and in vitro evaluation of 3D composite scaffold based on collagen/hyaluronic acid sponge and electrospun polycaprolactone nanofibers for peripheral nerve regeneration. *J. Biomed. Mater. Res. A* **2021**, *109*, 300–312. [CrossRef]
18. Yardimci, A.I. Comparative Study of the Structural, Mechanical and Electrochemical Properties of Polyacrylonitrile (PAN)-Based Polypyrrole (PPy) and Polyvinylidene Fluoride (PVDF) Electrospun Nanofibers. *J. Macromol. Sci. B* **2022**, *61*, 1103–1115. [CrossRef]
19. Zhang, H.N.; Zhang, T.T.; Qiu, Q.H.; Qin, X.H. Quaternary ammonium salt–modified polyacrylonitrile/polycaprolactone electrospun nanofibers with enhanced antibacterial properties. *Text. Res. J.* **2021**, *91*, 2194–2203. [CrossRef]
20. Jirofti, N.; Golandi, M.; Movaffagh, J.; Ahmadi, F.S.; Kalalinia, F. Improvement of the Wound-Healing Process by Curcumin-Loaded Chitosan/Collagen Blend Electrospun Nanofibers: In Vitro and In Vivo Studies. *ACS Biomater. Sci. Eng.* **2021**, *7*, 3886–3897. [CrossRef]
21. Akbarzadeh, M.; Pezeshki-Modaress, M.; Zandi, M. Biphasic, tough composite core/shell PCL/PVA-GEL nanofibers for biomedical application. *J. Appl. Polym. Sci.* **2020**, *137*, 48713. [CrossRef]
22. Islam, M.A.; Begum, H.A.; Shahid, M.A.; Ali, Y. Antibacterial electrospun nanofibers from poly (vinyl alcohol) and *Mikania micrantha* with augmented moisture properties: Formation and evaluation. *J. Text. Inst.* **2021**, *112*, 1602–1610. [CrossRef]
23. Stella, S.M.; Vijayalakshi, U. Influence of chemically modified Luffa on the preparation of nanofiber and its biological evaluation for biomedical applications. *J. Biomed. Mater. Res. A* **2019**, *107*, 610–620. [CrossRef] [PubMed]
24. Du, Z.W.; Jia, S.W.; Xiong, P.; Cai, Z.J. Preparation of protein nanoparticle-coated poly(hydroxybutyrate) electrospun nanofiber based scaffold for biomedical applications. *Int. J. Polym. Mater. Polym. Biomater.* **2022**, *71*, 677–691. [CrossRef]
25. Malik, S.; Hussain, T.; Nazir, A.; Khenoussi, N.; Cheema, S.A. Oriented electrospun nanofibers on stand-alone multi-segmented cylindrical collectors. *J. Text. Inst.* **2021**, *112*, 955–964. [CrossRef]
26. Munawar, M.A.; Schubert, D.W. Highly Oriented Electrospun Conductive Nanofibers of Biodegradable Polymers-Revealing the Electrical Percolation Thresholds. *ACS Appl. Polym. Mater.* **2021**, *3*, 2889–2901. [CrossRef]
27. Han, Y.S.; Hong, H.J.; Park, S.M.; Kim, D.S. Metal–Electrolyte Solution Dual-Mode Electrospinning Process for In Situ Fabrication of Electrospun Bilayer Membrane. *Adv. Mater. Interfaces* **2020**, *7*, 2000571. [CrossRef]
28. Maurya, A.K.; Weidenbacher, L.; Spano, F.; Fortunato, G.; Rossi, R.M.; Frenz, M.; Dommann, A.; Neels, A.; Sadeghpour, A. Structural insights into semicrystalline states of electrospun nanofibers: A multiscale analytical approach. *Nanoscale* **2019**, *15*, 7176–7187. [CrossRef]
29. Shao, Z.G.; Chen, J.Y.; Ke, L.-J.; Wang, Q.F.; Wang, X.; Li, W.W.; Zheng, G.F. Directional Transportation in a Self-Pumping Dressing Based on a Melt Electrospinning Hydrophobic Mesh. *ACS Biomater. Sci. Eng.* **2021**, *7*, 5918–5926. [CrossRef]
30. Sengupta, P.; Ghosh, A.; Bose, N.; Mukherjee, S.; Chowdhury, A.R.; Datta, P. A comparative assessment of poly(vinylidene fluoride)/conducting polymer electrospun nanofiber membranes for biomedical applications. *J. Appl. Polym. Sci.* **2020**, *137*, 49115. [CrossRef]
31. Gwon, G.; Choi, H.J.; Bae, J.H.; Binti Zulkifli, N.A.; Jeong, W.; Yoo, S.S.; Hyun, D.C.; Lee, S.W. An All-Nanofiber-Based Substrate-Less, Extremely Conformal, and Breathable Organic Field Effect Transistor for Biomedical Applications. *Adv. Funct. Mater.* **2022**, *32*, 2204645. [CrossRef]
32. Shi, S.; Si, Y.F.; Han, Y.T.; Wu, T.; Irfan Iqbal, M.; Fei, B.; Li, R.K.Y.; Hu, J.L.; Qu, J.P. Non-Toxic Crosslinking of Electrospun Gelatin Nanofibers for Tissue Engineering and Biomedicine—A Review. *Adv. Mater.* **2022**, *34*, 2107938. [CrossRef]
33. Kim, H.K.; Chung, H.J.; Park, T.G. Biodegradable polymeric microspheres with "open/closed" pores for sustained release of human growth hormone. *J. Control. Release* **2006**, *112*, 167–174. [CrossRef]
34. Kim, Y.J.; Kang, J.H.; Shen, B.W.; Wang, Y.Q.; He, Y.; Lee, M.S. Open–closed switching of synthetic tubular pores. *Nat. Commun.* **2015**, *6*, 8650. [CrossRef]
35. Liu, H.; Zhao, X.P. Thermal Conductivity Analysis of High Porosity Structures with Open and Closed Pores. *Int. J. Heat Mass Transf.* **2022**, *183A*, 122089. [CrossRef]
36. Yang, J.; Shi, G.X.; Bei, J.Z.; Wang, S.G.; Cao, Y.L.; Shang, Q.X.; Yang, G.H.; Wang, W.J. Fabrication and surface modification of macroporous poly(L-lactic acid) and poly(L-lactic-*co*-glycolic acid) (70/30) cell scaffolds for human skin fibroblast cell culture. *J. Biomed. Mater. Res.* **2002**, *62*, 438–446. [CrossRef]
37. Pati, F.; Adhikari, B.; Dhara, S. Development of chitosan-tripolyphosphate non-woven fibrous scaffolds for tissue engineering application. *J. Mater. Sci. Mater. Med.* **2012**, *23*, 1085–1096. [CrossRef]
38. Safari, S.; Ehsani, M.; Zandi, M. Stimuli-responsive electrospun nanofibers based on PNVCL-PVAc copolymer in biomedical applications. *Prog. Biomater.* **2021**, *10*, 245–258. [CrossRef]

39. Kahdim, Q.S.; Abdelmoula, N.; Al-Karagoly, H.; Albukhaty, S.; Al-Saaidi, J. Fabrication of a Polycaprolactone/Chitosan Nano fibrous Scaffold Loaded with Nigella sativa Extract for Biomedical Applications. *BioTech* **2023**, *12*, 19. [CrossRef]
40. Wang, N.; Yang, Y.J.; Al-Deyab, S.S.; El-Newehy, M.; Yu, J.Y.; Ding, B. Ultra-light 3D nanofibre-nets binary structured nylon 6–polyacrylonitrile membranes for efficient filtration of fine particulate matter. *J. Mater. Chem. A* **2015**, 3, 23946–23954. [CrossRef]
41. Chen, Y.J.; Mensah, A.; Wang, Q.Q.; Li, D.W.; Qiu, Y.Y.; Wie, Q.F. Hierarchical porous nanofibers containing thymol/beta-cyclodextrin: Physico-chemical characterization and potential biomedical applications. *Mater. Sci. Eng. C* **2020**, *115*, 111155. [CrossRef]
42. Salehi, M.; Niyakan, M.; Ehterami, A.; Haghi-Daredeh, S.; Nazarnezhad, S.; Abbaszadeh-Goudarzi, G.; Vaez, A.; Hashemi, S.F.; Rezaei, N.; Mousavi, S.R. Porous electrospun poly(ε-caprolactone)/gelatin nanofibrous mat containing cinnamon for wound healing application: In vitro and in vivo study. *Biomed. Eng. Lett.* **2020**, *10*, 149–161. [CrossRef] [PubMed]
43. Ghaee, A.; Bagheri-Khoulenjani, S.; Afshar, H.A.; Bogheiri, H. Biomimetic nanocomposite scaffolds based on surface modified PCL-nanofibers containing curcumin embedded in chitosan/gelatin for skin regeneration. *Comp. B Eng.* **2019**, *177*, 107339. [CrossRef]
44. Esmaeili, E.; Eslami-Arshaghi, T.; Hosseinzadeh, S.; Elahirad, E.; Jamalpoor, Z.; Hatamie, S.; Soleimani, M. The biomedical potential of cellulose acetate/polyurethane nanofibrous mats containing reduced graphene oxide/silver nanocomposites and curcumin: Antimicrobial performance and cutaneous wound healing. *Int. J. Biol. Macromol.* **2020**, *152*, 418–427. [CrossRef] [PubMed]
45. Chen, J.; Zhang, T.H.; Hua, W.K.; Li, P.Y.; Wang, X.F. 3D Porous poly(lactic acid)/regenerated cellulose composite scaffolds based on electrospun nanofibers for biomineralization. *Colloids Surf. A Physicochem. Eng. Asp.* **2020**, *585*, 124048. [CrossRef]
46. Ahmed, M.K.; Mansour, S.F.; Al-Wafi, R.; Abdel-Fattah, E. Nanofibers scaffolds of co-doped Bi/Sr-hydroxyapatite encapsulated into polycaprolactone for biomedical applications. *J. Mater. Res. Technol.* **2021**, *13*, 2297–2309. [CrossRef]
47. Tamari, S. Optimum design of the constant-volume gas pycnometer for determining the volume of solid particles. *Meas. Sci. Technol.* **2004**, *15*, 549. [CrossRef]
48. Khataei, S.; Al-Musawi, M.H.; Asadi, K.; Ramezani, S.; Abbasian, M.; Ghorbani, M. Effect of molecular weight and content of polyvinylpyrrolidone on cell proliferation, loading capacity and properties of electrospun green tea essential oil-incorporated polyamide-6/polyvinylpyrrolidone nanofibers. *J. Drug Deliv. Sci. Technol.* **2023**, *82*, 104310. [CrossRef]
49. Zadeh, Z.E.; Solouk, A.; Shafieian, M.; Nazarpak, M.H. Electrospun polyurethane/carbon nanotube composites with different amounts of carbon nanotubes and almost the same fiber diameter for biomedical applications. *Mater. Sci. Eng. C* **2021**, *118*, 111403. [CrossRef]
50. He, W.; Ma, Z.W.; Yong, T.; Teo, W.E.; Ramakrishna, S. Fabrication of collagen-coated biodegradable polymer nanofiber mesh and its potential for endothelial cells growth. *Biomaterials* **2005**, *26*, 7606–7615. [CrossRef]
51. Hotaling, N.A.; Bharti, K.; Kriel, H.; Simon, C.G., Jr. DiameterJ: A validated open source nanofiber diameter measurement tool. *Biomaterials* **2015**, *61*, 327–338. [CrossRef]
52. Bouchet, M.; Gauthier, M.; Maire, M.; Ajji, A.; Lerouge, S. Towards compliant small-diameter vascular grafts: Predictive analytical model and experiments. *Mater. Sci. Eng. C* **2019**, *100*, 715–723. [CrossRef]
53. Nejad, M.R.; Yousefzadeh, M.; Solouk, A. Electrospun PET/PCL small diameter nanofibrous conduit for biomedical application. *Mater. Sci. Eng. C* **2020**, *110*, 110692. [CrossRef]
54. Liu, Y.; Chaparro, F.J.; Gray, Z.; Gaumer, J.; Cybyk, D.B.; Ross, L.; Gosser, J.; Tian, Z.; Jia, Y.; Dull, T.; et al. 3D reconstruction of bias effects on porosity, alignment and mesoscale structure in electrospun tubular polycaprolactone. *Polymer* **2021**, *232*, 124120. [CrossRef]
55. Liu, Y.-X.; Chaparro, F.J.; Tian, Z.T.; Jia, Y.Z.; Gosser, J.; Gaumer, J.; Ross, L.; Tafreshi, H.; Lannutti, J.J. Visualization of porosity and pore size gradients in electrospun scaffolds using laser metrology. *PLoS ONE* **2023**, *18*, e0282903. [CrossRef]
56. Agueda, J.R.S.; Madrid, J.; Mondragon, J.M.; Lim, J.; Tan, A.; Wang, I.; Duguran, N.; Bondoc, A. Synthesis and Characterization of Electrospun Polyvinylidene Fluoride-based (PVDF) Scaffolds for Renal Bioengineering. *J. Phys. Conf. Ser.* **2021**, *2071*, 012005. [CrossRef]
57. Liu, W.Y.; Walker, G.; Price, S.; Yang, X.D.; Li, J.; Bunt, C. Electrospun Membranes as a Porous Barrier for Molecular Transport: Membrane Characterization and Release Assessment. *Pharmaceutics* **2021**, *13*, 916. [CrossRef]
58. Tahami, S.R.; Nemati, N.H.; Keshvari, H.; Khorasani, M.T. Effect of Electrical Potential on the Morphology of Polyvinyl Alcohol/ Sodium Alginate Electrospun Nanofibers, Containing Herbal Extracts of Calendula Officinalis for Using in Biomedical Applications. *J. Mod. Process. Manuf. Prod.* **2020**, *9*, 43–46.
59. Stella, S.M.; Sridhar, T.M.; Ramprasath, R.; Gimbun, J.; Vijayalakshmi, U. Physio-Chemical and Biological Characterization of Novel HPC (Hydroxypropylcellulose):HAP (Hydroxyapatite):PLA (Poly Lactic Acid) Electrospun Nanofibers as Implantable Material for Bone Regenerative Application. *Polymers* **2023**, *15*, 155. [CrossRef]
60. Zhang, Q.C.; Rudolph, T.; Benitez, A.J.; Gould, O.E.C.; Behl, M.; Kratz, K.; Lendlein, A. Temperature-controlled reversible pore size change of electrospun fibrous shape-memory polymer actuator based meshes. *Smart Mater. Struct.* **2019**, *28*, 055037. [CrossRef]
61. Havlícek, K.; Svobodová, L.; Bakalova, T.; Lederer, T. Influence of electrospinning methods on characteristics of polyvinyl butyral and polyurethane nanofibres essential for biological applications. *Mater. Des.* **2020**, *194*, 108898. [CrossRef]

62. Krysiak, Z.J.; Szewczyk, P.K.; Berniak, K.; Sroczyk, E.A.; Boratyn, E.; Stachewicz, U. Stretchable skin hydrating PVB patches with controlled pores' size and shape for deliberate evening primrose oil spreading, transport and release. *Biomater. Adv.* **2022**, *136*, 212786. [CrossRef] [PubMed]
63. Chen, Y.J.; Jia, Z.H.; Shafiq, M.; Xie, X.R.; Xiao, X.H.; Castro, R.; Rodrigues, J.; Wu, J.L.; Zhou, G.D.; Mo, X.M. Gas foaming of electrospun poly(L-lactide-co-caprolactone)/silk fibroin nanofiber scaffolds to promote cellular infiltration and tissue regeneration. *Coll. Surf. B Biointerfaces* **2021**, *201*, 111637. [CrossRef] [PubMed]
64. McLaren, R.L.; Laycock, C.J.; Brousseau, E.; Owen, G.R. Examining slit pore widths within plasma-exfoliated graphitic material utilising Barrett–Joyner–Halenda analysis. *New J. Chem.* **2021**, *45*, 12071–12080. [CrossRef]
65. Yang, H.M.; Song, X.L.; Zhang, X.C.; Ao, W.Q.; Qiu, G.H. Synthesis of vanadium-doped $SnO_2$ nanoparticles by chemical co-precipitation method. *Mater. Lett.* **2003**, *57*, 3124–3127. [CrossRef]
66. Sing, K.S.W.; Williams, R.T. Physisorption Hysteresis Loops and the Characterization of Nanoporous Materials. *Absorpt. Sci. Technol.* **2004**, *22*, 773–782. [CrossRef]
67. Kim, C.-Y.; Lee, J.-K.; Kim, B.-I. Synthesis and pore analysis of aerogel–glass fiber composites by ambient drying method. *Colloids Surf. A Physicochem. Eng. Asp.* **2008**, *313–314*, 179–182. [CrossRef]
68. Bazzi, M.; Shabani, I.; Mohandesi, J.A. Enhanced mechanical properties and electrical conductivity of Chitosan/Polyvinyl Alcohol electrospun nanofibers by incorporation of graphene nanoplatelets. *J. Mech. Behav. Biomed. Mater.* **2022**, *125*, 104975. [CrossRef]
69. Choma, J.; Jaroniec, M.; Burakiewicz-Mortka, W.; Kloske, M. Critical appraisal of classical methods for determination of meso pore size distributions of MCM-41 materials. *Appl. Surf. Sci.* **2002**, *196*, 216–223. [CrossRef]
70. Zhang, S.Y.; Yan, D.; Zhao, L.F.; Lin, J.Y. Composite fibrous membrane comprising PLA and PCL fibers for biomedical application. *Compos. Commun.* **2022**, *34*, 101268. [CrossRef]
71. Peinador, R.I.; Calvo, J.I.; Aim, R.B. Comparison of Capillary Flow Porometry (CFP) and Liquid Extrusion Porometry (LEP) Techniques for the Characterization of Porous and Face Mask Membranes. *Appl. Sci.* **2020**, *10*, 5703. [CrossRef]
72. Jena, A.; Gupta, K. Pore Volume of Nanofiber Nonwovens. *Int. Nonwovens J.* **2005**, *2*. [CrossRef]
73. Yunok, T.; Matsumoto, K.; Nakamura, K. Pore Size Distribution Measurements of Nonwoven Fibrous Filter by Differential Flow Method. *Membrane* **2004**, *29*, 227–235. [CrossRef]
74. Fatema, N.; Bhatia, S.K. Comparisons between geotextile pore sizes obtained from capillary flow and dry sieving tests. *Geotech. Test. J.* **2019**, *43*, 853–876. [CrossRef]
75. Kolb, H.E.; Schmitt, R.; Dittler, A.; Kasper, G. On the accuracy of capillary flow porometry for fibrous filter media. *Sep. Purif. Technol.* **2018**, *199*, 198–205. [CrossRef]
76. He, X.; Wang, Y.-n.; Zhou, J.F.; Wang, H.B.; Ding, W.; Shi, B. Suitability of Pore Measurement Methods for Characterizing the Hierarchical Pore Structure of Leather. *J. Am. Leather Chem. Assoc.* **2019**, *114*, 41–47.
77. Liu, Y.; Lannutti, J.J. Characterization of electrospun porosities: Current techniques. In Proceedings of the Nanofiber, Applications and Related Technologies NART 2021, Istanbul, Turkey, 8–10 September 2021; pp. 54–63.
78. Appell, M.; Jackson, M.A. Applications of Nanoporous Materials in Agriculture. *Adv. Appl. Nanotechnol. Agric.* **2013**, *1143*, 167–176.
79. Orsolini, P.; Michen, B.; Huch, A.; Tingaut, P.; Caseri, W.R.; Zimmermann, T. Characterization of Pores in Dense Nanopapers and Nanofibrillated Cellulose Membranes: A Critical Assessment of Established Methods. *ACS Appl. Mater. Interfaces* **2015**, *7*, 25884–25897. [CrossRef]
80. Raja, I.S.; Fathima, N.N. Gelatin–Cerium Oxide Nanocomposite for Enhanced Excisional Wound Healing. *ACS Appl. Bio Mater.* **2018**, *1*, 487–495. [CrossRef]
81. Hao, J.J.; Lu, C.X.; Zhou, P.C.; Li, D.H. Pore structure development of polyacrylonitrile nascent fibers in water stretching process. *Thermochim. Acta* **2013**, *569*, 42–47. [CrossRef]
82. Kanungo, I.; Fathima, N.N.; Rao, J.R.; Nair, B.U. Influence of PCL on the material properties of collagen based biocomposites and in vitro evaluation of drug release. *Mater. Sci. Eng. C* **2013**, *33*, 4651–4659. [CrossRef]
83. Landry, M.R. Thermoporometry by differential scanning calorimetry: Experimental considerations and applications. *Thermochim. Acta* **2005**, *433*, 27–50. [CrossRef]
84. Ishikiriyama, K.; Todoki, M. Evaluation of water in silica pores using differential scanning calorimetry. *Thermochim. Acta* **1995**, *256*, 213–226. [CrossRef]
85. Abolhasani, M.M.; Naebe, M.; Amiri, M.H.; Shirvanimoghaddam, K.; Anwar, S.; Michels, J.J.; Asadi, K. Hierarchically Structured Porous Piezoelectric Polymer Nanofibers for Energy Harvesting. *Adv. Sci.* **2020**, *7*, 2000517. [CrossRef] [PubMed]
86. Gustafsson, S.; Westermann, F.; Hanrieder, T.; Jung, L.; Ruppach, H.; Mihranyan, A. Comparative Analysis of Dry and Wet Porometry Methods for Characterization of Regular and Cross-Linked Virus Removal Filter Papers. *Membranes* **2019**, *9*, 1. [CrossRef]
87. Fashandi, H.; Karimi, M. Characterization of porosity of polystyrene fibers electrospun at humid atmosphere. *Thermochim. Acta* **2012**, *547*, 38–46. [CrossRef]
88. Balasubramaniam, B.; Kumar, S.A.; Singh, K.A.; Bhunia, S.; Verma, K.; Tian, L.M.; Gupta, R.K.; Gaharwar, A.K. Electrically Conductive $MoS_2$ Reinforced Polyacrylonitrile Nanofibers for Biomedical Applications. *Adv. NanoBiomed Res.* **2022**, *2*, 2100105. [CrossRef]

89. Jalalah, M.; Ahmad, A.; Saleem, A.; Bilal Qadir, M.; Khaliq, Z.; Khan, M.Q.; Nazir, A.; Faisal, M.; Alsaiari, M.; Irfan, M.; et al. Electrospun Nanofiber/Textile Supported Composite Membranes with Improved Mechanical Performance for Biomedical Applications. *Membranes* **2022**, *12*, 1158. [CrossRef]
90. Chen, Y.J.; Qiu, Y.Y.; Chen, W.B.F.; Wei, Q.F. Electrospun thymol-loaded porous cellulose acetate fibers with potential biomedical applications. *Mater. Sci. Eng. C* **2020**, *109*, 110536. [CrossRef]
91. Chen, S.; Shen, L.L.; Huang, D.; Du, J.; Fan, X.X.; Wie, A.L.; Chen, W.Y. Facile synthesis, microstructure, formation mechanism, in vitro biocompatibility, and drug delivery property of novel dendritic $TiO_2$ nanofibers with ultrahigh surface area. *Mater. Sci. Eng. C* **2020**, *115*, 111100. [CrossRef]
92. Cheng, H.; Li, X.N.; Li, T.H.; Qin, D.F.; Tang, T.F.; Li, Y.P.; Wang, G.X. Electrospun Nanofibers with High Specific Surface Area to Prepare Modified Electrodes for Electrochemiluminescence Detection of Azithromycin. *J. Nanomater.* **2021**, *2021*, 9961663. [CrossRef]
93. Li, W.Y.; Chao, S.; Li, Y.M.; Bai, F.Q.; Teng, Y.K.; Li, X.; Li, L.J.; Wang, C. Dual-layered composite nanofiber membrane with Cu-BTC-modified electrospun nanofibers and biopolymeric nanofibers for the removal of uremic toxins and its application in hemodialysis. *J. Membr. Sci.* **2022**, *642*, 119964. [CrossRef]
94. Arabpour, Z.; Baradaran-Rafii, A.; Bakhshaiesh, N.L.; Ai, J.; Ebrahimi-Barough, S.; Malekabadi, H.E.; Nazeri, N.; Vaez, A.; Salehi, M.; Sefat, F.; et al. Design and characterization of biodegradable multi layered electrospun nanofibers for corneal tissue engineering applications. *J. Biomed. Mater. Res.* **2019**, *107*, 2340–2349. [CrossRef]
95. Lim, S.K.; Hwang, S.-H.; Chang, D.I.; Kim, S.H. Preparation of mesoporous $In_2O_3$ nanofibers by electrospinning and their application as a CO gas sensor. *Sens. Actuators B Chem.* **2010**, *149*, 28–33. [CrossRef]
96. Prajapati, Y.N.; Verma, N. Adsorptive desulfurization of diesel oil using nickel nanoparticle-doped activated carbon beads with/without carbon nanofibers: Effects of adsorbate size and adsorbent texture. *Fuel* **2017**, *189*, 186–194. [CrossRef]
97. Othman, F.E.C.; Yusof, N.; Petru, M.; Md Nordin, N.A.H.; Hamid, M.F.; Ismail, A.F.; Rushdan, A.I.; Hassan, S.A. Polyethy leneimine-impregnated activated carbon nanofiber composited graphene-derived rice husk char for efficient post-combustion $CO_2$ capture. *Nanotechnol. Rev.* **2022**, *11*, 926–944. [CrossRef]
98. Al-Ghouti, M.A.; Da'ana, D.A. Guidelines for the use and interpretation of adsorption isotherm models: A review. *J. Hazard. Mater.* **2020**, *393*, 122383. [CrossRef]
99. Filimon, A.; Olaru, N.; Doroftei, F.; Coroaba, A.; Dunca, S. Processing of quaternized polysulfones solutions as tool in design of electrospun nanofibers: Microstructural characteristics and antimicrobial activity. *J. Mol. Liq.* **2021**, *330*, 115664. [CrossRef]
100. Filimon, A.; Stoica, I.; Onofrei, M.D.; Bargan, A.; Dunca, S. Quaternized polysulfones-based blends: Surface properties and performance in life quality and environmental applications. *Polym. Test.* **2018**, *71*, 285–295. [CrossRef]
101. Chaiarwut, S.; Ekabutr, P.; Chuysinuan, P.; Chanamuangkon, T.; Supaphol, P. Surface immobilization of PCL electrospun nanofibers with pexiganan for wound dressing. *J. Polym. Res.* **2021**, *28*, 344. [CrossRef]
102. Scaffaro, R.; Lopresti, F.; Maio, A.; Botta, L.; Rigogliuso, S.; Ghersi, G. Electrospun PCL/GO-g-PEG structures: Processing-morphology-properties relationships. *Comp. A Appl. Sci. Manuf.* **2017**, *92*, 97–107. [CrossRef]
103. Jia, X.W.; Qin, Z.Y.; Xu, J.X.; Kong, B.H.; Liu, Q.; Wang, H. Preparation and characterization of pea protein isolate-pullulan blend electrospun nanofiber films. *Int. J. Biol. Macromol.* **2020**, *157*, 641–647. [CrossRef] [PubMed]
104. McCarthy, A.; Saldana, L.; McGoldrick, D.; John, J.V.; Kuss, M.; Chen, S.X.; Duan, B.; Carlson, M.A.; Xie, J.W. Large-scale synthesis of compressible and re-expandable three-dimensional nanofiber matrices. *Nano Sel.* **2021**, *2*, 1566–1579. [CrossRef]
105. Dorati, R.; Chiesa, E.; Pisani, S.; Genta, I.; Modena, T.; Bruni, G.; Brambilla, C.R.M.; Benazzo, M.; Conti, B. The Effect of Process Parameters on Alignment of Tubular Electrospun Nanofibers for Tissue Regeneration Purposes. *J. Drug Deliv. Sci. Technol.* **2020**, *58*, 101781. [CrossRef]
106. Murphy, R.; Turcott, A.; Banuelos, L.; Dowey, E.; Goodwin, B.; O'Halloran Cardinal, K. SIMPoly: A Matlab-Based Image Analysis Tool to Measure Electrospun Polymer Scaffold Fiber Diameter. *Tissue Eng. C Methods* **2020**, *26*, 628–636. [CrossRef]
107. Götz, A.; Senz, V.; Schmidt, W.; Huling, J.; Grabow, N.; Illner, S. General image fiber tool: A concept for automated evaluation of fiber diameters in SEM images. *Measurement* **2021**, *177*, 109265. [CrossRef]
108. Li, Y.; Shen, Q.; Shen, J.; Ding, X.B.; Liu, T.; He, J.H.; Zhu, C.Y.; Zhao, D.; Zhu, J.D. Multifunctional Fibroblasts Enhanced via Thermal and Freeze-Drying Post-treatments of Aligned Electrospun Nanofiber Membranes. *Adv. Fiber Mater.* **2021**, *3*, 26–37. [CrossRef]
109. Cai, Z.J.; Xiong, P.; He, S.Q.; Zhu, C. Improved piezoelectric performances of highly orientated poly(β-hydroxybutyrate) electrospun nanofiber membrane scaffold blended with multiwalled carbon nanotubes. *Mater. Lett.* **2019**, *240*, 213–216. [CrossRef]
110. Hellert, C.; Wortmann, M.; Frese, N.; Grötsch, G.; Cornelißen, C.; Ehrmann, A. Adhesion of Electrospun Poly(acrylonitrile) Nanofibers on Conductive and Isolating Foil Substrates. *Coatings* **2021**, *11*, 249. [CrossRef]
111. Storck, J.L.; Grothe, T.; Mamun, A.; Sabantina, L.; Klöcker, M.; Blachowicz, T.; Ehrmann, A. Orientation of electrospun magnetic nanofibers near conductive areas. *Materials* **2020**, *13*, 47. [CrossRef]
112. Bazrafshan, Z.; Stylios, G.K. Custom-built electrostatics and supplementary bonding in the design of reinforced Collagen-g-P (methyl methacrylate-co-ethyl acrylate)/nylon 66 core-shell fibers. *J. Mech. Behav. Biomed. Mater.* **2018**, *87*, 19–29. [CrossRef]
113. He, H.J.; Wang, Y.M.; Farkas, B.; Nagy, Z.K.; Molnar, K. Analysis and prediction of the diameter and orientation of AC electrospun nanofibers by response surface methodology. *Mater. Des.* **2020**, *194*, 108902. [CrossRef]

114. Shahverdi, F.; Barati, A.; Salehi, E.; Arjomandzadegan, M. Biaxial electrospun nanofibers based on chitosan-poly (vinyl alcohol) and poly (ε-caprolactone) modified with $CeAlO_3$ nanoparticles as potential wound dressing materials. *Int. J. Biol. Macromol.* **2022**, *221*, 736–750. [CrossRef]
115. El-Morsy, M.A.; Afifi, M.; Ahmed, M.K.; Awwad, N.S.; Ibrahium, H.A.; Alqahtani, M.S. Electrospun nanofibrous scaffolds of polycaprolactone containing binary ions of Pd/vanadate doped hydroxyapatite for biomedical applications. *J. Drug Deliv. Sci. Technol.* **2022**, *70*, 103153. [CrossRef]
116. El-Naggar, M.E.; Shalaby, E.S.; Abd-Al-Aleem, A.H.; Abu-Saied, M.A.; Youssef, A.M. Synthesis of environmentally benign antimicrobial dressing nanofibers based on polycaprolactone blended with gold nanoparticles and spearmint oil nanoemulsion. *J. Mater. Res. Technol.* **2021**, *15*, 3447–3460. [CrossRef]
117. Teaima, M.H.; Abdelnaby, F.A.; Fadel, M.; El-Nabarawi, M.A.; Shoueir, K.R. Synthesis of Biocompatible and Environmentally Nanofibrous Mats Loaded with Moxifloxacin as a Model Drug for Biomedical Applications. *Pharmaceutics* **2020**, *12*, 1029. [CrossRef]
118. Sambaer, W.; Zatloukal, M.; Kimmer, D. 3D air filtration modeling for nanofiber based filters in the ultrafine particle size range. *Chem. Eng. Sci.* **2012**, *82*, 299–311. [CrossRef]
119. Joshi, J.; Homburg, S.V.; Ehrmann, A. Atomic force microscopy (AFM) on biopolymers and hydrogels for biotechnological applications—Possibilities and limits. *Polymers* **2022**, *14*, 1267. [CrossRef]
120. Beigmoradi, R.; Samimi, A.; Mohebbi-Kalhori, D. Controllability of the hydrophilic or hydrophobic behavior of the modified polysulfone electrospun nanofiber mats. *Polym. Test.* **2021**, *93*, 106970. [CrossRef]
121. Sharma, D.; Dhingra, S.; Banerjee, A.; Saha, S.; Bhattacharyya, J.; Satapathy, B.K. Designing suture-proof cell-attachable copolymer-mediated and curcumin- β-cyclodextrin inclusion complex loaded aliphatic polyester-based electrospun antibacterial constructs. *Int. J. Biol. Macromol.* **2022**, *216*, 397–413. [CrossRef]
122. Arumugam, M.; Murugesan, B.; Sivakumar, P.M.; Pandiyan, N.; Chinnalagu, D.K.; Rangasamy, G.; Mahalingam, S. Electrospun silk fibroin and gelatin blended nanofibers functionalized with noble metal nanoparticles for enhanced biomedical applications. *Process Biochem.* **2023**, *124*, 221–234. [CrossRef]
123. Kichi, M.K.; Torkaman, R.; Mohammadi, H.; Toutounchi, A.; Kharaziha, M.; Alihosseini, F. Electrochemical and in vitro bioactivity behavior of poly (ε-caprolactone) (PCL)-gelatin-forsterite nano coating on titanium for biomedical application. *Mater. Today Commun.* **2020**, *24*, 101326. [CrossRef]
124. Drobota, M.; Gradinaru, L.M.; Vlad, S.; Bargan, A.; Butnaru, M.; Angheloiu, M.; Afori, M. Preparation and Characterization of Electrospun Collagen Based Composites for Biomedical Applications. *Materials* **2020**, *13*, 3961. [CrossRef] [PubMed]
125. Lasenko, I.; Sanchaniya, J.V.; Kanukuntla, S.P.; Ladani, Y.; Viluma-Gudmona, A.; Kononova, O.; Lusis, V.; Tipans, I.; Selga, T. The Mechanical Properties of Nanocomposites Reinforced with PA6 Electrospun Nanofibers. *Polymers* **2023**, *15*, 673. [CrossRef] [PubMed]
126. Pakolpakcil, A.; Draczynski, Z.; Szulc, J.; Stawski, D.; Tarzynska, N.; Bednarowicz, A.; Sikorski, D.; Hernandez, C.; Sztajnowski, S.; Krucinska, I.; et al. An In Vitro Study of Antibacterial Properties of Electrospun Hypericum perforatum Oil-Loaded Poly(lactic Acid) Nonwovens for Potential Biomedical Applications. *Appl. Sci.* **2021**, *11*, 8219. [CrossRef]
127. Ryu, H.I.; Koo, M.S.; Kim, S.J.; Kim, S.K.; Park, Y.-A.; Park, S.M. Uniform-thickness electrospun nanofiber mat production system based on real-time thickness measurement. *Sci. Rep.* **2020**, *10*, 20847. [CrossRef]
128. Adhikari, U.; An, X.X.; Rijal, N.; Hopkins, T.; Khanal, S.; Chavez, T.; Tatu, R.; Sankar, J.; Little, K.J.; Horn, D.B.; et al. Embedding magnesium metallic particles in polycaprolactone nanofiber mesh improves applicability for biomedical applications. *Acta Biomater.* **2019**, *98*, 215–234. [CrossRef]
129. Sordini, L.; Silva, J.C.; Garrudo, F.F.F.; Rodrigues, C.A.V.; Marques, A.C.; Linhardt, R.J.; Cabral, J.M.S.; Morgado, J.; Castelo Ferreira, F. PEDOT:PSS-Coated Polybenzimidazole Electroconductive Nanofibers for Biomedical Applications. *Polymers* **2021**, *13*, 2786. [CrossRef]
130. Zarei, M.; Samimi, A.; Khorram, M.; Abdi, M.M.; Golestaneh, S.I. Fabrication and characterization of conductive polypyrrole/chitosan/collagen electrospun nanofiber scaffold for tissue engineering application. *Int. J. Biol. Macromol.* **2021**, *168*, 175–186. [CrossRef]
131. Conte, A.A.; Sun, K.; Hu, X.; Beachley, V.Z. Effects of Fiber Density and Strain Rate on the Mechanical Properties of Electrospun Polycaprolactone Nanofiber Mats. *Front. Chem.* **2020**, *8*, 610. [CrossRef]
132. Mozaffari, A.; Gashti, M.P. Air Plasma Functionalization of Electrospun Nanofibers for Skin Tissue Engineering. *Biomedicines* **2022**, *10*, 617. [CrossRef]
133. Chen, S.X.; John, J.V.; McCarthy, A.; Carlson, M.A.; Li, X.W.; Xie, J.W. Fast transformation of 2D nanofiber membranes into pre-molded 3D scaffolds with biomimetic and oriented porous structure for biomedical applications. *Appl. Phys. Rev.* **2020**, *7*, 021406. [CrossRef]
134. Movahedi, M.; Salehi, A.O.M.; Hajipour, F.P.; Etemad, S. Casein release and characterization of electrospun nanofibres for cartilage tissue engineering. *Bull. Mater. Sci.* **2022**, *45*, 76. [CrossRef]
135. Karim, A.M.; Kavehpour, H.P. Effect of viscous force on dynamic contact angle measurement using Wilhelmy plate method. *Colloids Surf. A Physicochem. Eng. Asp.* **2018**, *548*, 54–60. [CrossRef]

136. Zefirov, V.V.; Lubimtsev, N.A.; Stakhanov, A.I.; Elmanovich, I.V.; Kondratenko, M.S.; Lokshin, B.V.; Gallyamov, M.O.; Khokhlov, A.R. Durable crosslinked omniphobic coatings on textiles via supercritical carbon dioxide deposition. *J. Supercrit. Fluids* **2018**, *133*, 30–37. [CrossRef]
137. Mahltig, B.; Fischer, A. Inorganic/organic polymer coatings for textiles to realize water repellent and antimicrobial properties—A study with respect to textile comfort. *J. Polym. Sci. B Polym. Phys.* **2010**, *48*, 1562–1568. [CrossRef]
138. Liu, Y.Y.; Chen, X.Q.; Xin, J.H. Hydrophobic duck feathers and their simulation on textile substrates for water repellent treatment. *Bioinspiration Biomim.* **2008**, *3*, 046007. [CrossRef]
139. Gashti, M.P.; Dehdast, S.A.; Berenjian, A.; Shabani, M.; Zarinabadi, E.; Fard, G.C. PDDA/Honey Antibacterial Nanofiber Composites for Diabetic Wound-Healing: Preparation, Characterization, and In Vivo Studies. *Gels* **2023**, *9*, 173. [CrossRef]
140. Merin, D.D.; Jose, R.A.; Arulananth, T.S.; Sundarraj, A.A.; Inbamalar, T.M.; Meharie, M.G. Nanoclay-Incorporated Polycaprolactone Matrix via Electrospinning Techniques-Enriched Spectroscopic Responses. *J. Nanomater.* **2023**, *2023*, 1194158.
141. Serbezeanu, D.; Vlad-Bubulac, T.; Rusu, D.; Gradisteanu Pircalabioru, G.; Samoila, I.; Dinescu, S.; Aflori, M. Functional Polyimide-Based Electrospun Fibers for Biomedical Application. *Materials* **2019**, *12*, 3201. [CrossRef]
142. Sazegar, M.; Bazgir, S.; Katbab, A.A. Preparation and characterization of water-absorbing gas-assisted electrospun nanofibers based on poly(vinyl alcohol)/chitosan. *Mater. Today Commun.* **2020**, *25*, 101489. [CrossRef]
143. Abdolbaghian, H.; Bazgir, S. Fabrication and characterization of gas-assisted core-shell hydrogel nanofibers as a drug release system with antibacterial activity. *Eur. Polym. J.* **2022**, *174*, 111302. [CrossRef]
144. Munawar, M.A.; Schubert, D.W. Revealing Electrical and Mechanical Performances of Highly Oriented Electrospun Conductive Nanofibers of Biopolymers with Tunable Diameter. *Int. J. Mol. Sci.* **2021**, *22*, 10295. [CrossRef] [PubMed]
145. Pedrotty, D.M.; Koh, J.; Davis, B.H.; Taylor, D.A.; Wolf, P.; Niklason, L.E. Engineering skeletal myoblasts: Roles of threedimensional culture and electrical stimulation. *Am. J. Physiol. Heart Circ. Physiol.* **2005**, *288*, H1620–H1626. [CrossRef] [PubMed]
146. Schwarz-Pfeiffer, A.; Obermann, M.; Weber, M.O.; Ehrmann, A. Smarten up garments through knitting. *IOP Conf. Ser. Mater. Sci. Eng.* **2016**, *141*, 012008. [CrossRef]
147. Tyurin, I.N.; Getmantseva, V.V.; Andreeva, E.G. Van der Pauw Method for Measuring the Electrical Conductivity of Smart Textiles. *Fibre Chem.* **2019**, *51*, 139–146. [CrossRef]
148. Blachowicz, T.; Ehrmann, G.; Ehrmann, A. Recent Developments in Additive Manufacturing of Conductive Polymer Composites. *Macromol. Mater. Eng.* 2023, 2200692; *early view*. [CrossRef]
149. Simsek, M.; von Kruechten, L.; Buchner, M.; Duerkop, A.; Baeumner, A.J.; Wongkaew, N. An efficient post-doping strategy creating electrospun conductive nanofibers with multi-functionalities for biomedical applications. *J. Mater. Chem. C* **2019**, *7*, 9316–9325. [CrossRef]
150. Archana, D.; Dutta, J.; Dutta, P.K. Evaluation of chitosan nano dressing for wound healing: Characterization, in vitro and in vivo studies. *Int. J. Biol. Macromol.* **2013**, *57*, 193–203. [CrossRef]
151. Mi, F.-L.; Shyu, S.-S.; Wu, Y.-B.; Lee, S.-T.; Shyong, J.-Y.; Huang, R.-N. Fabrication and characterization of a sponge-like asymmetric chitosan membrane as a wound dressing. *Biomaterials* **2001**, *22*, 165–173. [CrossRef]
152. Chen, X.; Wang, X.; Wang, S.; Zhang, X.; Yu, J.; Wang, C. Mussel-inspired polydopamine-assisted bromelain immobilization onto electrospun fibrous membrane for potential application as wound dressing. *Mater. Sci. Eng. C. Mater. Biol. Appl.* **2020**, *110*, 110624. [CrossRef]
153. Zhong, G.F.; Qiu, M.Y.; Zhang, J.B.; Jiang, F.C.; Yue, X.; Huang, C.; Zhao, S.Y.; Zeng, R.; Zhang, C.; Qu, Y. Fabrication and characterization of PVA@PLA electrospinning nanofibers embedded with *Bletilla striata polysaccharide* and *Rosmarinic acid* to promote wound healing. *Int. J. Biol. Macromol.* **2023**, *234*, 123693. [CrossRef]
154. Gu, S.-Y.; Wang, Z.-M.; Ren, J.; Zhang, C.-Y. Electrospinning of gelatin and gelatin/poly(l-lactide) blend and its characteristics for wound dressing. *Mater. Sci. Eng. C* **2009**, *29*, 1822–1828. [CrossRef]
155. Naseri-Nosar, M.; Farzamfar, S.; Sahrapeyma, H.; Ghorbani, S.; Bastami, F.; Vaez, A.; Salehi, M. Cerium oxide nanoparticle-containing poly (ε-caprolactone)/gelatin electrospun film as a potential wound dressing material: In vitro and in vivo evaluation. *Mater. Sci. Eng. C* **2017**, *81*, 366–372. [CrossRef]
156. Samadian, H.; Zamiri, S.; Ehterami, A.; Farzamfar, S.; Vaez, A.; Khastar, H.; Alam, M.; Ai, A.; Derakhshankhah, H.; Allahyari, Z.; et al. Electrospun cellulose acetate/gelatin nanofibrous wound dressing containing berberine for diabetic foot ulcer healing: In vitro and in vivo studies. *Sci. Rep.* **2020**, *10*, 8312. [CrossRef]
157. Zheng, J.; Yang, Y.W.; Shi, X.Q.; Xie, Z.G.; Hu, J.L.; Liu, Y.C. Effects of preparation parameters on the properties of the crosslinked pectin nanofiber mats. *Carbohydr. Polym.* **2021**, *269*, 118314. [CrossRef]
158. Salemi, M.S.; Bahrami, G.; Arkan, E.; Izadi, Z.; Miraghaee, S.; Samadian, H. Co-electrospun nanofibrous mats loaded with bitter gourd (*Momordica charantia*) extract as the wound dressing materials: In vitro and in vivo study. *BMC Complement. Med. Ther.* **2021**, *21*, 111. [CrossRef]
159. Xia, Y.; He, L.F.; Feng, J.D.; Xu, S.J.; Yao, L.R.; Pan, G.W. Waterproof and Moisture-Permeable Polyurethane Nanofiber Membrane with High Strength, Launderability, and Durable Antimicrobial Properties. *Nanomaterials* **2022**, *12*, 1813. [CrossRef]
160. Mustapha, R.; Zoughaib, A.; Ghaddar, N.; Ghali, K. Modified upright cup method for testing water vapor permeability in porous membranes. *Energy* **2020**, *195*, 117057. [CrossRef]
161. Beristain-Bauza, S.C.; Mani-López, E.; Palou, E.; López-Malo, A. Antimicrobial activity and physical properties of protein films added with cell-free supernatant of *Lactobacillus rhamnosus*. *Food Control* **2016**, *62*, 44–51. [CrossRef]

162. Shekarabi, A.S.; Oromiehie, A.R.; Vaziri, A.; Ardjmand, M.; Safekordi, A.A. Investigation of the effect of nanoclay on the properties of quince seed mucilage edible films. *Food Sci. Nutr.* **2014**, *2*, 821–827. [CrossRef]
163. Sabantina, L.; Hes, L.; Mirasol, J.R.; Cordero, T.; Ehrmann, A. Water Vapor Permeability through PAN Nanofiber Mat with Varying Membrane-Like Areas. *Fibres Text. East. Eur.* **2019**, *27*, 12–15. [CrossRef]
164. Huang, J.H.; Qian, X.M. Comparison of Test Methods for Measuring Water Vapor Permeability of Fabrics. *Text. Res. J.* **2008**, *78*, 342–352. [CrossRef]
165. Sun, N.; Wang, G.-G.; Zhao, H.-X.; Cai, Y.-W.; Li, J.-Z.; Li, G.-Z.; Zhang, X.-N.; Wang, B.-L.; Han, J.-C.; Wang, Y.H.; et al. Waterproof, breathable and washable triboelectric nanogenerator based on electrospun nanofiber films for wearable electronics. *Nano Energy* **2021**, *90*, 106639. [CrossRef]
166. Yardimci, A.I.; Durmus, A.; Kayhan, M.; Tarhan, O. Antibacterial Activity of AgNO3 Incorporated Polyacrylonitrile/Polyvinylidene Fluoride (PAN/PVDF) Electrospun Nanofibrous Membranes and Their Air Permeability Properties. *J. Macromol. Sci. B* **2022**, *61*, 749–762. [CrossRef]
167. Kim, H.J.; Park, S.H. Reinforced tensile strength and wettability of nanofibrous electrospun cellulose acetate by coating with waterborne polyurethane and graphene oxide. *J. Eng. Fibers Fabr.* **2022**, *17*, 15589250221127353. [CrossRef]
168. Sarwar, M.N.; Ali, H.G.; Ullah, S.; Yamashita, K.; Shahbaz, A.; Nisar, U.; Hashmi, M.; Kim, I.-S. Electrospun PVA/CuONPs/Bitter Gourd Nanofibers with Improved Cytocompatibility and Antibacterial Properties: Application as Antibacterial Wound Dressing. *Polymers* **2022**, *14*, 1361. [CrossRef]
169. Kawabata, S. Method and Apparatus for Measuring Air Permeability of Fiber Material Such as Cloth or Nonwoven Fabric of Every Kind. Patent No. JPH056133B2, 14 December 1987.
170. Wang, Y.W.; Pan, J.Q.; Liu, H.Y.; Liu, K.; Huang, F.H.; Niu, S.X.; Cheng, G.Y.; Wang, D.D. Test Equipment Used for Air Circulation Performance of Diesel Soot Particulate Filter. Patent CN108019263A, 11 May 2018.
171. Wagner, C.G.; Cain, D.E. Method and Apparatus for Determining Permeability and Thickness of Refractory Coatings on Foundry Molds and Cores. U.S. Patent US4366703A, 4 January 1983.
172. Lyu, L.X.; Daichi, K.; Yang, Y.; Xu, T. Gas Permeability Detecting Device and Determination Method for Tissue Engineering Porous Scaffold. Patent CN106596374A, 26 April 2017.
173. Sabantina, L.; Ehrmann, A. New testing device for air permeability. *Commun. Dev. Assem. Text. Prod.* 2023, in print.
174. Lv, X.H.; Tang, Y.; Tian, Q.F.; Wang, Y.P.; Ding, T. Ultra-stretchable membrane with high electrical and thermal conductivity via electrospinning and in-situ nanosilver deposition. *Compos. Sci. Technol.* **2020**, *200*, 108414. [CrossRef]
175. Zhang, X.S.; Wang, B.; Wu, N.; Han, C.; Wu, C.Z.; Wang, Y.D. Flexible and thermal-stable SiZrOC nanofiber membranes with low thermal conductivity at high-temperature. *J. Europ. Ceram. Soc.* **2020**, *40*, 1877–1885. [CrossRef]
176. Li, Z.J.; Cheng, B.; Ju, J.G.; Kang, W.M.; Liu, Y. Development of a novel multi-scale structured superhydrophobic nanofiber membrane with enhanced thermal efficiency and high flux for membrane distillation. *Desalination* **2021**, *501*, 114834. [CrossRef]
177. Han, Z.Y.; Cheng, Z.Q.; Chen, Y.; Liang, Z.W.; Li, H.F.; Ma, Y.J.; Feng, X. Fabrication of highly pressure-sensitive, hydrophobic, and flexible 3D carbon nanofiber networks by electrospinning for human physiological signal monitoring. *Nanoscale* **2019**, *11*, 5942–5950. [CrossRef]
178. Datsyuk, V.; Trotsenko, S.; Trakakis, G.; Boden, A.; Vyzas-Asimakopoulos, K.; Parthenios, J.; Galiotis, C.; Reich, S.; Papagelis, K. Thermal properties enhancement of epoxy resins by incorporating polybenzimidazole nanofibers filled with graphene and carbon nanotubes as reinforcing material. *Polym. Test.* **2020**, *82*, 106317. [CrossRef]
179. Yang, G.; Zhang, X.D.; Shang, Y.; Xu, P.H.; Pan, D.; Su, F.M.; Ji, Y.X.; Feng, Y.Z.; Liu, Y.Z.; Liu, C.T. Highly thermally conductive polyvinyl alcohol/boron nitride nanocomposites with interconnection oriented boron nitride nanoplatelets. *Compos. Sci. Technol.* **2021**, *201*, 108521. [CrossRef]
180. Yin, C.-G.; Ma, Y.; Liu, Z.-J.; Fan, J.-C.; Shi, P.-H.; Xu, Q.-J.; Min, Y.-L. Multifunctional boron nitride nanosheet/polymer composite nanofiber membranes. *Polymer* **2019**, *162*, 100–107. [CrossRef]

membranes

*Article*

# Physico-Chemical, Mechanical, and Biological Properties of Polylactide/*Portulaca oleracea* Extract Electrospun Fibers

**Nikoleta Stoyanova [1], Mariya Spasova [1,*], Nevena Manolova [1], Iliya Rashkov [1], Sabina Taneva [2], Svetlana Momchilova [2] and Ani Georgieva [3]**

[1] Laboratory of Bioactive Polymers, Institute of Polymers, Bulgarian Academy of Sciences, Acad. G. Bonchev Street, bl. 103, BG-1113 Sofia, Bulgaria
[2] Department of Lipid Chemistry, Institute of Organic Chemistry with Centre of Phytochemistry, Bulgarian Academy of Sciences, Acad. G. Bonchev Street, bl. 9, BG-1113 Sofia, Bulgaria
[3] Institute of Experimental Morphology, Pathology and Anthropology with Museum, Bulgarian Academy of Sciences, Acad. G. Bonchev Street, bl. 25, BG-1113 Sofia, Bulgaria
* Correspondence: mspasova@polymer.bas.bg; Fax: +359-(0)2-8700309

**Abstract:** Electrospinning was used to create fibrous polylactide (PLA) materials loaded with *Portulaca oleracea* (*P. oleracea*) plant extract obtained by supercritical carbon dioxide. Morphological, physico-chemical, mechanical, and biological characteristics of the fibers were studied. According to the SEM results, the diameters of smooth and defect-free fibers fabricated by a one-pot electrospinning method were at micron scale. All the obtained materials possess good mechanical properties. Additionally, it was found that the composite fibers exhibited considerable antioxidant activity. The antimicrobial activity of the fibrous materials against Gram-positive and Gram-negative bacteria was determined as well. In vitro studies showed that the electrospun biomaterials had no cytotoxic effects and that the combination of PLA and the *P. oleracea* extract in the fiber structure promoted cell survival and proliferation of normal mouse fibroblasts. The obtained results reveal that microfibrous mats containing the polyester—PLA and the plant extract—*P. oleracea* can be suitable for applications in wound healing.

**Keywords:** plant extract; *Portulaca oleracea*; polyester; electrospinning; normal fibroblasts; wound healing

**Citation:** Stoyanova, N.; Spasova, M.; Manolova, N.; Rashkov, I.; Taneva, S.; Momchilova, S.; Georgieva, A. Physico-Chemical, Mechanical, and Biological Properties of Polylactide/*Portulaca oleracea* Extract Electrospun Fibers. *Membranes* **2023**, 13, 298. https://doi.org/10.3390/membranes13030298

Academic Editor: Andrea Ehrmann

Received: 21 January 2023
Revised: 25 February 2023
Accepted: 28 February 2023
Published: 2 March 2023



## 1. Introduction

*Portulaca oleracea* (known as purslane) is a herbaceous annual that is distributed all over the world [1]. Purslane is grown as a specialty crop recognized for its dietary and therapeutic benefits, especially in Asia and in Mediterranean countries [2]. This herb is a rich source of essential nutrients, mainly minerals [3,4], vitamins A, C, E, and B, and omega-3 fatty acids [5,6], and contains bioactive phytochemicals such as carotenoids and phenolic antioxidants with proven health benefits [7,8]. Various parts of this herb possess antioxidant, anti-inflammatory, antitumor, antidiabetic, hepatoprotective, anti-insomnia, analgesic, skeletal muscle-relaxant, gastroprotective, neuroprotective, wound-healing, and antiseptic properties [9–12].

Biodegradable polymers (synthetic or natural) have attracted considerable interest in recent years. They find diverse applications in packaging, agriculture, medicine, etc. [13]. Due to their significant diversity and synthetic versatility, aliphatic polyesters are the most extensively investigated class of biodegradable polymers [14]. Polylactide (PLA) is a polymer derived from renewable resources that belongs to the class of biodegradable aliphatic polyesters. This polymer and its copolymers are extensively used as advanced drug delivery carriers [15].

In recent years, the electrospinning method has gained much attention due to the fact that it allows facile fabrication of continuous fibers with diameters ranging from tens of nanometers to several micrometers. This electrohydrodynamic process uses the application of a high voltage to generate a jet from polymer solution or melt which is highly stretched and elongated to generate nano- and microfibers [16]. Because of their beneficial characteristics such as a large surface-area-to-volume ratio and a high porosity with small pore size, the electrospun fibrous materials find applications in drug delivery [17], tissue engineering [18], cosmetics [19], filtration [20], protective clothing [21], food packaging [22], etc. The use of electrospun nano- and microfibers in the pharmaceutical industry has increased recently [23]. Antibiotics [24], proteins [25], extracts [26], DNA [27], RNA, and anticancer drugs [28,29], along with other bioactive substances [30], have been loaded in electrospun nanofibers to treat diverse diseases.

The preparation of PLA-based nanofibrous materials by the electrospinning technique has received growing attention [31]. The morphology of the PLA mats and their properties are affected by a variety of factors such as viscosity, conductivity, polymer concentration, surface tension, and solvent system [32]. Electrospun PLA nano- and microfibers possess biodegradability, biocompatibility, non-toxicity, and good thermo-mechanical properties, combined with high specific surface area, and therefore have been widely investigated for biomedical applications as wound dressings [33], drug carriers [34], and tissue engineering scaffolds [35], as well as membranes for separation [36]. However, studies reporting the incorporation of plant extracts in the PLA-based electrospun fibrous materials are scarce.

Plant species have been used medicinally since ancient times because of the better patient tolerance and acceptance. There are four main natural product sources: plants, animals, sea species, and microbes [37]. They show a remarkable variety in chemical composition and structure. Electrospinning can increase the therapeutic potential of plant extracts by encapsulating them in suitable polymer matrixes and in this way improving their bioavailability and maintaining the needed concentration of bioactive compound at the target area [38].

To our knowledge, there is only one study in the literature, which was carried out by us, reporting the incorporation of the *P. oleracea* plant extract in electrospun polymer fibers. Recently, we have performed experiments to find the optimal process conditions for the encapsulation of the *P. oleracea* extract in a suitable polymer matrix [39]. The effect of the extract concentration on the morphology and some properties of the obtained materials were determined. The novelty of the present study consists in revealing the physico-chemical, thermal, mechanical, and biological properties of recently developed PLA/*P. oleracea* fibrous materials, and in studying their potential for their application in the biomedical field.

## 2. Materials and Methods

### 2.1. Used Materials

The following materials have been used in the present study without further purification: PLA (Ingeo™ Biopolymer 4032D, NatureWorks, Minnetonka, MN, USA; Mw = 259,000 g $mol^{-1}$, Mw/Mn = 1.94, as determined using size-exclusion chromatography with polystyrene standards), methylene chloride (DCM; Merck, Darmstadt, Germany), and ethanol absolute (Merck, Darmstadt, Germany).

Purslane plant material was collected in the Novo Zhelezare area (Plovdiv region, Bulgaria). Using a SEPAREX (France) high-pressure extractor with a 2 L extraction vessel and operating at a pressure of up to 1000 bar, 800 g of air-dried plant leaves and stems were milled and extracted with supercritical $CO_2$. The supercritical extraction was conducted for 2 h at a pressure of 400 bar and a temperature of 80 °C.

DPPH (2,2-diphenyl-1-picrylhydrazyl) and a reference mixture of fatty acid methyl esters from Sigma-Aldrich (Darmstadt, Germany), ethidium bromide (EtBr; Sigma Chemical, Balcatta, WA, Australia), and acridine orange (AO; Sigma Chemical, Balcatta, WA, Australia) were used without additional purification as they were of analytical grade. Penicillin and streptomycin (LONZA, Cologne, Germany) antibiotics and fetal calf serum (FCS; Gibco, Wien, Austria) were added to Dulbecco Modified Eagle's Medium (DMEM; Sigma-Aldrich, Darmstadt, Germany).

The mouse BALB/3T3 clone A31 cell line (ATCC, CCL-163) was obtained from the American Type Cultures Collection (ATCC, Rockville, MD, USA).

*Staphylococcus aureus* (*S. aureus*) 749 and *Escherichia coli* 3588 (*E. coli*) 74 were purchased from the National Bank for Industrial Microorganisms and Cell Cultures (NBIMCC), Sofia, Bulgaria.

### 2.2. Electrospinning for Preparation of Fibrous Mats

Fibrous PLA and PLA/*P. oleracea* materials were obtained by electrospinning, as described in detail elsewhere [39]. In brief, prior to electrospinning, the following solutions were prepared: (i) PLA (10 wt%) and (ii) PLA (10 wt%)/*P. oleracea* (7.5 wt% with respect to PLA weight). A mixed solution of DCM and EtOH (90/10 *w*/*w*) was used for the dissolution of the polymer and the extract.

The obtained PLA or PLA/*P. oleracea* solutions were put in a syringe (5 mL) fitted with a metal needle (gauge size: 20GX1$\frac{1}{2}$") connected to the positively charged electrode of a high-voltage power supply (up to 30 kV). A grounded drum with a diameter of 45 mm was positioned 15 cm away from the needle tip, rotating at a constant speed of 1000 rpm. An infusion pump (NE-300 Just InfusionTM syringe pump, New Era Pump Systems Inc., Farmingdale, NY, USA) was used to supply the spinning solution at a regulated feed rate of 3 mL/h at a constant applied voltage of 25 kV, at a room temperature of 21 °C, and at a relative humidity of 52%.

### 2.3. Characterization of the Materials

The spinning solutions' dynamic viscosity measurements were performed via a Brookfield DV-II+ Pro programmable viscometer equipped with a sample thermostatic cup and a cone spindle for the one/plate option, at room temperature (25 °C).

Scanning electron microscopy (SEM) was used to evaluate the fibers' morphology. Before the sample observation, the fibrous materials were vacuum-coated with a fine gold layer and analyzed using a Jeol JSM-5510 scanning electron microscope (JEOL Co., Ltd., Tokyo, Japan).

Using the ImageJ software [40], at least 50 fibers from SEM micrographs were evaluated in order to determine the average fiber diameter, fiber distribution, and morphology according to the previously reported criteria for the complex evaluation of electrospun mats [41].

Surface wettability analysis was used in order to evaluate the static contact angle via a DSA 10-MK2 drop shape analyzer system (Krüss, Hamburg, Germany) at 20 ± 0.2 °C. Contact angles of the fibrous materials were measured by dropping a deionized water droplet (10 μL) controlled by a computer dosing system on the surface. The droplet's temporal photographs were captured. Computer analysis of the obtained pictures was used in order to determine the contact angles. The final results represented an average of 20 measurements taken on various regions of the mat surfaces.

Thermogravimetric analysis (TGA) was carried out on the Perkin Elmer TGA 4000 (Waltham, MA, USA) at a 10 °C/min heating rate under argon at a flow of 60 mL/min. Pyris v. 11.0.0.0449 software was used for instrument control, data collecting, and data processing.

The mechanical properties of the fibrous materials were determined via tensile measurements performed in a single column system for mechanical testing, the INSTRON 3344, equipped with a loading cell of 50 N and the Bluehill universal software. The initial length between the clamps was 40 mm and the used stretching rate was 10 mm/min. The fibrous samples were cut in the direction of the collector rotation with dimensions of 20 × 60 mm$^2$. A Digital Thickness Gauge FD 50 (Kafer GmbH, München, Germany) was used to determine the thickness of the fibrous materials. The average thickness was ca. 300 μm ± 20 nm. For the sake of statistical significance, 10 specimens of each sample were tested, after which the average values of Young's modulus, the ultimate stress, and maximum deformation at break were determined.

The antioxidative properties of the materials was evaluated using the 2,2-diphenyl-1-picryl-hydrazyl-hydrate (DPPH) free radical method. A volume of 0.5 mL of an ethanol solution of *P. oleracea* (0.375 mg) was mixed with 2.5 mL of an ethanol solution of DPPH at a concentration of $1 \times 10^4$ mol $L^{-1}$. Three milliliters of DPPH solution in ethanol was added to PLA (5 mg mat) or PLA/*P. oleracea* (5 mg mat containing 0.375 mg of *P. oleracea*) fibrous mats. The as-prepared blended solutions were incubated in the dark for half an hour at room temperature (21 °C). Using a DU 800 UV-visible spectrophotometer (Beckman Coulter, Brea, CA, USA), the amount of DPPH radicals left in the solution was determined. The following equation was used to assess the antioxidant activity (AA):

$$\text{Inhibition, AA, \%} = \left[ \frac{\left(A_{DPPH} - A_{sample}\right)}{A_{DPPH}} \right] \times 100\,, \quad (1)$$

where $A_{sample}$ is the absorbance for the DPPH• solution at 517 nm after the addition of the solution containing plant extract or fibrous materials, and $A_{DPPH}$ is the absorbance for the DPPH• solution at 517 nm. Experiments were performed in triplicate.

### 2.4. Determination of Fatty Acids Composition and Acid Value

Gas chromatography (GC) with a flame ionization detector was used for the determination of fatty acids composition after acid-catalyzed transesterification of plant supercritical $CO_2$ extract to methyl esters [42]. Prior to GC analysis, fatty acid methyl esters (FAME) were purified by preparative thin-layer chromatography on a silica gel plate using a mobile phase of hexane-acetone (100:6 *v*/*v*). GC was conducted on a Shimadzu GC 2030 chromatograph equipped with a flame ionization detector and a Simplicity Wax capillary column (30 m × 0.32 mm × 0.25 µm, Supelco). The operating temperature program was a gradient from 170 °C to 260 °C at 2 °C/min and 5 min held at the final temperature. The injector and detector temperatures were 260 °C and 280 °C, respectively. Nitrogen was used as a carrier gas at 0.6 mL/min flow rate, with a split ratio of 1:50. Peak identification was done according to retention times and compared to that of a standard FAME mixture. The acid value (AV) was estimated by titration with ethanolic KOH.

### 2.5. Antibacterial Activity Assessment

The antibacterial activities of the electrospun mats were determined against the Gram-positive bacteria *S. aureus* 749 and Gram-negative bacteria *E. coli* 3588 by applying the disk diffusion assay. For that purpose, in vitro studies were performed using the Tryptone glucose extract agar (DIFCO Laboratories, Detroit, MI, USA) solid medium. The surface of the solid agar was inoculated with a suspension of cell culture with a cell concentration of $1 \times 10^5$ cells/mL. Within 5–10 min after inoculation, PLA mat and PLA/*P. oleracea* were placed on the inoculated surface (one disc with a diameter of 17 mm and weight of 5.0 mg per Petri dish). The Petri dishes were incubated for 24 h at 37 °C. Subsequently, the diameters of the inhibition zones around the disks were observed.

### 2.6. Dual Staining Using AO and EtBr

Dual staining of mouse BALB/c 3T3 fibroblast cells was performed using acridine-orange (AO) and ethidium bromide (EtBr) in order to evaluate cell death. The cells were plated at a concentration of $2 \times 10^5$ cells $\times$ $mL^{-1}$ Dulbecco's Modified Eagle Medium (DMEM) supplemented with 10% fetal bovine serum (FBS) on glass lamellas, placed at the bottom of 24-well plates, and incubated at 37 °C for 24 h in a $CO_2$ incubator to form a monolayer. PLA and PLA/*P. oleracea* fibrous mats were then sterilized using UV light and placed in the 24-well plates for additional 24 h of incubation. After that, the electrospun mats were removed and glass lamellas were washed twice with phosphate-buffered saline (PBS, pH 7.4) to remove unattached cells. Subsequently, the lamellas were stained with AO and EtBr at a ratio of 1:1 (10 µg/mL), and were observed on a fluorescence microscope (Leika DM 5000B, Wetzlar, Germany).

### 2.7. Mouse Fibroblast Adhesion on the Fibrous Surface

The adhesion of mouse BALB/3T3 cells on the surface of PLA and a PLA/*P. oleracea* fibrous mat was monitored by direct SEM observation. For that purpose, the fibrous samples were placed in 24-well tissue culture plates (Falcon Becton Dickinson, USA). A total of $2 \times 10^5$ cells per well were seeded on each sample and cultured for 72 h in 1 mL DMEM with 10% FBS. In order to prevent the mats from floating, thin Teflon rings adapted to the inner diameter of the wells were used. The samples were fixed after 72 h of incubation with 2.5 wt% glutaraldehyde solution in saline at 4 °C for 4 h. Then, the samples were carefully washed three times with saline and, prior to freeze-drying, with distilled water. Before the SEM observations, the specimens were vacuum-coated with gold under vacuum.

## 3. Results and Discussion

### 3.1. Morphology and Physico-Chemical Properties of the Fibrous Mats

In the present study, the morphology and properties of fibrous materials based on biocompatible and biodegradable polyester—PLLA and a natural plant extract of *P. oleracea* prepared by one-pot electrospinning were investigated. Electrospinning is a versatile technique for the fabrication of nanofibrous materials and the final morphology strongly depends on the intrinsic properties of the solution itself, its viscosity, and conductivity. It is known that electrospinning of low-viscosity solutions results in discontinuous fiber formation. Therefore, the dynamic viscosities of PLA and PLA/*P. oleracea* spinning solutions were measured prior to conducting the one-pot electrospinning. The determined viscosity value of the PLA solution (10 wt%) was 1180 ± 10 cP. The addition of the crude extract to the PLA solution resulted in a change of the solution color from transparent to saturated green (Figure 1) and led to a significant increase of the measured value of the dynamic viscosity to 4350 ± 15 cP. We assume that some components of the *P. oleracea* extract restricted the possibility of the polymer chains to move which resulted in the significant increase of the mixed solution's viscosity.

**(a)** **(b)**

**Figure 1.** Digital images of spinning solutions of: (**a**) PLA and (**b**) PLA/*P. oleracea*.

Subsequently, after the spinning solutions preparation and measuring of their dynamic viscosities, the solutions were subjected to electrospinning. The morphology of the obtained electrospun fibrous materials was evaluated by using scanning electron microscopy. Representative SEM images of the PLA and PLA/*P. oleracea* mats were shown in Figure 2. The shown SEM micrographs at different magnifications reveal the morphology of the obtained fibrous materials. As can be easily seen, the electrospinning of the PLA solution with concentration 10 wt% reproducibly resulted in the fabrication of fibers with mean fiber diameter of 1100 ± 200 nm. The prepared PLA fibers are continuous, defect-free, and with smooth surface. The addition of the plant extract in the PLA spinning solution and its consequent subjection to electrospinning resulted in the fabrication of fibers with larger diameters compared to PLA alone. The mean diameter of the PLA/*P. oleracea* fibers was 2200 ± 550 nm. The detected increase in the diameters of the composite fibers is most probably due to the significant increase in the dynamic viscosity of the PLA/*P. oleracea* spinning solution (4350 ± 15 cP) compared to the viscosity value of the PLA solution (1180 cP).

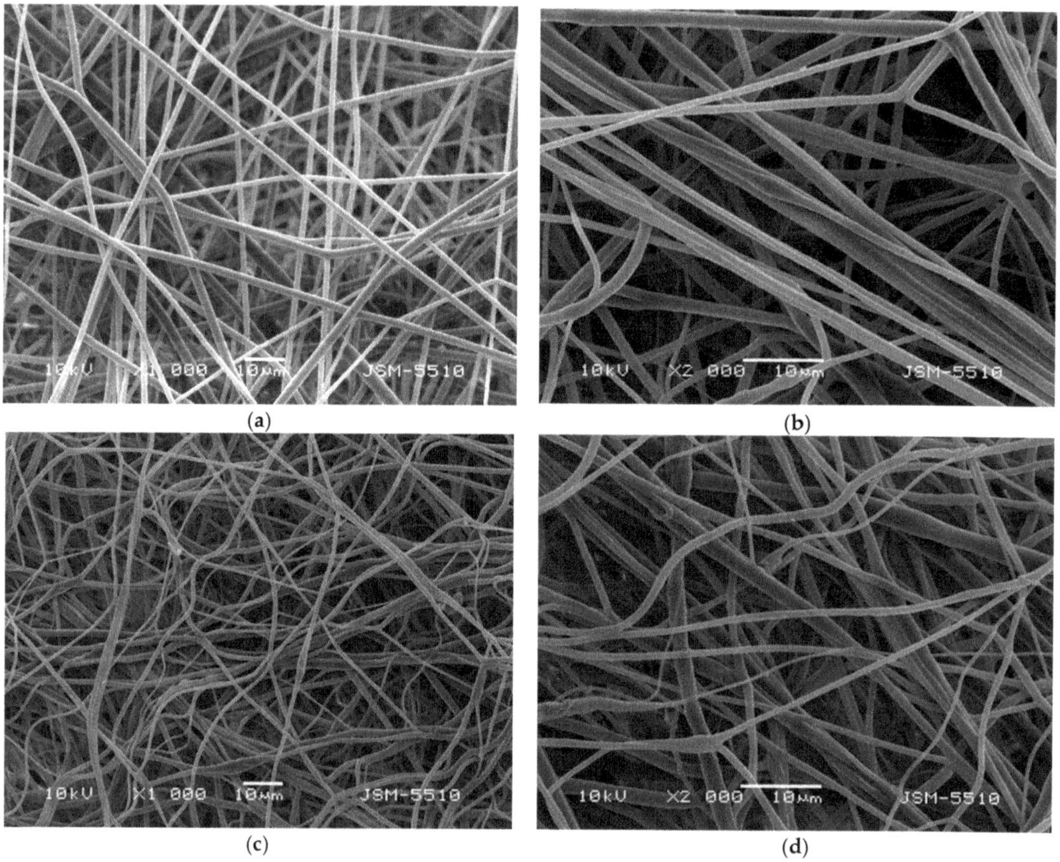


**Figure 2.** SEM images of: (**a**,**b**) PLA fibrous mat and (**c**,**d**) PLA/*P. oleracea* fibrous mat at magnifications ×1000 (**a**,**c**) and ×2000 (**b**,**d**).

The wettability of electrospun fibers is one of the most important factors for their applications, especially in biomedicine, pharmacy, agriculture, etc. Therefore, the surface wettability of the produced fibrous materials was assessed. The results from the water contact angle measurements showed that the PLA mats were hydrophobic since a water drop placed on them maintained its spherical form at a water contact angle of 110 ± 3.5°. The presence of the extract in the fibers in the case of the PLA/*P. oleracea* mat resulted in a slight reduction in the water contact angle value. The 94 ± 2.5° contact angle value shows that the fabricated PLA/*P. oleracea* mats are hydrophobic as well.

The effect of the plant extract in the PLA matrix on the thermal characteristics of the hybrid fibrous mats was studied by thermogravimetric analyses. The thermograms of electrospun PLA materials and hybrid electrospun PLA/*P. oleracea* mats are shown in Figure 3. The TGA of the pristine natural plant extract of *P. oleracea* was performed as well. Thermal stability of the neat extract showed a continuous mass loss in a one-degradation step, being almost fully degraded at 800 °C. The weight loss (ca. 3–4%) started to occur at approximately 150 °C, which can be ascribed to the evaporation of moisture (desorption of water) or some extract volatiles. The main mass loss started at around 300 °C.

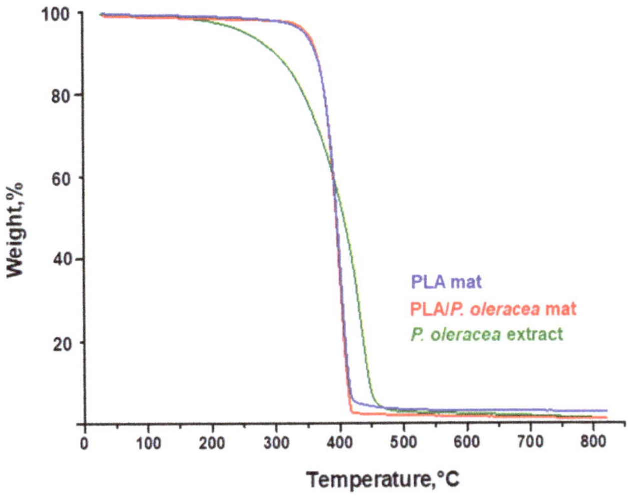


**Figure 3.** TG thermograms of crude *P. oleracea* extract, PLA, and PLA/*P. oleracea* mats.

As seen in Figure 3, the electrospun PLA and PLA/*P. oleracea* mats showed one decomposition peak. The thermal decomposition of electrospun PLA mat started at 330 °C and ended at 425 °C due to the decomposition of the polyester. Additionally, the presence of the natural extract in a concentration of 7.5 wt% does not alter the thermal behavior of the composite mat. The thermal degradation of electrospun PLA/*P. oleracea* mats began at 325 °C and ended at 430 °C. The residual mass at 800 °C was 1.20%, 1.20%, and 2.83% for the neat extract, the PLA/*P. oleracea* mat, and the PLA mat, respectively.

The mechanical properties are one of the most important properties of the electrospun fibers. They play an important role in determining the fibers' applications. The mechanical characteristics of the fibrous mats depend strongly on measurement technique, conditions of fiber fabrication, fiber orientation, point bonding, crosslinking, etc. The addition of a second component to the spinning solution might have a significant effect on the mechanical behavior of the resulting composite fibers. Therefore, it is crucial to study the influence of the extract on the mechanical properties of the hybrid PLA/*P. oleracea* mats. The mechanical characteristics of the obtained electrospun mats were determined using a single-column tensile testing machine. The typical stress–strain curves of PLA and PLA/*P. oleracea* mats are shown in Figure 4. The tensile strength values of the PLA mat and the hybrid mat containing the natural extract were very similar. The tensile strength of the PLA/*P. oleracea* mat was ca. 3.78 MPa, while the tensile strength of the PLA fibrous material reaches 3.9 MPa. This finding proved that the incorporation of the *P. oleracea* extract (at a concentration of 7.5 wt% with respect to the polymer weight) in the polymer matrix does not lead to a decrease in the mechanical properties of the composite material, thus preserving its good mechanical properties.

Various parts of purslane are known for medicinal and pharmacological uses because of its antioxidant activity [9]. The antioxidant activity of plants is due to their antioxidants, the majority of which are phenolic compounds such as phenolic acids and flavonoids, along with organic acids such as rosmarinic, caffeic, chlorogenic, p-coumaric, ferulic acids, quercetin, rutin, kaempferol, fumaric, oxalic, citric, acotinic, and malic acids. These compounds are capable of reducing oxidative stress by scavenging free radical species, and many of them have been identified in *P. oleracea* [43]. There are no data in the literature about the composition of the supercritical $CO_2$ *P. oleracea* extract. Our initial analyses revealed significant content of waxes, chlorophyll, about 40% lipids of neutral and polar classes in comparable amounts, some presence of terpenes, and other compounds. As for the fatty acids composition of this extract, lignoceric acid (24:0) was predominant at

30%, followed by behenic acid (22:0) at 18%, linoleic and linolenic acids (18:2 and 18:3, respectively) at 10%, palmitic (16:0), arachidic (20:0), and cerotic (26:0) acids at 7%, stearic (18:0) and palmitoleic (16:1) acids at 3%, oleic acid (9–18:1) at 2%, and other four acids (12:0, 14:0, 11–18:1 and 20:1) at below 1%. A high acid value (14 mg KOH/g) had been expected most probably because of the oxalic acid presence in such plants.

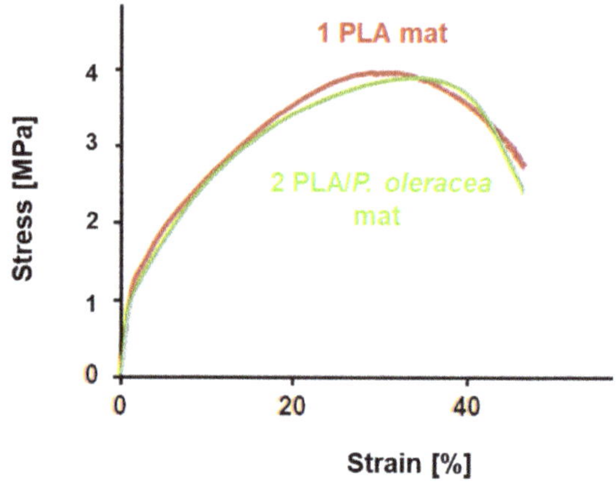


**Figure 4.** Stress–strain curves of: electrospun PLA mat and PLA/*P. oleracea* mat.

The antioxidant activity of the fibrous mats obtained in the present study was evaluated using the DPPH radical scavenging assay. This method was used to screen the radical scavenging activity of the *P. oleracea* crude extract as well. It is known that DPPH creates a violet color in ethanol or methanol solution; however, in the presence of antioxidants, the color fades to yellowish hues. The results of the antioxidant activity as well as the digital images of the DPPH solution in the presence of different samples were presented in Figure 5. As it can be easily seen, the color of the DPPH solution in contact with the fibrous PLA mat was deep violet and the absorbance of the radical decreased by approximately 4.1%. In contrast, the color of the DPPH solution in contact with PLA/*P. oleracea* mat changed its color to yellowish and the DPPH absorbance decreased by approximately 78%, revealing high antioxidant activity of the extract-containing fibrous material. For the sake of comparison, the change in absorbance of the DPPH solution upon contact with an ethanol solution of *P. oleracea* was measured and it was 88.2%. The obtained results revealed that the incorporated *P. oleracea* extract preserves it strong antioxidant activity into the polymer fibrous mat and imparted antioxidant properties to the hybrid material.

### 3.2. Antibacterial Activity of the Fibrous Materials

There are few reports showing the antimicrobial potential of purslane [44]. It has been reported that the *Portulaca elatior* root contains a trehalose-binding lectin possessing antibacterial and antifungal activities [45]. Further, Du et al. reported that *Portulaca oleracea* L. exhibited different levels of antibacterial activities against Gram-positive and Gram-negative bacteria, showing antibacterial activity in Gram-positive bacteria [46]. The authors suggest that this might be due to the several structural differences between Gram-positive and Gram-negative bacterial cell walls, as the latter has an outer membrane and a unique periplasmic space. Knowing this literature data, it was of interest to us to determine the antibacterial potential of the electrospun fibrous mats obtained in this study. Thus, discs with diameters of 17 mm were cut and placed in contact for 24 h with Gram-positive bacteria—*S. aureus* and Gram-negative bacteria—*E. coli*. The digital images of the Petri dishes are shown in Figure 6. As seen in Figure 6a, the bacteria cells grow normally (control). As expected, the PLA fibrous mats do not show any antibacterial activity against the tested pathogenic bacteria. However, no zones of inhibition were detected around the fibrous PLA/*P. oleracea* discs either.

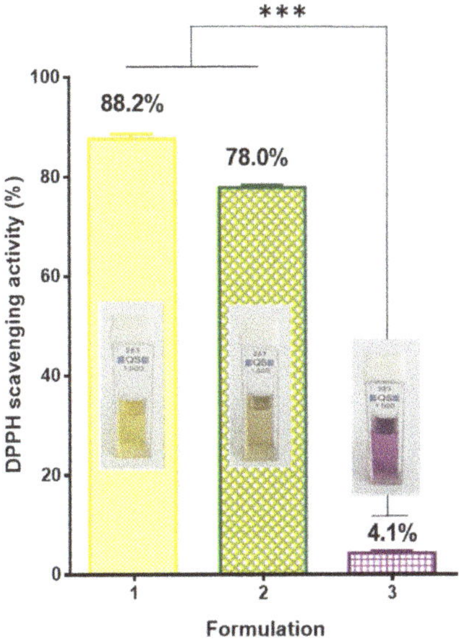


**Figure 5.** Antioxidant activity: 1—ethanol solution of the *P. oleracea* crude extract; 2—PLA/*P. oleracea* mat; 3—PLA mat and inset—digital images of 1—DPPH solution in the presence of the *P. oleracea* crude extract; 2—DPPH solution in the presence of a PLA/*P. oleracea* mat; and 3—DPPH solution in the presence of a PLA mat. *** $p < 0.001$.

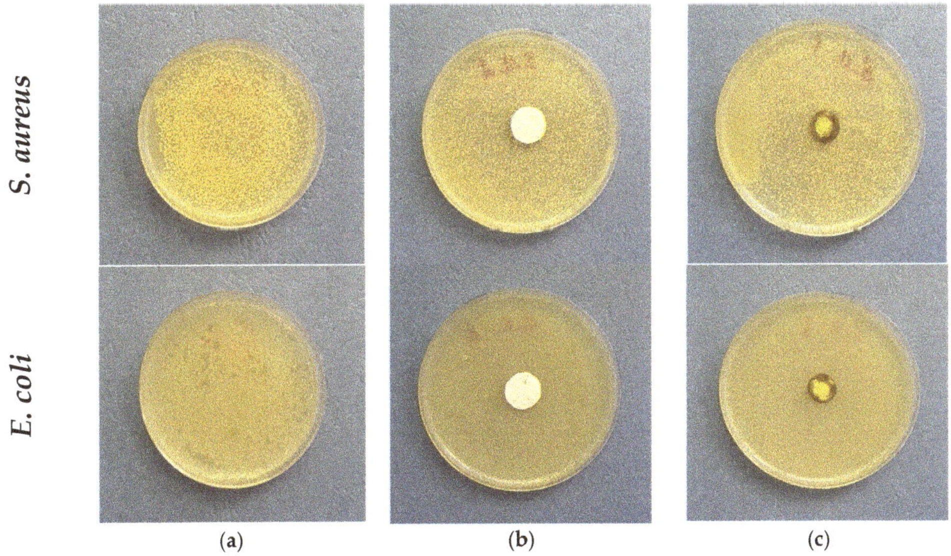


**Figure 6.** Digital photographs of (**a**) control bacteria; (**b**) PLA mat; and (**c**) PLA/*P. oleracea* mat after 24 h with *S. aureus* and *E. coli*. The cell type is marked in the left of each row.

### 3.3. Cytotoxicity Assay and Cell Staining

Nano- and microfibrous scaffolds created by electrospinning exhibit multiscale functionalities, including the ability to release locally specific bioactive molecules from synthetic or natural origin to targeted cell types [47]. This feature is highly desirable in regulating appropriate cell phenotypes for tissue engineering and wound healing applications [48,49]. Furthermore, the electrospun mats resemble the extracellular matrix of human body tissues and facilitate tissue regeneration. *P. oleracea* extract possesses a wide spectrum of pharmacological properties due to the presence of active components such as flavonoids, terpenoids, and vitamins that contribute to epithelialization and promote skin renewal. The biocompatibility of the obtained materials is a main factor for their future biomedical applications. The in vitro compatibility of the PLA and PLA/*P. oleracea* mats was assessed by MTT assay [33], while the cellular morphology of mouse fibroblasts was observed on a fluorescence microscope after staining cells that were cultivated in contact with fibrous materials. Fibroblasts play a key role in restoring the integrity of injured tissue. Therefore, mouse BALB/c 3T3 fibroblasts were cultured for 24 h with electrospun PLA and PLA/*P. oleracea* mats, and were then stained using fluorescent dyes (AO and EtBr). This staining method allows to discriminate between dead and viable cells. Acridine orange stains both live and dead cells, emitting green fluorescence as a result of its intercalation in the double-stranded DNA. As opposed to AO, EtBr is not able to pass through the membrane of viable cells and stains only dead and late apoptotic cells with poor membrane integrity, generating red fluorescence. Figure 7 presents the fluorescence micrographs showing the cells' morphology. Untreated fibroblast cells are characterized by a normal morphological structure with pale green nuclei and bright green nucleoli. No change was observed in the staining of the nuclei and cytoplasm in cells after their treatment with PLA and PLA/*P. oleracea* mats. The cell morphology remained normal. Furthermore, the previously obtained results from the MMT test as well as the cell staining reveal that the number of fibroblasts increased after being in contact with the fibrous mats containing the plant extract. This result shows that the electrospun PLA/*P. oleracea* mat is a prospective biomaterial with no toxicity supporting fibroblast attachment and proliferation in vitro.

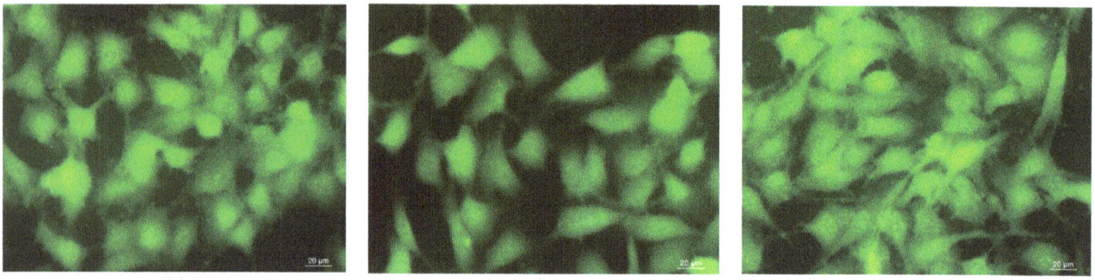

(a) (b) (c)

**Figure 7.** Fluorescence micrographs of AO and EtBr double-stained mouse BALB/c 3T3 fibroblast cells incubated for 24 h: (**a**) untreated cells; (**b**) PLA mat; and (**c**) PLA/*P. oleracea* mat. Live cells are shown in green. Bar: 20 μm.

The mouse BALB/3T3 cell line is suitable for preliminary assessment of cell viability, proliferation, and adhesion to the obtained fibrous materials. The adhesion of fibroblasts to the surface of the prepared PLA and PLA/*P. oleracea* fibrous materials was assessed by SEM analysis. As shown in Figure 8, the fibroblasts attach well to all fibrous materials, particularly when the plant extract of *P. oleracea* is loaded into the PLA fibers (Figure 8c). Moreover, considerable cell spreading is observed. The cells gain a specific shape following the fibers and spreading between them. This result revealed that the mouse fibroblasts adhere and spread well on the fibrous PLA and PLA/*P. oleracea* mats. This finding confirms the cell compatibility of the created novel materials and their potential use for wound healing and tissue engineering applications.

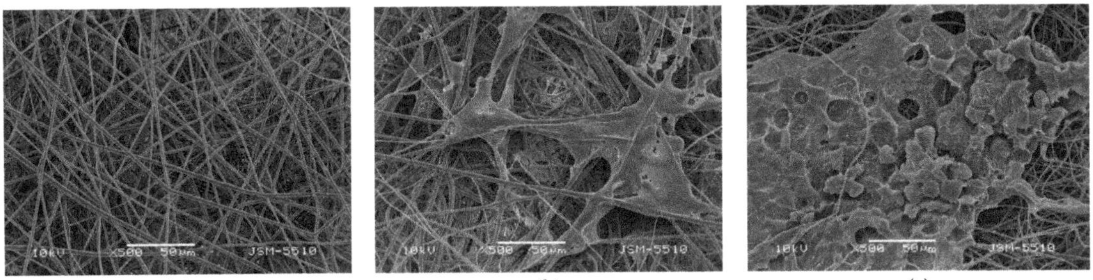


(a) (b) (c)

**Figure 8.** SEM micrographs of: (**a**) PLA fibers; (**b**) mouse fibroblasts on PLA fibers; and (**c**) mouse fibroblasts on PLA/*P. oleracea* after 72 h contact.

## 4. Conclusions

In the present study, the physico-chemical, mechanical, and biological properties of electrospun PLA and PLA/*P. oleracea* materials were characterized. The incorporation of a crude *P. oleracea* extract, obtained by using supercritical $CO_2$, imparted to the fibrous mat strong antioxidant activity while preserving the good mechanical properties of the fibers. Moreover, the in vitro tests with normal fibroblasts reveal that the electrospun PLA/*P. oleracea* material is biocompatible, promoting fibroblast adhesion, attachment, and proliferation. Thus, the created fibrous materials containing the plant extract could be suitable candidates for wound dressing applications.

**Author Contributions:** Conceptualization and supervision, M.S., N.M. and I.R.; methodology, N.S., M.S., S.T., S.M. and A.G.; formal analysis, investigation, and data curation, N.S., M.S, S.T. and A.G., original draft preparation, N.S., M.S., N.M., I.R., S.M. and A.G.; supervision, I.R.; funding acquisition, M.S.; writing—review and editing, M.S., N.M. and I.R. All authors have read and agreed to the published version of the manuscript.

**Funding:** This research was funded by Operational Program 'Science and Education for Smart Growth' 2014–2020, co-financed by the European Union through the European Structural and Investment Funds, grant BG05M2OP001-1.002-0012 'Sustainable utilization of bio-resources and waste of medicinal and aromatic plants for innovative bioactive products'.

**Institutional Review Board Statement:** Not applicable.

**Data Availability Statement:** The data presented in this study are available on request from the corresponding author.

**Acknowledgments:** Financial support from the Operational Program 'Science and Education for Smart Growth' 2014–2020, co-financed by the European Union through the European Structural and Investment Funds, grant BG05M2OP001-1.002-0012 'Sustainable utilization of bio-resources and waste of medicinal and aromatic plants for innovative bioactive products' is gratefully acknowledged. Research equipment of Distributed Research Infrastructure INFRAMAT, part of the Bulgarian National Roadmap for Research Infrastructures, supported by the Bulgarian Ministry of Education and Science, was partially used in this study.

**Conflicts of Interest:** The authors declare no conflict of interest.

## References

1. Movahedian, A.; Ghannadi, A.; Vashirnia, M. Hypocholesterolemic effects of purslane extracts on serum lipids in rabbits fed with high cholesterol levels. *Int. J. Pharmacol.* **2007**, *3*, 285–289.
2. Chan, K.; Islam, M.; Kamil, M.; Radhakrishnan, R.; Zakaria, M.; Habibullah, M.; Attas, A. The analgesic and anti-inflammatory effects of *Portulaca oleracea* L. subsp.sativa (Haw.) Celak. *J. Ethnopharmacol.* **2000**, *73*, 445–451. [CrossRef]
3. Uddin, M.; Juraimi, A.; Ali, M.; Ismail, M. Evaluation of antioxidant properties and mineral composition of Purslane (*Portulaca oleracea* L.) at different growth stages. *Int. J. Mol. Sci.* **2012**, *13*, 10257–10267. [CrossRef] [PubMed]
4. Okafor, I.; Ayalokunrin, M.; Orachu, L. A review on *Portulaca oleracea* (Purslane) plant its nature and biomedical benefits. *Int. J. Biomed. Res.* **2014**, *5*, 75–80. [CrossRef]
5. Petropoulos, S.; Karkanis, A.; Fernandes, A.; Barros, L.; Ferreira, I.; Ntatsi, G.; Petrotos, K.; Lykas, E.; Khah, L. Chemical composition and yield of six genotypes of common purslane (*Portulaca oleracea* L.): An alternative source of omega-3 fatty acids. *Plant Foods Hum. Nutr.* **2015**, *70*, 420–426. [CrossRef] [PubMed]
6. Simopoulos, A. Omega-3 fatty acids and antioxidants in edible wild plants. *Biol. Res.* **2004**, *37*, 263–277. [CrossRef]
7. Alam, M.; Juraimi, A.; Rafii, M.; Hamid, A.; Aslani, F.; Hasan, M.; Zainudin, M.; Uddin, M. Evaluation of antioxidant compounds, antioxidant activities, and mineral composition of 13 collected purslane (*Portulaca oleracea* L.) accessions. *Biomed. Res. Int.* **2014**, *2014*, 296063. [CrossRef] [PubMed]
8. Zhou, Y.-X.; Xin, H.-L.; Rahman, K.; Wang, S.-J.; Peng, C.; Zhang, H. *Portulaca oleracea* L.: A Review of phytochemistry and pharmacological effects. *Biomed. Res. Int.* **2015**, *925631*, 1–11.
9. Kumar, A.; Sreedharan, S.; Kashyap, A.; Singh, P.; Ramchiary, N. A review on bioactive phytochemicals and ethnopharmacological potential of purslane (*Portulaca oleracea* L.). *Heliyon* **2022**, *8*, e08669. [CrossRef]
10. Chugh, V.; Mishra, V.; Sharma, K. Purslane (*Portulaca oleracea* L.): An underutilized wonder plant with potential pharmacological value. *Pharm. J.* **2019**, *8*, 236–246.
11. Zhao, R.; Shao, X.; Jia, G.; Huang, Y.; Liu, Z.; Song, B.; Hou, J. Anti-cervical carcinoma effect of *Portulaca oleracea* L. polysaccharides by oral administration on intestinal dendritic cells. *BMC Complement. Altern. Med.* **2019**, *19*, 161. [CrossRef]
12. Gallo, M.; Conte, E.; Naviglio, D. Analysis and comparison of the antioxidant component of *Portulaca oleracea* leaves obtained by different solid-liquid extraction techniques. *Antioxidants* **2017**, *6*, 64. [CrossRef]
13. Vroman, I.; Tighzert, L. Biodegradable polymers. *Materials* **2009**, *2*, 307–344. [CrossRef]
14. Camerona, D.; Shaver, M. Aliphatic polyester polymer stars: Synthesis, properties and applications in biomedicine and nanotechnology. *Chem. Soc. Rev.* **2011**, *40*, 1761–1776. [CrossRef] [PubMed]
15. Tyler, B.; Gullotti, D.; Mangraviti, A.; Utsuki, T.; Brem, H. Polylactic acid (PLA) controlled delivery carriers for biomedical applications. *Adv. Drug Deliv. Rev.* **2016**, *107*, 163–175. [CrossRef]
16. Xue, J.; Wu, T.; Dai, Y.; Xia, Y. Electrospinning and electrospun nanofibers: Methods, materials, and applications. *Chem. Rev.* **2019**, *119*, 5298–5415. [CrossRef] [PubMed]
17. Sill, T.; von Recum, H. Electrospinning: Applications in drug delivery and tissue engineering. *Biomaterials* **2008**, *29*, 1989–2006. [CrossRef]
18. Hong, J.; Yeo, M.; Yang, G.; Kim, G. Cell-electrospinning and its application for tissue engineering. *Int. J. Mol. Sci.* **2019**, *20*, 6208. [CrossRef]
19. Xu, H.; Wu, Z.; Zhao, D.; Liang, H.; Yuan, H.; Wang, C. Preparation and characterization of electrospun nanofibers-based facial mask containing hyaluronic acid as a moisturizing component and huangshui polysaccharide as an antioxidant component. *Int. J. Biol. Macromol.* **2022**, *214*, 212–219. [CrossRef] [PubMed]
20. Bonfim, D.; Cruz, F.; Guerra, V.; Aguiar, M. Development of filter media by electrospinning for air filtration of nanoparticles from PET bottles. *Membranes* **2021**, *11*, 293. [CrossRef]
21. Baji, A.; Agarwal, K.; Oopath, S. Emerging developments in the use of electrospun fibers and membranes for protective clothing applications. *Polymers* **2020**, *12*, 492. [CrossRef]
22. Zhang, C.; Li, Y.; Wang, P.; Zhang, H. Electrospinning of nanofibers: Potentials and perspectives for active food packaging. *Compr. Rev. Food Sci. Food Saf.* **2020**, *19*, 479–502. [CrossRef]
23. Shitole, M.; Dugam, S.; Desai, N.; Tade, R.; Nangare, S. Pharmaceutical applications of electrospun nanofibers: A state-of-the-art review. *Asian J. Pharm. Technol.* **2020**, *10*, 187–201. [CrossRef]
24. Toncheva, A.; Paneva, D.; Maximova, V.; Manolova, N.; Rashkov, I. Antibacterial fluoroquinolone antibiotic-containing fibrous materials from poly(l-lactide-co-d,l-lactide) prepared by electrospinning. *Eur. J. Pharm. Sci.* **2012**, *47*, 642–651. [CrossRef]
25. Akhmetova, A.; Heinz, A. Electrospinning proteins for wound healing purposes: Opportunities and challenges. *Pharmaceutics* **2021**, *13*, 4. [CrossRef] [PubMed]
26. Yang, S.; Kim, E.; Kim, S.; Kim, Y.; Oh, W.; Lee, J.; Jang, Y.; Sabina, Y.; Ji, B.; Yeum, J. Electrospinning fabrication of poly(vinylalcohol)/*Coptis chinensis* extract nanofibers for antimicrobial exploits. *Nanomaterials* **2018**, *8*, 734. [CrossRef] [PubMed]
27. Matthew, R.; Gopi, M.; Menon, P.; Jayakumar, R.; Vijayachandran, L. Synthesis of electrospun silica nanofibers for protein/DNA binding. *Mater. Lett.* **2016**, *184*, 5–8. [CrossRef]
28. Darbasizadeh, B.; Mortazavi, S.; Kobarfard, F.; Jaafari, M.; Hashemi, A.; Farhadnejad, H.; Feyzi-barnaji, B. Electrospun doxorubicin-loaded PEO/PCL core/sheath nanofibers for chemopreventive action against breast cancer cells. *J. Drug Deliv. Sci. Technol.* **2021**, *64*, 102576. [CrossRef]

29. Abid, S.; Hussain, T.; Raza, Z.; Nazir, A. Current applications of electrospun polymeric nanofibers in cancer therapy. *Mater. Sci. Eng. C* **2019**, *97*, 966–977. [CrossRef]
30. Hermosilla, J.; Pastene-Navarrete, E.; Acevedo, F. Electrospun fibers loaded with natural bioactive compounds as a biomedical system for skin burn treatment. A Review. *Pharmaceutics* **2021**, *13*, 2054. [CrossRef]
31. Maleki, H.; Azimi, B.; Ismaeilimoghadam, S.; Danti, S. Poly(lactic acid)-based electrospun fibrous structures for biomedical applications. *Appl. Sci.* **2022**, *12*, 3192. [CrossRef]
32. Maleki, H.; Gharehaghaji, A.A.; Criscenti, G.; Moroni, L.; Dijkstra, P.J. The influence of process parameters on the properties of electrospun PLLA yarns studied by the response surface methodology. *J. Appl. Polym. Sci.* **2015**, *132*, 41388. [CrossRef]
33. Spasova, M.; Paneva, D.; Manolova, N.; Radenkov, P.; Rashkov, I. Electrospun chitosan-coated fibers of poly(L-lactide) and poly(L-lactide)/poly(ethylene glycol): Preparation and characterization. *Macromol. Biosci.* **2008**, *8*, 153–162. [CrossRef] [PubMed]
34. Toncheva, A.; Spasova, M.; Paneva, D.; Manolova, N.; Rashkov, I. Polylactide (PLA)-based electrospun fibrous materials containing ionic drugs as wound dressing materials: A review. *Int. J. Polym. Mater.* **2014**, *63*, 657–671. [CrossRef]
35. Spasova, M.; Stoilova, O.; Manolova, N.; Rashkov, I.; Altankov, G. Preparation of PLLA/PEG nanofibers by electrospinning and potential applications. *J. Bioact. Compat. Polym.* **2007**, *22*, 62–76. [CrossRef]
36. Iulianelli, A.; Russo, F.; Galiano, F.; Desiderio, G.; Basile, A.; Figoli, A. PLA easy fil—white-based membranes for $CO_2$ separation. *Greenh. Gases Sci. Technol.* **2019**, *9*, 360–369. [CrossRef]
37. Chin, Y.; Balunas, M.; Chai, H.; Kinghorn, A. Drug discovery from natural sources. *AAPS J.* **2006**, *8*, E239–E253. [CrossRef]
38. Khan, A.; Xiangyang, S.; Ahmad, A.; Mo, X. Electrospinning of crude plant extracts for antibacterial and wound healing applications: A review. *SM J. Biomed. Eng.* **2018**, *4*, 1024.
39. Spasova, M.; Stoyanova, N.; Manolova, N.; Rashkov, I.; Taneva, S.; Momchilova, S.; Georgieva, G. Facile preparation of novel antioxidant fibrous material based on natural plant extract from *Portulaca oleracea* and polylactide by electrospinning for biomedical applications. *Polym. Int.* **2022**, *71*, 689–696. [CrossRef]
40. Rasband, W.S. ImageJ, US National Institutes of Health, Bethesda, MD (1997–2021). Available online: http://imagej.nih.gov/ij (accessed on 3 January 2016).
41. Spasova, M.; Mincheva, R.; Paneva, D.; Manolova, N.; Rashkov, I. Perspectives on: Criteria for complex evaluation of the morphology and alignment of electrospun polymer nanofibers. *J. Bioact. Compat. Polym.* **2006**, *21*, 465–479. [CrossRef]
42. Christie, W.W. *Lipid Analysis: Isolation, Separation, Identification, and Structural Analysis of Lipids*, 3rd ed.; The Oily Press: Bridgwater, UK, 2003; pp. 205–224.
43. Erkan, N. Antioxidant activity and phenolic compounds of fractions from *Portulaca oleracea* L. *Food Chem.* **2012**, *133*, 775–781. [CrossRef]
44. Elkhayat, E.S.; Ibrahim, S.R.M.; Aziz, M.A. Portulene, a new diterpene from *Portulaca oleracea* L. *J. Asian Nat. Prod. Res.* **2008**, *10*, 1039–1043. [CrossRef]
45. Ferreira da Silva, J.D.; Pedrosa da Silva, S.; Michelle da Silva, P.; Vieira, A.M. Portulaca elatior root contains a trehalose-binding lectin with antibacterial and antifungal activities. *Int. J. Biol. Macromol.* **2019**, *126*, 291–297. [CrossRef] [PubMed]
46. Du, Y.-K.; Liu, J.; Li, X.-M.; Pan, F.-F.; Zhi-Guo Wen, Z.-G.; Zhang, T.-C.; Yang, P.-L. Flavonoids extract from *Portulaca oleracea* L. induce Staphylococcus aureus death by apoptosis-like pathway. *Int. J. Food Prop.* **2017**, *20*, S534–S542. [CrossRef]
47. Toncheva, A.; Paneva, D.; Manolova, N.; Rashkov, I.; Mita, L.; Crispi, S.; Mita, D. Dual vs. single spinneret electrospinning for the preparation of dual drug containing non-woven fibrous materials. *Colloids Surf. A Physicochem. Eng. Asp.* **2013**, *439*, 176–183. [CrossRef]
48. Ramier, J.; Bouderlique, T.; Stoilova, O.; Manolova, N.; Rashkov, I.; Langlois, V.; Renard, E.; Albanese, P.; Grande, D. Biocomposite scaffolds based on electrospun poly(3-hydroxybutyrate) nanofibers and electrosprayed hydroxyapatite nanoparticles for bone tissue engineering applications. *Mater. Sci. Eng. C* **2014**, *38*, 161–169. [CrossRef] [PubMed]
49. Ramier, J.; Grande, D.; Bouderlique, T.; Stoilova, O.; Manolova, N.; Rashkov, I.; Langlois, V.; Albanese, P.; Renard, E. From design of bio-based biocomposite electrospun scaffolds to osteogenic differentiation of human mesenchymal stromal cells. *J. Mater. Sci. Mater. Med.* **2014**, *25*, 1563–1575. [CrossRef] [PubMed]

*membranes* MDPI

*Article*

# Electrochemical and Electroconductive Behavior of Silk Fibroin Electrospun Membrane Coated with Gold or Silver Nanoparticles

**Wilson Agudelo [1], Yuliet Montoya [1,2,*], Alejandra Garcia-Garcia [3], Adriana Restrepo-Osorio [4] and John Bustamante [1,2]**

1 Grupo de Dinámica Cardiovascular, Línea Ingeniería de Tejidos y Protésica Cardiovascular, Universidad Pontificia Bolivariana, Medellín 050031, Colombia
2 Comité de Trabajo de Bioingeniería Cardiovascular, Sociedad Colombiana de Cardiología y Cirugía Cardiovascular, Bogotá 110121, Colombia
3 Grupo de Síntesis y Modificación de Nanoestructuras y Materiales Bidimensionales, Centro de Investigación en Materiales Avanzados S.C., Parque PIIT, Km 10, Autopista Monterrey-Aeropuerto, Apodaca 66628, Mexico
4 Grupo de Investigación sobre Nuevos Materiales, Universidad Pontificia Bolivariana, Medellín 050031, Colombia
* Correspondence: yuliet.montoya@upb.edu.co

**Abstract:** The surface modification of materials obtained from natural polymers, such as silk fibroin with metal nanoparticles that exhibit intrinsic electrical characteristics, allows the obtaining of biocomposite materials capable of favoring the propagation and conduction of electrical impulses, acting as communicating structures in electrically isolated areas. On that basis, this investigation determined the electrochemical and electroconductive behavior through electrochemical impedance spectroscopy of a silk fibroin electrospun membrane from silk fibrous waste functionalized with gold or silver nanoparticles synthetized by green chemical reduction methodologies. Based on the results obtained, we found that silk fibroin from silk fibrous waste ($SF_W$) favored the formation of gold (AuNPs-$SF_W$) and silver (AgNPs-$SF_W$) nanoparticles, acting as a reducing agent and surfactant, forming a micellar structure around the individual nanoparticle. Moreover, different electrospinning conditions influenced the morphological properties of the fibers, in the presence or absence of beads and the amount of sample collected. Furthermore, treated $SF_W$ electrospun membranes, functionalized with AuNPs-$SF_W$ or AgNPS-$SF_W$, allowed the conduction of electrical stimuli, acting as stimulators and modulators of electric current.

**Keywords:** electrochemical impedance spectroscopy; silk fibroin; metal nanoparticles; electrospun membranes; impregnation coatings

**Citation:** Agudelo, W.; Montoya, Y.; Garcia-Garcia, A.; Restrepo-Osorio, A.; Bustamante, J. Electrochemical and Electroconductive Behavior of Silk Fibroin Electrospun Membrane Coated with Gold or Silver Nanoparticles. *Membranes* **2022**, *12*, 1154. https://doi.org/10.3390/membranes12111154

Academic Editor: Andrea Ehrmann

Received: 4 October 2022
Accepted: 3 November 2022
Published: 16 November 2022



## 1. Introduction

The recent integration of nanotechnology into the biomedical field has made it possible to make use of the physicochemical properties inherent to materials on a nanometric scale to propose new therapeutic methodologies, such as the controlled release of drugs [1], tissue repair [2], techniques for cell diagnosis, and therapy [3]. Moreover, it is necessary to carry out processes that allow the obtaining of nanostructures with controlled size and shape, high purity in environmentally friendly conditions, and with low levels of toxicity [2,3]. To achieve this, the use of functionalization processes is required, generally with elements of natural origin, giving rise to biocomposite materials.

In the search for a biomaterial that mimics functional behavior and has a favorable biocompatible response, different polymers of natural origin have been studied, among which is silk fibroin (SF), a protein extracted from different families of silkworms such as Bombyx mori and of some arthropods such as spiders. This protein has exhibited biocompatibility, controllable biodegradability, and superior mechanical properties to those of other natural biopolymers such as collagen [4,5], so SF is considered a material of interest for use in the field of tissue engineering. Moreover, it can be electrospun under specific conditions.

Nowadays, the incorporation of inorganic nanoparticles with biocompatible and electroconductive characteristics in the functionalization of materials with application to tissue engineering seeks to provide controllable electrical properties that do not interfere with the electrical conductance of the action potential, and that in turn allows for modulating the geometry and topography of the material to mimic the morphological characteristics of the native extracellular matrix [6,7]. In view of this, it is necessary to carry out processes that allow obtaining nanostructures with controlled size and shape, high purity, in environmentally friendly conditions, and with low levels of cytotoxicity [2,3]. To achieve this, the use of functionalization processes is required, generally with elements of natural origin, giving rise to biocomposite materials [8,9].

On the other hand, to determine the electro-conductive characteristics of biocomposite materials, several techniques can be used, among which is electrochemical impedance spectroscopy, which allows the electrical properties of materials to be analyzed at different conditions, obtaining frequency spectra of impedance, conductivity, and permittivity. These are adjusted by mathematical models that allow for elucidating the electrical behavior (resistor, inductor, capacitor, constant phase elements, among others) [10].

In this context, and in order to propose an electrospun membrane that exhibits electrochemical and electroconductive characteristics, the research purpose was the development of a biocomposite electrospun membrane with electro-conductive properties based on a natural polymeric structure of silk fibroin from silk fibrous waste, coated with gold or silver nanoparticles, which were synthesized from a chemical reduction with the same $SF_w$ in one step. This biocomposite could be used as a therapeutic strategy for tissue engineering, like an electroconductive scaffold that facilitates ionic interaction between isolated electric areas.

## 2. Materials and Methods

For this investigation, gold trichloride hydrochloride ($HAuCl_4$ $3H_2O$), lithium bromide (LiBr, ReagentPlus, ≥99%), sodium carbonate ($Na_2CO_3$, ReagentPlus), and polyethylene oxide (PEO, Mv ~ 900,000) were obtained from (Sigma Aldrich, St. Louis, MO, USA). Sodium hydroxide (NaOH pellets), methanol for analysis (EMSURE ACS, ISO, Reag. Ph Eur), potassium chloride (EMSURE), potassium dihydrogen phosphate (EMSURE ISO), sodium phosphate, dibasic, heptahydrate (EMSURE ACS), and sodium bicarbonate (EMSURE ISO) were obtained from (Merck Millipore, Darmstadt, Germany). Silver nitrate ($AgNO_3$ pure, pharma grade) and sodium chloride (NaCl pure, pharma grade) were obtained from (PanReac Applichem, Darmstadt, Germany). The NIH/3T3 fibroblast cell line was purchased from (American Type Culture Collection (ATCC), Manassas, VA, USA) and Dulbecco's Modified Eagle's Medium was obtained from (Lonza Bioscience, Durham, NC, USA). Fetal bovine serum was purchased from Microgen, trypan blue was obtained from (Loba Chemie Mumbai, India), and 3-(4,5-dimethylthiazol-2-yl)-2,5-diphenyltetrazolium bromide (MTT) was acquired from (Alfa Aesar, Haverhill, MA, USA).

### 2.1. Silk Fibroin from Silk Fibrous Waste Extraction ($SF_w$)

Silk fibrous wastes from the Bombyx mori silkworm acquired from Corporación para el Desarrollo de la Sericultura del Cauca—CORSEDA (Popayán, Cauca, Colombia) were used. The silk fibers were degummed by immersion in aqueous $Na_2CO_3$ solution. Then, it was dissolved in an aqueous LiBr solution and dialyzed until reaching a stable conductivity, following the protocol reported by Jaramillo-Quiceno et al. [11].

### 2.2. Synthesis of Gold and Silver Nanoparticles

$SF_w$ to 0.5% $v/v$ was mixed with each of the precursor solutions of 2.5 mM $HAuCl_4$ and 10 mg/mL $AgNO_3$. Then, each of the solutions was incubated in light at 60 W for 24 h.

### 2.3. Preparation and Treatment of Silk Fibroin Electrospun Membranes

A $SF_w$/PEO homogeneous solution was electrospun using a flow of 0.7 mL/h and voltage of 16 kV, volumetric ratio of $SF_w$:PEO of 50:50, and a needle-collector distance between 10 cm to 25 cm, with a flat plate type static collector. Then, electrospun $SF_w$/PEO membranes were treated in methanol for 10 min and 15 min and taken under vacuum for 24 h to promote the evaporation of the solvent. Finally, they were washed with deionized water at 37 °C for 48 h to remove the PEO, and finally, they were dried at room temperature for 24 h.

### 2.4. Impregnation Coatings of Treated Electrospun Membranes with AuNPs-$SF_w$ o AgNPs-$SF_w$

Treated membranes were coated with AuNPs-$SF_w$ or AgNPs-$SF_w$ solutions by impregnation coatings for periods of 4 h and 24 h. Then, one sample for each immersion period and type of nanoparticle was washed with deionized water, while the other samples were not washed. Finally, all membranes were dried at room temperature.

### 2.5. UV-Visible Spectrophotometry and FTIR Spectroscopy Analysis

Gold and silver nanoparticles, as well as the treated and functionalized membranes, were analyzed in a UV-visible spectrophotometer (UV-Vis-NIR Cary 5000, Agilent Technologies, Santa Clara, CA, USA) and (Lambda Bio 10, Perkin Elmer, Waltham, MA, USA) in a wavelength range between (1100–200) nm. As controls, 0.5% $SF_w$ solutions and membranes treated without functionalizing with the nanoparticles were used. To determine changes in the functional groups of untreated, treated, and functionalized membranes, a (Nicolet iS50 FTIR-ATR spectrometer, Thermo Fisher Scientific, Waltham, MA, USA) at a resolution of 4 $cm^{-1}$ and 32 scans was used.

### 2.6. Scanning Electron (SEM) and Field Emission (FESEM) Microscopy Analysis

Gold and silver nanoparticles, untreated, treated, and functionalized membranes were analyzed on a (Nova NanoSEM 200, Hillsboro, OR, USA) field emission scanning electron microscope and on a scanning electron microscope (NeoScope JCM-6000 Plus, Jeol, Akishima, Japan) operated at 15 kV. The micrographs were analyzed with Fiji® software developed by Schindelin et al. [12] for the diameters and distribution histograms' particles and fibers.

### 2.7. Electrochemical Impedance Spectroscopy Analysis

The electrochemical impedance spectroscopy (EIS) configuration consisted of an arrangement of three electrodes where the electrospun membranes were arranged on a graphite bar, which acted as a working electrode ($WE_G$) with an exposed area of 5.8 $cm^2$. Moreover, a graphite rod with a diameter of 5.2 mm was used as a counter electrode ($CE_G$), an Ag/AgCl electrode as a reference (RE) and Hank's Balanced Salt solution as a working solution. For linear sweep voltammetry measurements, a potentiostat (Ivium CompactStat.h, Ivium Technologies BV, Eindhoven, Netherlands) with scanning rates of 100 mV/s in a potential range between (0–1) V was used, using an Ag/AgCl reference electrode with saturated KCl solution. For electrochemical impedance measurements, these were performed at a sweep frequency of $1 \times 10^6$ Hz to 0.1 Hz with 10 points for each decade, with an AC potential with 0.08 V amplitude and a range of DC potentials between 0 V and 1 V with a step of 400 mV between measurements.

### 2.8. MTT Assay

The cytotoxicity of the membranes without or functionalized with the gold or silver nanoparticles was measured with the MTT test, which allows for determining cell viability based on the mitochondrial activity of the cells from the direct interaction with the membranes. For this assay, $4.5 \times 10^3$ cells/well of 3T3 fibroblasts were seeded in 96-well plates. Then, the membranes interacted with the 3T3 fibroblasts for 24 h. After MTT was added during (4–5) h, then it was treated with isopropanol to dissolve the formazan crystals. Later, it was left in incubation at 37 °C and the absorbance was measured at 570 nm using a spectrophotometer microplate reader (Multiskan FC, Thermo Fisher Scientific, Waltham, MA, USA).

### *2.9. Statistical Analysis*

The synthesis of the gold and silver nanoparticles, the UV-visible measurements, and the electrochemical impedance spectroscopy were performed in triplicate at three independent times. In addition, a multivariate statistical analysis was performed using the Statgraphics Centurion XVI software (Statgraphics Technologies, The plains, VA, USA) which allowed obtaining the population mean, the standard deviation, and the 95% confidence intervals. For the statistical analysis of the particle and fiber size of the SEM and STEM micrographs, the Fiji® software was used, obtaining the mean of the population, the standard deviation, and the 95% confidence intervals.

## 3. Results and Discussion

### *3.1. Synthesis of Gold and Silver Nanoparticles*

In the analysis by UV-Vis spectrophotometry (Figure 1a), the presence of absorption bands corresponding to the surface plasmon resonance (SPR) of each metal ion at synthesized solutions were located at wavelengths of 526 nm for AuNPs-$SF_w$ and 439 nm for AgNPs-$SF_w$.

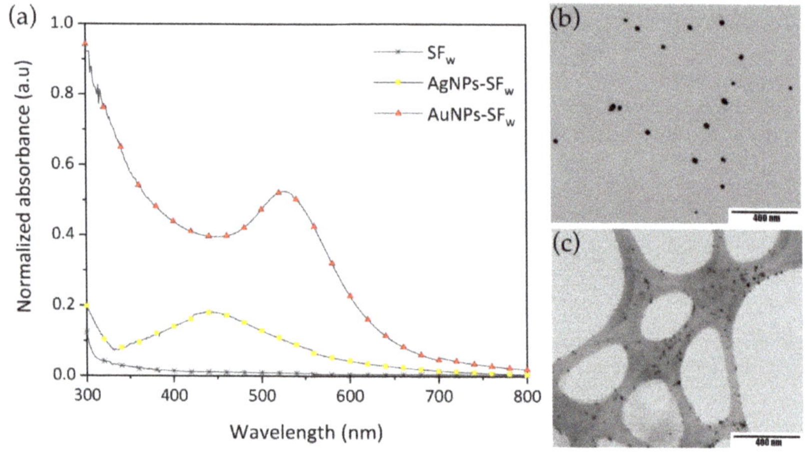


**Figure 1.** (**a**) Normalized UV-Vis absorption spectra of gold (line with red triangles) and silver (line with yellow squares) nanoparticles. STEM micrographs of (**b**) AuNPs-$SF_w$ and (**c**) AgNPs-$SF_w$. Scale bar, 400 nm.

The presence of tyrosine residuals in the $SF_w$ structure are responsible for the process of formation of gold and silver nanoparticles, which, based on the pH conditions in which the synthesis reaction takes place, establishes the mechanisms of particle formation and stability. In the case of gold nanoparticles, when presenting pH adjustment, the reaction mechanism is mediated by the electrons transfer through the deprotonation of oxygen and the formation of tyrosinase ion. However, for silver nanoparticles, as they do not present a pH adjustment, a reaction mediated by the transfer of electrons coupled to protons is generated [13–16].

From the analysis of the STEM micrographs of the metal nanoparticles (Figure 1b,c), it was found that the AuNPs-$SF_w$ present particles with an average size (23 ± 6) nm, while AgNPs-$SF_w$ exhibited average sizes of (14 ± 8) nm. The changes in particle size for each type of nanostructure are related to the final value of the pH and the concentration of the reducing agent, which in each case results in the reaction pathway that each synthesis follows and the way in which gold or silver ions are coupled in the nucleation and growth processes [15,17,18].

### *3.2. Preparation of Silk Fibroin Electrospun Membranes*

To obtain the silk fibroin electrospun membranes, PEO was used as a modifying element of the electrospinning conditions [19]. From the results obtained, for the 50:50 $SF_W$/PEO ratio (Figure 2a–d), the decrease in protein concentration caused an increase in viscosity that favored the polymer chains to overcome the surface tension, resulting in uniform fibers without the presence of defects [20].

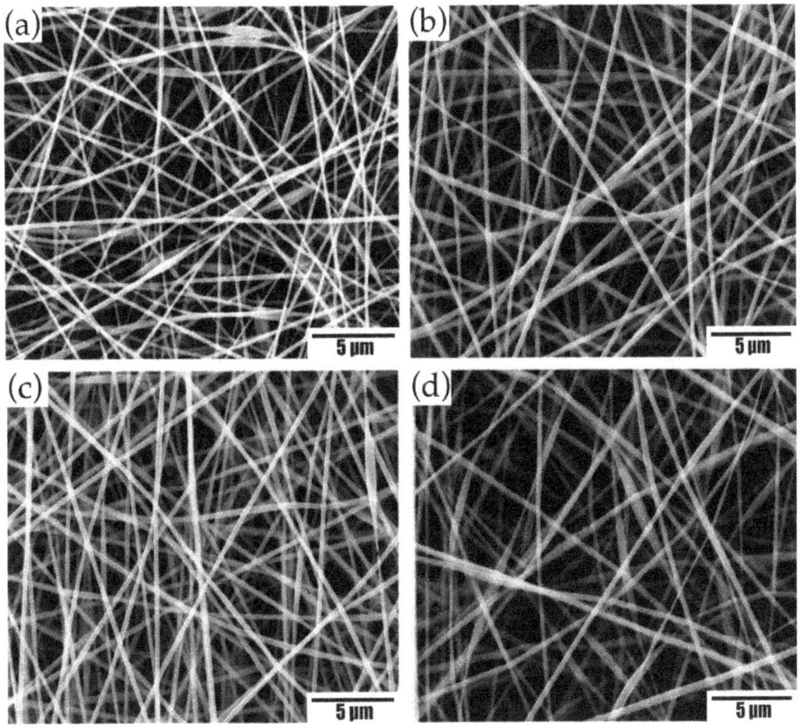


**Figure 2.** SEM micrographs of SFw/PEO electrospun membranes obtained at a volumetric ratio of 50:50, at distances of (**a**) 10 cm; (**b**) 15 cm; (**c**) 20 cm; (**d**) 25 cm. Scale bar, 5 µm.

The effect of the collection distance was evaluated. The membranes obtained at 50:50 $SF_W$/PEO had fiber sizes of (194 ± 39) nm, (296 ± 47) nm, (264 ± 30) nm, and (281 ± 57) nm for distances of 10 cm, 15 cm, 20 cm, and 25 cm, respectively.

Furthermore, it was observed that the electrospinning distance has an influence on the sample collection area, where at distances of 10 cm and 15 cm, a circular geometry was observed without completely covering the collector, while, at distances of 20 cm and 25 cm, the collector was completely covered. In view of this, the difference in the collection area of the fibers is related to the projection of these on the collector, which decreases as the distance between the tip of the needle and the substrate decreases, forming a circular collection area [21].

According to the results, $SF_W$/PEO electrospun membranes at a volumetric ratio of 50:50 and (20–25) cm were used, conditions that favored obtaining membranes with continuous fibers, without beads and a larger collection area.

### 3.3. Treatment of Electrospun Membranes

The use of organic solvents, such as ethanol and methanol as treatments in the silk fibroin electrospun membranes, reduces the percentage of non-crystalline structures of the amorphous phase present in the protein and decrease the solubility of the membrane in an aqueous medium. The use of these solvents favors the transition of fibroin macromolecules and their non-crystalline secondary structures to crystalline ones in the form of β-bends, increasing the crystalline phase of the membranes and making them hydrophobic [22,23].

To avoid the loss of material due to the solubilization of the membranes in aqueous media, the dehydration due to long periods of immersion, and remove the higher content of synthetic polymer, $SF_w$/PEO electrospun membranes at a volumetric ratio of 50:50 and a distance of (20–25) cm were treated in methanol immersion for periods of 10 min and 15 min with subsequent washing in deionized water for 48 h. Based on the results obtained, changes in fiber sizes and rough morphologies were evidenced in the micrographs (Figure 3a–d), caused by the phase separation between $SF_w$ and PEO extraction [24]. It was observed that the apparent porosity of the membranes is reduced by the increase in the size and thickening of the fibers, which could contribute to greater connectivity between fibers.

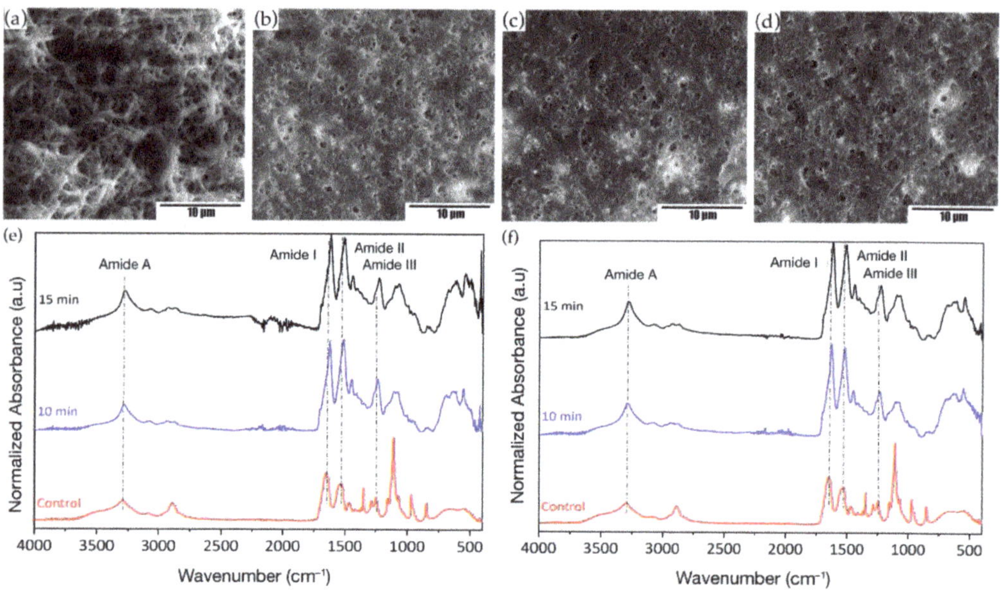


**Figure 3.** SEM micrographs of $SF_w$/PEO electrospun membranes at a volumetric ratio of 50:50 treated with methanol at distances of 20 cm, for (**a**) 10 min; (**b**) 15 min. The distance of 25 cm, during (**c**) 10 min; (**d**) 15 min. Scale bar, 10 μm. Infrared absorption spectra of electrospun $SF_w$/PEO membranes at a volumetric ratio of 50:50 treated with methanol for 10 min and 15 min. Where (**e**) 20 cm; (**f**) 25 cm. Untreated $SF_w$/PEO electrospun membranes were used as controls.

Concerning the functional changes that the electrospun membranes suffered from methanol, the analysis by infrared spectroscopy was performed, which were compared with the control spectrum of pristine $SF_w$ and membranes without treatment. From the results obtained (see Figure 3e–f), it was found that the $SF_w$ spectrum showed the characteristic peaks of the vibrational modes of the amides A (N-H stretching), I (C=O stretching), II (N-H bending and C-N stretching), and III (C-C stretch, C-N, and C-H bend) at wavenumbers of 3274 $cm^{-1}$, 1635 $cm^{-1}$, 1515 $cm^{-1}$, and 1230 $cm^{-1}$, respectively. In the untreated $SF_w$/PEO membranes, these bands were located at wavenumbers of 3282 $cm^{-1}$, 1640 $cm^{-1}$, 1528 $cm^{-1}$, and 1241 cm$^{-1}$, which could indicate that electrospinning processes enhance the ability of the protein to self-organize its structure, similarly to native silk I (Table 1) [24].

**Table 1.** Wavenumbers for the main amide vibrational modes of the pristine SF$_W$ spectra, electrospun membranes without and with methanol treatment for 10 min and 15 min.

| Sample | Vibrational Modes | | |
|---|---|---|---|
| | Amide I (cm$^{-1}$) | Amide II (cm$^{-1}$) | Amide III (cm$^{-1}$) |
| SF$_W$ | 1635 | 1515 | 1230 |
| F50 [1]d20 [2]—control | 1640 | 1528 | 1241 |
| F50d20—10 min | 1620 | 1509 | 1230 |
| F50d20—15 min | 1621 | 1515 | 1227 |
| F50d25—control | 1640 | 1528 | 1241 |
| F50d25—10 min | 1621 | 1513 | 1230 |
| F50d25—15 min | 1620 | 1511 | 1229 |

[1] FXX—Ratio of SF$_W$ used in the solution. [2] dXX—Fiber collection distance.

Concerning the bands located at wavenumbers at 1100 cm$^{-1}$, 962 cm$^{-1}$, and 841 cm$^{-1}$ in the absorption spectrum of the untreated SF$_W$/PEO electrospun membranes, these corresponded to the stretching vibrations of the CO and bending of the CH out of the plane for polyethylene oxide [25,26], which after washing reduced in absorption intensity, indicating the removal of PEO from the electrospun membrane.

Regarding the changes in the fibroin structure, it was observed in the infrared absorption spectra of the treated electrospun membranes with methanol, the displacement of the bands associated with the characteristic vibrational modes of the protein, which were located at ~1620 cm$^{-1}$, ~1511 cm$^{-1}$, and 1228 cm$^{-1}$ (Table 1). These displacements indicate the transition of the structures corresponding to random spirals to secondary structures in the form of β-sheets and β-turns due to the rearrangement induced by methanol to the SF$_W$ chains and the formation of different hydrogen bonds [27,28].

Regarding the difference in the percentage of secondary crystalline structures in the form of β-sheets and β-turns varying the treatment times with methanol, it was found that immersion periods of 15 min increased the number of crystalline structures for both membrane types compared to being treated at 10 min immersion periods and untreated controls (Table 2).

**Table 2.** Secondary structures present in the vibrational mode of amide I for treated and untreated electrospun membranes with methanol for 10 min and 15 min. SF$_W$ films were used as control.

| Structure Type | SF$_W$ | SF$_W$/PEO (Untreated) | | F50d20 | | F50d25 | |
|---|---|---|---|---|---|---|---|
| | | d20 | d25 | 10 min | 15 min | 10 min | 15 min |
| β-sheets (strong, weak, intra and intermolecular) | 28% | 18% | 12% | 56% | 64% | 46% | 55% |
| β-turns | - | - | - | 33% | 2% | - | 32% |
| Side chains (Tyr) | 5% | 2% | 2% | 7% | 8% | 7% | 7% |
| Random coils | 24% | 45% | 51% | - | 8% | 37% | - |
| α-helix | 39% | - | - | - | 18% | - | - |
| turns | 4% | 35% | 35% | 4% | 1% | 10% | 5% |

Based on the above, it was found when that performing treatments with organic solvents such as methanol, the crystalline phase of SF$_W$ electrospun membranes increases due to the transition of macromolecules and non-crystalline to crystalline secondary structures, favoring the membrane's hydrophobicity with aqueous solutions [22,23].

### 3.4. Functionalization of Treated Membranes with Metal Nanoparticles

For the functionalization of treated membranes with gold or silver nanoparticles using impregnation coating, the membranes treated with methanol for 15 min were used, since these were those that had a higher content of crystalline structures. The samples were immersed in solutions with AuNPs-$SF_W$ and AgNPs-$SF_W$ for 4 h and 24 h. From the obtained results, a change in color of the membranes was evidenced, going from white to an intense red in the case of those functionalized with AuNPs-$SF_W$ and to a light yellow-brown for the membranes functionalized with AgNPs-$SF_W$. This change in tonality is related to the coupling of the different nanostructures on the surface of the electrospun material.

Analysis of the unwashed membranes was performed by field emission microscopy, where the micrographs showed that the membranes functionalized with AuNPs-$SF_W$ or AgNPs-$SF_W$ without washing after the interaction time presented a greater deposition of nanoparticles in the fibers, in comparison with the membranes that were washed after the exposure time with the metal nanoparticles, which was evidenced by the presence of bright areas, red for the AuNPs-$SF_W$, and yellow for the AgNPs-$SF_W$ (Figure 4a–h). Such behavior is due to the solubility of AuNPs-$SF_W$ or AgNPs-$SF_W$ in aqueous solvents, which, when washed after the functionalization process, were removed from the membrane surface. On the other hand, and related to the functionalization times, it was evidenced that at longer immersion times there was a greater deposition of nanoparticles on the surface of the treated membranes, which could be related to a favoring of the coupling nanoparticle-membrane product of electrostatic bonding between amino groups present on the surface of electrospun membranes and functional groups exposed in the coating of gold and silver nanostructures by silk fibroin [29].

Moreover, the analysis by UV-Vis spectrophotometry of the $SF_W$ membranes functionalized with the gold or silver nanoparticles was performed to detect the characteristic resonance plasmon for each type of nanostructure present on the surface of the membranes, where it was found that for the case of the membranes functionalized with AuNPs-SFw, absorption bands were presented at an approximate wavelength of 530 nm (Figure 4i), while for the membranes functionalized with AgNPs-SFw it was not possible to obtain the characteristic resonance band for silver nanoparticles (Figure 4j), which could indicate that there was a greater deposition of gold nanoparticles than silver on the membrane surface.

Regarding the changes in the functional groups of the membranes with gold or silver nanoparticles, it was evidenced that from the IR absorption spectra for each of the samples (Figure 4k,l) that no significant changes occurred with respect to the treated membranes, in the position of the characteristic bands of the vibrational modes of amides I, II, and III located at wavenumbers between (1623–1619) $cm^{-1}$, (1520–1511) $cm^{-1}$, and (1232–1227) $cm^{-1}$, respectively. When determining the percentage of crystalline and non-crystalline secondary structures, an increase in random coils, turns, and $\alpha$-helix was found in the functionalized membranes with AuNPs-$SF_W$ or AgNPs-$SF_W$ compared to the treated membranes without functionalization (Table 3), which is related to the contribution of amorphous secondary structures present in the protein used in the synthesis of the metal nanoparticles.

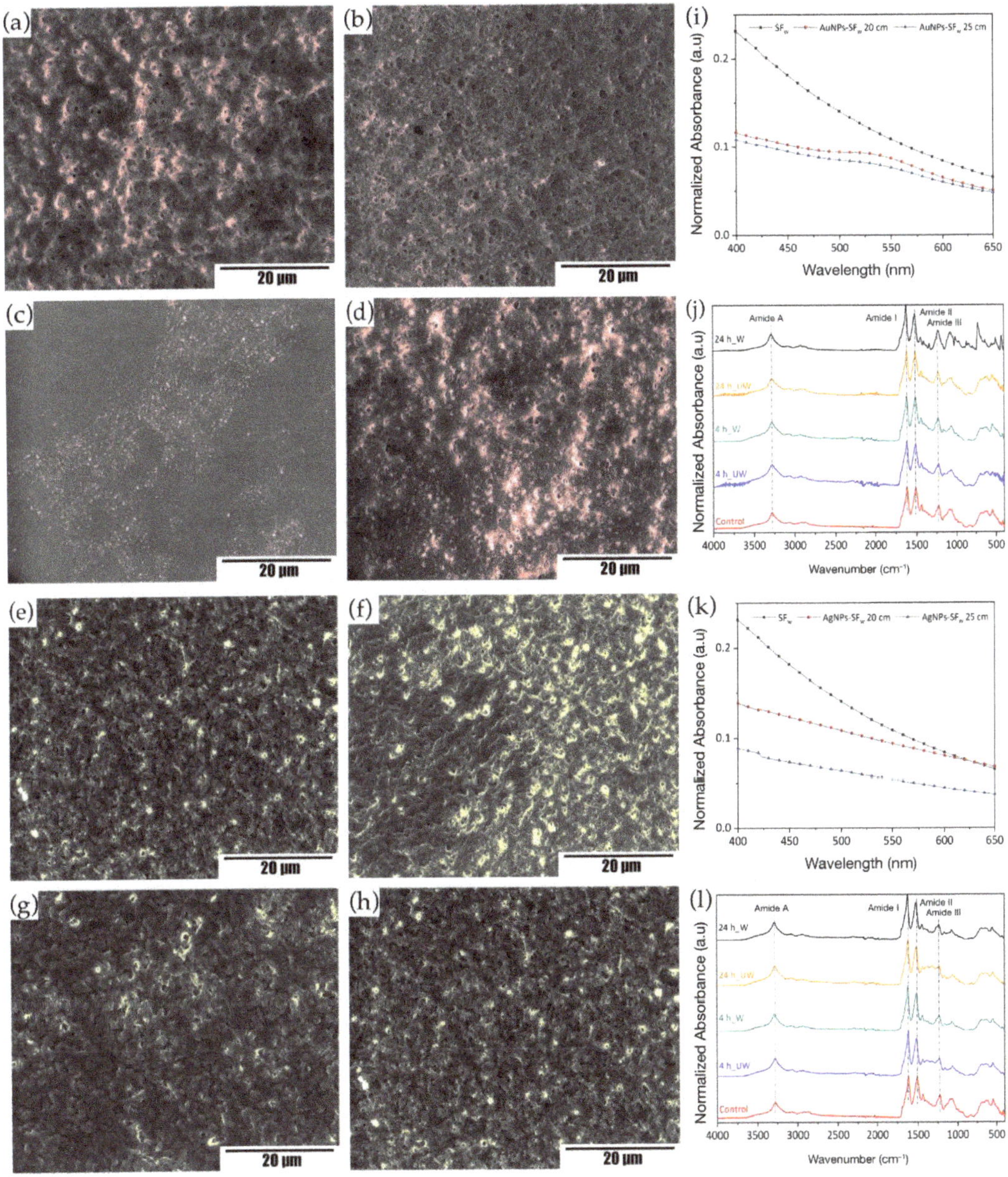


**Figure 4.** STEM micrographs of treated $SF_W$ membranes and functionalized by impregnation coating for 4 h and 24 h with and without washing. Where (**a**,**b**) with AuNPs-$SF_W$ with washing; (**c**,**d**) with AuNPs-$SF_W$ without washing; (**e**,**f**) with AgNPs-$SF_W$ with washing; (**g**,**h**) with AgNPs-$SF_W$ without washing. Scale bar, 20 µm. UV-Visible absorption spectra of treated and functionalized $SF_W$ membranes. Where (**i**) AuNPs-$SF_W$ and (**k**) AgNPs-$SF_W$. A $SF_W$ membrane without impregnation coating was used as a control. FTIR-ATR absorption spectra obtained from treated and functionalized $SF_W$ electrospun membranes. Where (**j**) AuNPs-$SF_W$ and (**l**) AgNPs-$SF_W$. Non-functionalized treated $SF_W$ membranes were used as controls. Where UW—unwashed and W—washed.

**Table 3.** Secondary structures present in the vibrational mode of amide I for the treated membranes and functionalized with AuNPs-$SF_w$ and AgNPs-$SF_w$ for 4 h and 24 h. Non-functionalized treated $SF_w$ membranes were used as controls.

| Structure Type | $SF_w$ (Control) | AuNPs-$SF_w$ | | | | AgNPs-$SF_w$ | | | |
|---|---|---|---|---|---|---|---|---|---|
| | | UW | | W | | UW | | W | |
| | | 4 h | 24 h | 4 h | 24 h | 4 h | 24 h | 4 h | 24 h |
| β-sheets (strong, weak, intra and intermolecular) | 55% | 52% | 60% | 52% | 55% | 48% | 47% | 46% | 46% |
| β-turns | 32% | 0% | 2% | 38% | 1% | - | - | - | - |
| Side chains (Tyr) | 7% | 4% | 8% | 5% | 2% | 5% | 4% | 4% | 4% |
| Random coils | - | 4% | 11% | - | 26% | 39% | 34% | 37% | 38% |
| α-helix | - | 35% | 12% | - | 1% | - | - | - | - |
| Turns | 5% | 4% | 8% | 5% | 16% | 8% | 14% | 13% | 12% |

Where UW—unwashed and W—washed.

### 3.5. Linear Sweep Voltammetry

Linear sweep voltammetry consists of the measurement of the faradic current between the $WE_G$ and RE as a result of the oxidation-reduction processes of an analyte by applying a change in the electric potential at the working electrode in relation to the fixed potential of the reference electrode [30]. The treated non-functionalized and functionalized with gold nanoparticles or silver membranes were positioned on the $WE_G$, and a graphite counter electrode ($CE_G$) was used to complete the electrical circuit.

From the results obtained, it was found that, for the treated membranes that were electrospun at 20 cm, the magnitude of the electric current was lower compared to that obtained for the $WE_G$. On the other hand, the membranes functionalized with AgNPs-$SF_w$ or AuNPs-$SF_w$ presented a decrease in the value of the faradic current with respect to that of the non-functionalized membranes and when comparing the functionalized membranes, it was found that those containing AgNPS-$SF_w$ exhibited a higher current compared to those functionalized with AuNPs-$SF_w$ (Figure 5a). Such behavior is related to the number of species that were oxidized on the surface of the electrode, which is controlled by the standard oxidation potentials ($E^0$) of the ionic species of $Cl^-$ and $H_2O$ molecules present in the electrophysiological solution (Equations (1) and (2)). However, the potential increased when approaching the $E^0$ of the species present in equilibrium, and the current recorded for the graphite electrode, and the product of the exchange of electrons from the oxidized species towards the surface of this, progressively increased. Furthermore, the absence of oxidation peaks for $Cl^-$ and $H_2O$ in the linear scanning voltammogram is due to the presence of unoxidized species in the vicinity of the electrode surface that can increase the value of the faradic current until reaching a maximum peak and subsequent decrease in current [31].

$$2Cl^-_{(aq)} \rightarrow Cl_2 + 2e^- (E^0 \cong +1.36 \text{ V}) \quad (1)$$

$$2H_2O_{(l)} \rightarrow O_{2(g)} + 4H^+ + 4e^- \left(E^0 \cong +1.23 \text{ V}\right) \quad (2)$$

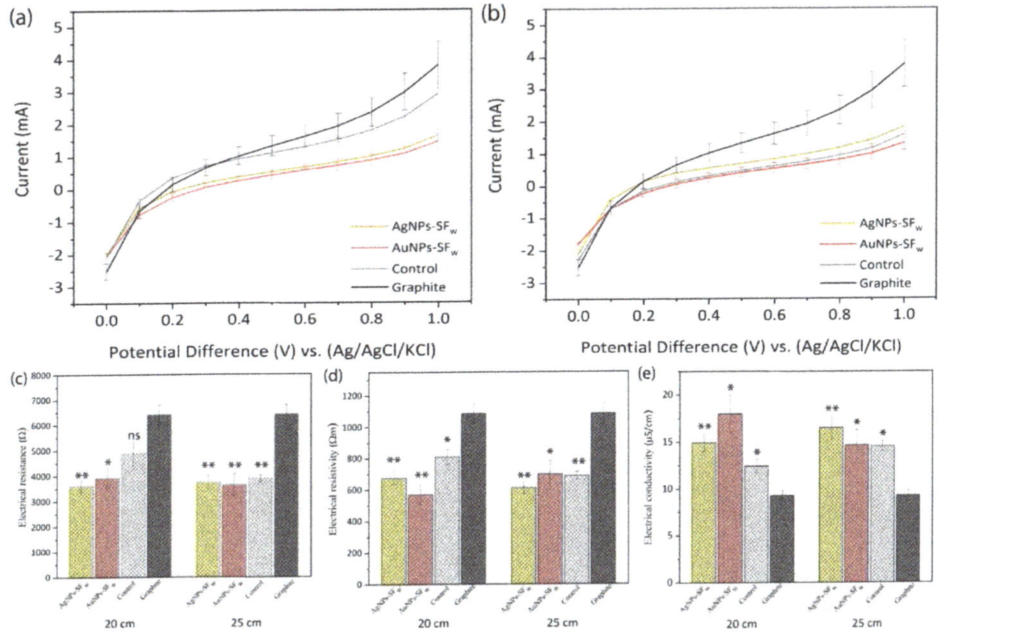


**Figure 5.** Linear scanning voltammograms were obtained in Hank's Balanced Salt solution at 100 mV/s. Where (**a**) $SF_W$ membranes obtained at 20 cm and functionalized with gold and silver nanoparticles and (**b**) $SF_W$ membranes obtained at 25 cm and functionalized with metal nanoparticles. Non-functionalized treated $SF_W$ membranes were used as controls. Electrical response of treated $SF_W$ membranes at distances of 20 cm and 25 cm with and without functionalization with AuNPs-$SF_W$ or AgNPs-$SF_W$. Where (**c**) resistivity; (**d**) resistance; (**e**) conductivity. * indicates statistical significance with $p$-values < 0.05 and ** $p$-values < 0.001.

On the other hand, the decrease in the faradic current registered for the $SF_W$-treated membranes is related to the lack of electrical coupling between the graphite electrode and the membrane, due to the absence of functional groups present on the surface of the $WE_G$, which favor the binding through electrostatic interactions with the functional groups exposed on the surface of the membranes that could facilitate a greater diffusion of electrons generated in oxidation reactions and the favorable response of the electric current [32,33]. Regarding the electrical behavior evidenced in the membranes functionalized with AuNPs-$SF_W$ or AgNPs-$SF_W$ apart from the oxidation reactions of the $Cl^-$ and $H_2O$ species, two additional reactions related to the oxidation of the metal nanoparticles were presented, which are mediated by the presence of chlorine ions in the solution that leads to the formation of silver chlorides for AgNPs-$SF_W$ (Equation (3)) and formation of Au (I) and Au (III) chlorides for AuNPs-$SF_W$ (Equations (4) and (5)) [34,35].

$$Ag_{(s)} + Cl^-{}_{(aq)} \rightarrow AgCl + e^- (E^0 \cong +0.22\text{ V}) \tag{3}$$

$$Au_{(s)} + 2Cl^-{}_{(aq)} \rightarrow AuCl_2{}^- + e^- (E^0 \cong +1.15\text{ V}) \tag{4}$$

$$Au_{(s)} + 4Cl^-{}_{(aq)} \rightarrow AuCl_4{}^- + 3e^- (E^0 \cong +1.00\text{ V}) \tag{5}$$

The decrease in the value of the electric current is related to the lack of electrical coupling between the metal nanoparticles and the surface of the electrospun membranes, where the functionalization mediated by the impregnation coating generates a thin film on the material, which does generate modifications in the exposed functional groups, which causes the electrons generated by the oxidation of the gold or silver nanoparticles arranged along the film to not diffuse through the material and reach the $WE_G$, generating an increase in the value of the registered current and obtaining an oxidation peak for the case of AgNPs-$SF_w$. Moreover, the difference in the faradic current registered for the treated membranes and functionalized with gold or silver nanoparticles correlates with the standard oxidation potential, which is lower for AgNPs-$SF_w$ compared to oxidation to Au (III) by AuNPs-$SF_w$, which favored the generation of more electrons due to greater oxidation of metallic silver ions and an increase in the faradic current captured by the electrode.

In relation to the membranes at 25 cm (Figure 5b), the electrical behavior was similar to that exhibited by the membranes obtained at 20 cm, where the $WE_G$ presented a greater magnitude of the electric current compared to the membranes treated with and without functionalization with metal gold or silver nanoparticles. The $SF_w$-treated membrane at 25 cm presented a lower current value compared to that obtained for the membrane at 20 cm, which is related to the fibrillar morphology, where the membranes at 25 cm presented a larger fiber size, which decreased the exposed surface area of the membranes and made it difficult for $Cl^-$ ions and/or $H_2O$ molecules to be oxidized on their surface and generate a greater current of electrons that were diffused to the surface of the $WE_G$.

On the other hand, the little variation in the faradic current obtained in the membranes at a distance of 25 cm and functionalized with AuNPs-$SF_w$ or AgNPS-$SF_w$ compared to the response of the membranes at a distance of 20 cm is associated with the formation of the thin film of nanoparticles on the surface of the membrane, which, as there was no modification in the functionalization procedure, resulted in a slight variation in the number of electrons diffused to the $WE_G$ product of the oxidation of AuNPs-$SF_w$ and AgNPs-$SF_w$ in the presence of $Cl^-$ ions.

When coating the graphite electrode with the $SF_w$ membranes treated with and without functionalization with the metal nanoparticles, a favorable response was found for each membrane evaluated in the electrochemical cell as a result of the different oxidation reactions involved according to the type of biomaterial of study, which was related to the concentration gradients of the ionic species present on the surface of the $WE_G$ and in the electrophysiological solution, which generated variations in the potential described by the Nernst (Equation (6)) [30].

$$E = E^0 + \frac{0.059}{n} \cdot \log \frac{\prod[Ox]^n}{\prod[Red]^n} \quad (6)$$

where $n$ is the number of electrons transferred, $\prod[Ox]^n$ y $\prod[Red]^n$ is the product operator of the concentrations of the oxidized and reduced species, raised to their stoichiometric coefficients in the vicinity of the $WE_G$.

The results presented a behavior similar to that exhibited by the cardiac action potential, where a reversible change in the membrane potential is generated by means of a bioelectric stimulus in the form of an electric current, produced by the sequential activation of various ionic currents generated by the diffusion of ions through the membrane in favor of its electrochemical gradient. This membrane potential is also described by the Nernst's Equation, which is mediated by the concentration gradients of the $Na^+$, $Ca^{2+}$, $Cl^-$ ions, and the $K^+$ concentration inside ($[K^+]_i$) and outside ($[K^+]_e$) of the cell membrane (Equation (7)) [36].

$$E_K = -61 \cdot \log \frac{[K^+]_i}{[K^+]_e} \quad (7)$$

Based on the above, the electrochemical response obtained from $SF_w$ electrospun treated membranes and functionalized with AuNPs-$SF_w$ or AgNPs-$SF_w$ could favor the diffusion of a bioelectric stimulus in the form of an electric current by modifying the resting potential and favoring the cardiac action potential. Furthermore, the functionalized membranes could exhibit favorable behavior when interacting in an in vitro model of functional cardiac cells. Therefore, these could be used as a therapeutic strategy for cardiac tissue engineering as an electroconductive bioactive scaffold.

On the other hand, Figure 5c–e show the electrical resistance, through the derivative of the voltage with respect to the current ($dV/dI$) at the place where the tangent touched the voltagram at a point, likewise, the resistivity ($\rho$) and conductivity ($\sigma$) (Equations (8) and (9)) of the treated membranes at 20 cm and 25 cm and functionalized with AuNPs-$SF_W$ and AgNPs-$SF_W$.

$$\rho = R \cdot \frac{S}{l} \quad (8)$$

$$\sigma = \frac{1}{\rho} \quad (9)$$

where $R$ is the resistance in ohms, $S$ is the exposed area in $m^2$, and $l$ is the separation between the working and reference electrode in m.

Considering each material as an individual working electrode evaluated in the electrochemical cell, it was found that the $SF_W$ membranes treated with and without functionalization with the metal nanoparticles generated less opposition to the flow of electric current compared to the graphite electrode from the results of resistance, resistivity, and electrical conductivity, which is associated with greater fiber connectivity in the treated membranes and the intrinsic electrical properties of the nanoparticles. Similarly, it could be noted that by increasing the distance from 20 cm to 25 cm, the $SF_W$-treated membranes presented an increase in electrical conductivity, which could be related to the arrangement of the fibers and their size, whereas the membranes functionalized with AugNPs-$SF_W$ or AgNPs-$SF_W$ showed variation in electrical conductivity, which is related to the number of nanoparticles arranged on the surface of the material.

### 3.6. Electrochemical Impedance Spectroscopy

Through electrochemical impedance spectroscopy, it is possible to find the electrochemical behavior of different biomaterials when interacting with the graphite working electrode by applying a small amplitude sinusoidal excitation signal and measuring the response in the form of current, voltage, or another signal of interest [37]. Therefore, impedance measurements were carried out at an amplitude of 86 mV, which emulated the transmembrane potential reached in the electrical depolarization of cardiac cells and at frequency sweeps between $1 \times 10^6$ to 0.1 Hz. Figure 6a presents the Argand plots or Nyquist plots for the graphite electrode without interacting with the membrane, where it presented a linear relationship between the imaginary impedance (Z") with respect to the real impedance (Z′), in which as the direct current electric potential applied to the system increased, a lower slope was presented, indicating that the kinematic reaction is limited by diffusion processes [37,38]

The behavior found for the graphite electrode could be related to the electrical double layer generated when interacting with Hank's Balanced Salt Solution, which under specific conditions can exhibit capacitive and conductive properties. Concerning to the elements that make up the Randles circuit, the parameter $R_s$ is associated with the resistance of the solution, the CPE1 corresponds to the non-ideal capacitive behavior of the electrode, which is strongly dependent on the frequency of the electrical conductivity, and CPE2 is due to the capacitance of the electrical double layer between the polarized WE and Hank's solution, while $R_{ct}$ represents the resistance to charge transfer between the solution and the electrode surface. The use of CPE constant phase elements instead of capacitors allowed us to achieve a better fit of the experimental data due to the lack of homogeneity in the system associated with the rough or porous surface of the electrode [39–41].

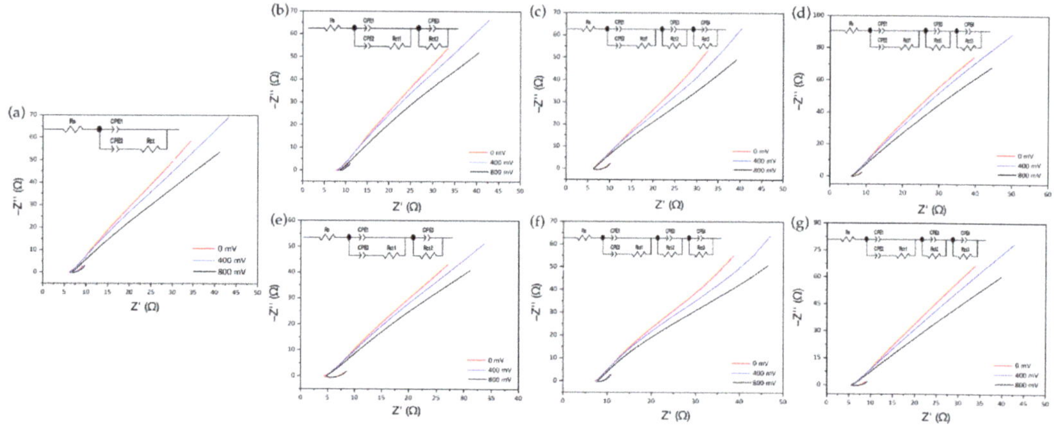


**Figure 6.** Nyquist plots between Z″ y Z′ for the graphite electrode, treated SFw membranes with and without functionalization in Hank's Balanced Salt solution at different DC electrical potentials. Where (**a**) graphite electrode; (**b**) treated SFw membrane at 20 cm; (**c**) AgNPs-SFw; (**d**) AuNPs-SFw; (**e**) treated SFw membrane at 25 cm; (**f**) AgNPs-SFw; (**g**) AuNPs-SFw. At the top left of each diagram, the Randles equivalent circuit is presented.

Regarding the Nyquist plots obtained for the treated $SF_w$ membranes with and without functionalization (Figure 6b–g), a similar behavior to that presented for the graphite electrode was observed, preserving the linear relationship between Z″ and Z′ and the slope value decrease as the DC electric potential increased. Based on these results, the Randles circuits were obtained, finding that the equivalent circuit obtained for the graphite electrode did not allow the adjustment of the experimental results for the treated membranes without or with functionalization. Thus, the circuit was complemented with elements composed of a CPE and $R_{ct}$ in parallel connected in series with the previous circuit, which corresponded to each of the additional bioactive electrochemical layers [42].

In Figure 7, an overview of the electrochemical behavior found and its corresponding equivalent circuit is presented, showing that at electric potentials of 0 mV (Figure 7a–d), an electrochemical equilibrium was presented between the ionic solution and working electrode with and without functionalization with the nanoparticles. However, as the electric potential increased to values of 400 mV (Figure 7e–h) and 800 mV (Figure 7i–l), the different oxidation reactions began to occur, which triggered the appearance of constant phase elements in equivalent circuits.

From the results, it was observed that when performing variations in the direct current electric potential, changes were generated in the $C_{CPE1}$ capacitance parameter and the exponent n corresponding to the constant phase element, which indicates the ability of a material to acquire and store energy in the form of an electric charge and its behavior as a capacitor or resistor, which vary as a result of the different oxidation reactions that occur on the surfaces of the treated membranes with or without functionalization with metal nanoparticles, generating electrons that are stored for short periods of time in the biomaterial or diffused through it [43].

On the other hand, regarding the change in the resistance $R_{ct1}$ due to the charge transfer between the solution and the surface of the type of membrane with which it interacts, this is strongly influenced by any modification of the electrode surface, finding that for the graphite electrode as electrical potential increases, resistance decreases. The opposite occurs for the different coatings of the working electrode with the treated membranes with or without functionalization with metal nanoparticles, whereas as the potential increased, the resistance increased due to the charge transfer processes. Such behavior could be related to (i) the electrostatic repulsion between the surface charge of both the treated $SF_w$ membranes, as well as those functionalized with AuNPs-$SF_W$ or AgNPs-$SF_W$ with ionic species carrying the same charge; and (ii) steric hindrances generated by structural modifications of the protein and/or by the presence of metal nanoparticles [39].

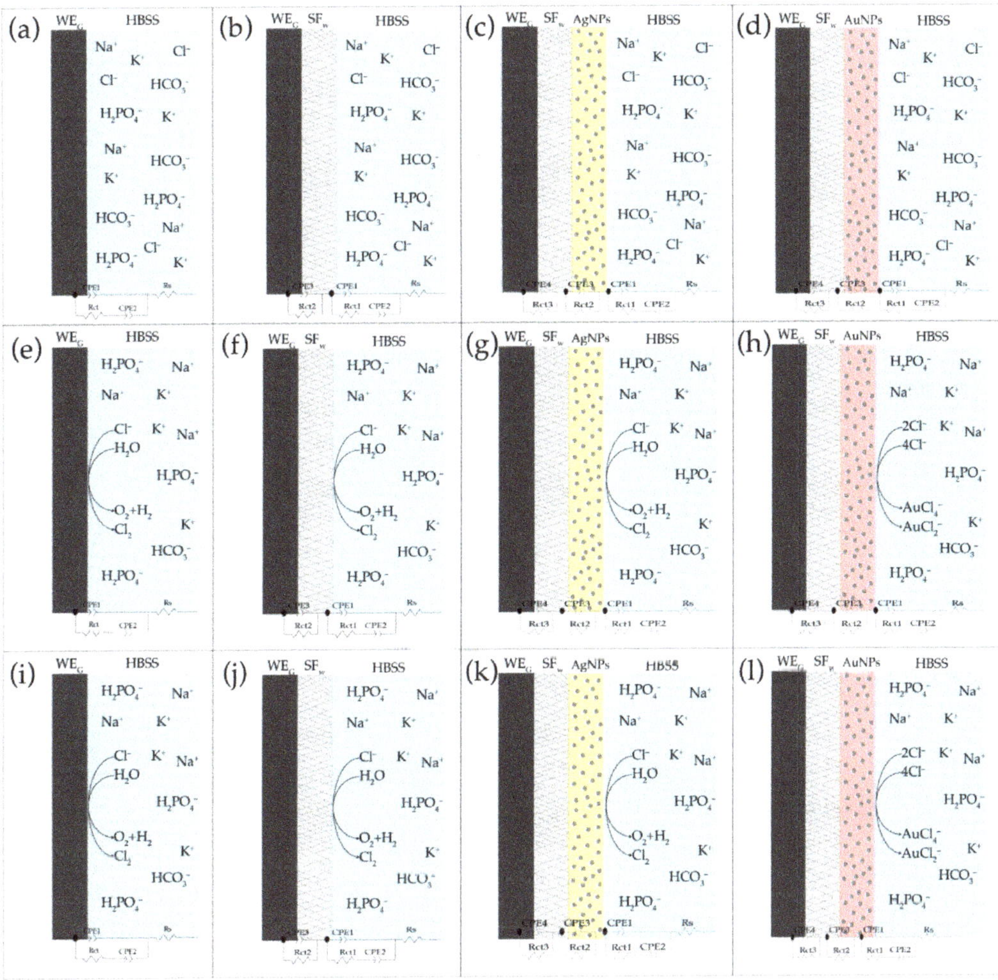


**Figure 7.** Schematic representation of the electrical behavior for the graphite electrode and the interaction with treated membranes with and without AuNPs-$SF_W$ or AgNPs-$SF_W$ functionalization, which describe the distribution of impedance components with their respective equivalent circuits to different DC electrical potentials. Where (**a–d**) 0 mV; (**e–h**) 400 mV; (**i–l**) 800 mV.

To determine the impedance for the different membranes evaluated by EIS, Bode plots were used to evaluate the change in the magnitude of the impedance with respect to the frequency. Finding that both for the graphite electrode (see Figure 8a) and for the treated $SF_W$ membranes obtained at distances of 20 cm and 25 cm with their respective surface modifications with AuNPs-$SF_W$ or AgNPs-$SF_W$ (see Figure 8b–g), it was determined that as the electric potential increased there was an increase in impedance. On the other hand, it was observed that, for the frequency range between $2 \times 10^5$ Hz to 100 Hz, an almost constant impedance value was presented, indicating a resistive behavior, while, at values lower than 100 Hz, a linear behavior was presented with a high slope, which could indicate capacitive behavior [44].

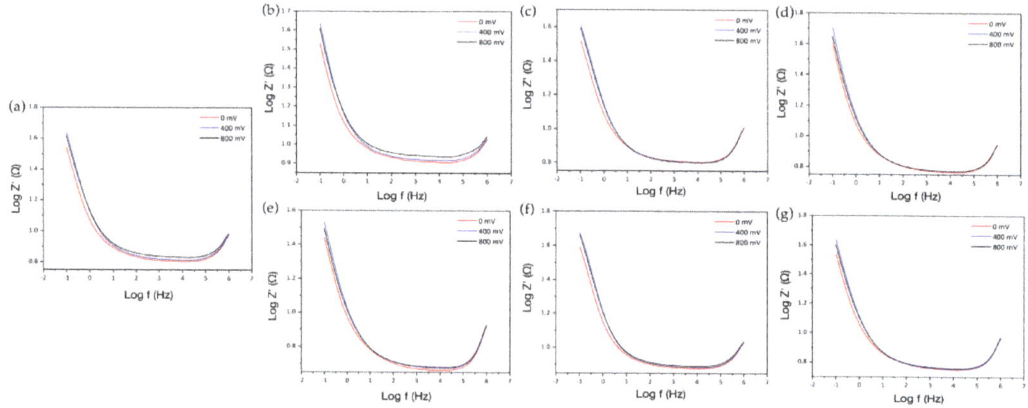


**Figure 8.** Bode plots between $\log_{10} |Z|$ y $\log_{10} f$ for the graphite electrode, treated $SF_w$ membranes with and without functionalization in Hank's Balanced Salt solution at different DC electrical potentials. Where (**a**) Graphite electrode; (**b**) $SF_w$ obtained at 20 cm; (**c**) AgNPS-$SF_w$; (**d**) AuNPs-$SF_w$; (**e**) $SF_w$ obtained at 25 cm; (**f**) AgNPS-$SF_w$; (**g**) AuNPs-$SF_w$.

To evaluate the change in the impedance value when coating the graphite electrode, the value of the impedance magnitude was determined at a frequency of 1,995 Hz close to a normal heart rate range, determined from 120 beats per minute, equivalent to 2 Hz. The results were not favorable as a function of the frequency close to that of the heart due to the lack of coupling through electrostatic interactions between the surface of the graphite electrode and the $SF_w$ membranes treated with and without functionalization (Table 4).

**Table 4.** Impedance values obtained at different DC electric potentials for the different membranes evaluated by EIS obtained at a frequency of 2 Hz.

| Sample | Electric Potential | | |
|---|---|---|---|
| | **0 mV** | **400 mV** | **800 mV** |
| Graphite | 9.80 | 10.72 | 10.95 |
| Treated $SF_w$ at 20 cm | 11.39 | 12.23 | 12.54 |
| AuNPs-$SF_w$ | 9.73 | 10.57 | 10.43 |
| AgNPs-$SF_w$ | 10.01 | 10.89 | 10.82 |
| Treatred $SF_w$ at 25 cm | 7.80 | 8.49 | 8.37 |
| AuNPs-$SF_w$ | 9.36 | 10.18 | 10.12 |
| AgNPs-$SF_w$ | 11.47 | 12.58 | 12.65 |

On the other hand, if only the first part of the equivalent circuit $[R_s \cdot (CPE1 \cdot (CPE2 \cdot R_{ct1}))]$ is considered, that is, the parameter $R_s$ associated with the resistance of the solution, $CPE1$ corresponding to the non-ideal capacitive behavior of the electrode, $CPE2$ is the capacitance of the electric double layer, and $R_{ct1}$ is the resistance to charge transfer between the solution and the electrode surface, and the impedance is determined at a frequency of 2 Hz using Equations (10)–(14), a favorable impedance behavior is obtained when coating the working electrode with each of the treated membranes and functionalized with metal nanoparticles [45].

$$Z_{Rs} = R_s \tag{10}$$

$$Z_{CPE} = \frac{1}{C_{CPE} \cdot (j \cdot \omega)^n} \tag{11}$$

$$Z_{CPE1,2} = \frac{e^{-2.528 \cdot n} \cdot Cos(1.571 \cdot n)}{C_{CPE1,2}} \tag{12}$$

$$Z_{Rct1} = R_{ct1} \tag{13}$$

$$Z_T = R_s + \frac{1}{\frac{1}{Z_{CPE1}} + \frac{1}{Z_{CPE2}+Z_{Rct1}}} \tag{14}$$

where $C_{CPE}$ corresponds to the capacitance in farads and $\omega$ is the angular frequency ($\omega = 2\pi f$). Equation (13) corresponds to the real part of Equation (12) evaluated at a frequency of 1995 Hz, which was obtained using (Maple$^{TM}$ 18 software, Maplesoft, Waterloo, Canada) [37].

The results showed that the increase in the total impedance values of the circuit $R_s \cdot (CPE1 \cdot (CPE2 \cdot R_{ct1}))$ for the treated $SF_w$ membranes, functionalized with AuNPs-$SF_w$ or AgNPs-$SF_w$ at direct current electric potentials of 400 mV and 800 mV, is related to the formation of Au (I) and Au (II) chloride species and AgCl during the oxidation processes of metal nanoparticles in the presence of $Cl^-$ ions, which modify the electrical properties of the composite layer-by-layer and increase the overall impedance of the electrical system.

### 3.7. Cell Viability of Membranes

The membranes at 20 cm and 25 cm that were not functionalized with nanoparticles presented viability of 60% and 25%, respectively. The membranes at 20 cm and 25 cm functionalized with gold nanoparticles caused viability of the fibroblasts of 45% and 19%, respectively (Figure 9). The results demonstrate that the microarchitecture of the membranes functionalized with gold nanoparticles favors cellular bioavailability and biocompatibility [46].

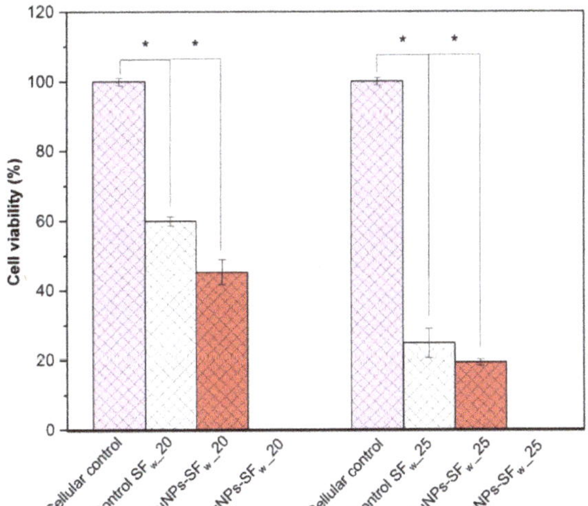


**Figure 9.** Cell viability of 3T3 fibroblasts on interaction with membranes without or functionalized with the gold or silver nanoparticles. Where _20 represents membranes manufactured at 20 cm and _25 at 25 cm. * indicates statistical significance with $p$-values < 0.05.

On the other hand, membranes functionalized with silver nanoparticles do not allow the generation of syncytium cells, and, on the contrary, generated cell toxicity and death. Therefore, this type of membrane could not be used as conductive structures for tissue engineering, due to their oxidative capacity, which causes a limiting barrier in cell–cell communication and could favor cell death processes and induce proinflammatory effects [47].

## 4. Conclusions

Based on the results of our research, it was possible to generate $SF_w$/PEO fibrillar membranes, modulating the morphological properties of the electrospun structure without the presence of defects. The treatment with organic solvents such as methanol contributed to these membranes, going from being soluble in aqueous solvents to insoluble in them by a crystalline transition of fibroin macromolecules.

Composed membranes with AuNPs-$SF_w$ or AgNPs-$SF_w$ were obtained layer-by-layer by impregnation techniques. The resulted membranes showed favorable electrical response with average conductivity values of 16.3 μS/cm (AuNPs-$SF_w$), 15.7 μS/cm (AgNPs-$SF_w$), and 13.5 μS/cm (Treated $SF_w$) compared to 9.3 μS/cm for graphite, with which this type of membrane allowed the diffusion of electrical stimuli behaving as stimulators and modulators of electrical current.

On the other hand, it was found that treated SFw electrospun membranes at distances of 20 cm and 25 cm, coated with AuNPs-SFw or AgNPs-SFw, reduce the impedance values, which could modulate charge transfer and modify the conduction rate of the electrical stimulus when evaluated in an electrophysiological solution.

Furthermore, it was found that electrospun membranes at 20 cm functionalized with AuNPs-SFw generate an increase in electrical conductivity when evaluated in an electrophysiological solution and favor of cell viability. Therefore, it is considered that this type of biomaterial composed of layers could be used as an electroconductive biomaterial for tissue engineering.

**Author Contributions:** Conceptualization, W.A., Y.M. and J.B.; methodology, W.A., Y.M. and J.B.; validation, W.A. and Y.M.; formal analysis, W.A., Y.M. and A.G.-G.; data curation, W.A. and Y.M.; writing—original draft preparation, W.A. and Y.M.; writing—review and editing, W.A., Y.M., A.G.-G., A.R.-O. (only review) and J.B.; supervision, Y.M. and J.B. All authors have read and agreed to the published version of the manuscript.

**Funding:** This work was supported by Universidad Pontificia Bolivariana, Ministry of Science, Technology and Innovation—MINCIENCIAS, Colombia (Grant No. 647 of 2014—national doctorate, Grant No. 812 of 2018—Young Researchers and Innovators, Grant No. 121084467592—project MINCIENCIAS).

**Institutional Review Board Statement:** Not Applicable.

**Data Availability Statement:** The data presented in this study are available on request from the corresponding author.

**Acknowledgments:** The authors are grateful to Nayeli Pineda Aguilar and Alejandro Arizpe Zapata from CIMAV for the technical support.

**Conflicts of Interest:** The authors declare that they have no relation, condition or circumstance that constitutes a potential conflict of interest.

## References

1. Capulli, A.; MacQueen, L.; Sheehy, S.P.; Parker, K. Fibrous scaffolds for building hearts and heart parts. *Adv. Drug Deliv. Rev.* **2016**, *96*, 83–102. [CrossRef] [PubMed]
2. Engelbrekt, C.; Sørensen, K.H.; Zhang, J.; Welinder, A.C.; Jensen, P.S.; Ulstrup, J. Green synthesis of gold nanoparticles with starch-glucose and application in bioelectrochemistry. *J. Mater. Chem.* **2009**, *19*, 7839. [CrossRef]
3. Iravani, S. Green synthesis of metal nanoparticles using plants. *Green Chem.* **2011**, *13*, 2638–2650. [CrossRef]
4. Nabieva, I.; Sadriddinov, B.; Khasanova, M.; Alimova, K.; Klemola, K.; Pearson, J. Preparation of Natural Silk Waste Solutions by High Frecuency Heating. *AUTEX Res. J.* **2004**, *4*, 143–146.
5. Ríos, A.D.; Sánchez, S.; Álvarez-López, L.J.C.C.; Restrepo-Osorio, A. Electrospinning of silk fibroin obtained from colombian silk fibrous waste and their potential applications as biomaterials. In Proceedings of the 6th Amazon & Pacific Green Materials Congress and Sustainable Construction Materials LAT-RILEM Conference, Cali, Colombia, 27–29 April 2016.
6. Cristallini, C.; Vitale, E.; Giachino, C.; Rastaldo, R. Nanoengineering in Cardiac Regeneration: Looking Back and Going Forward. *Nanomaterials* **2020**, *10*, 1587. [CrossRef]
7. Feng, J.; Shi, H.; Yang, X.; Xiao, S. Self-Adhesion Conductive Sub-micron Fiber Cardiac Patch from Shape Memory Polymers to Promote Electrical Signal Transduction Function. *ACS Appl. Mater. Interfaces* **2021**, *13*, 19593–19602. [CrossRef]

8. Fouad, H.; Khalil, K.A.; Alshammari, B.A.; Abdal-Hay, A.; El-Salam, N.M.A. Development of New Bio-Composite of PEO/Silk Fibroin Blends Loaded with Piezoelectric Material. *Polymers* **2022**, *14*, 4209. [CrossRef]
9. Wang, F.; Wu, H.; Venkataraman, V.; Hu, X. Silk fibroin-poly(lactic acid) biocomposites: Effect of protein-synthetic polymer interactions and miscibility on material properties and biological responses. *Mater. Sci. Eng. C* **2019**, *104*, 109890. [CrossRef]
10. Caicedo-Eraso, J.C.; Díaz-Arango, F.O.; Osorio-Alturo, A. Espectroscopia de impedancia eléctrica aplicada al control de la calidad en la industria alimentaria. *Cienc. Tecnol. Agropecu.* **2019**, *21*, 1–20. [CrossRef]
11. Jaramillo-Quiceno, N.; Restrepo-Osorio, A. Water-annealing treatment for edible silk fibroin coatings from fibrous waste. *J. Appl. Polym. Sci.* **2019**, *137*, 48505. [CrossRef]
12. Schindelin, J.; Arganda-Carreras, I.; Frise, E.; Kaynig, V.; Longair, M.; Pietzsch, T.; Preibisch, S.; Rueden, C.; Saalfeld, S.; Schmid, B.; et al. Fiji: An open-source platform for biological-image analysis. *Nat. Methods* **2012**, *9*, 676–682. [CrossRef] [PubMed]
13. Hwang, H.; McCaslin, T.G.; Hazel, A.; Pagba, C.V.; Nevin, C.M.; Pavlova, A.; Barry, B.A.; Gumbart, J.C. Redox-Driven Conformational Dynamics in a Photosystem-II-Inspired β-Hairpin Maquette Determined through Spectroscopy and Simulation. *J. Phys. Chem. B* **2017**, *121*, 3536–3545. [CrossRef] [PubMed]
14. Hammarström, L.; Styring, S. Proton-coupled electron transfer of tyrosines in Photosystem II and model systems for artificial photosynthesis: The role of a redox-active link between catalyst and photosensitizer. *Energy Environ. Sci.* **2011**, *4*, 2379–2388. [CrossRef]
15. Si, S.; Bhattacharjee, R.R.; Banerjee, A.; Mandal, T.K. A Mechanistic and Kinetic Study of the Formation of Metal Nanoparticles by Using Synthetic Tyrosine-Based Oligopeptides. *Chem.–A Eur. J.* **2006**, *12*, 1256–1265. [CrossRef] [PubMed]
16. Cheng, H.; Chen, C.; Zhang, S. Electrochemical Behavior and Sensitive Determination of L-Tyrosine with a Gold Nanoparticles Modified Glassy Carbon Electrode. *Anal. Sci.* **2009**, *25*, 1221–1225. [CrossRef]
17. Soltani, N.; Saion, E.; Erfani, M.; Rezaee, K.; Bahmanrokh, G.; Drummen, G.P.C.; Bahrami, A.; Hussein, M.Z. Influence of the Polyvinyl Pyrrolidone Concentration on Particle Size and Dispersion of ZnS Nanoparticles Synthesized by Microwave Irradiation. *Int. J. Mol. Sci.* **2012**, *13*, 12412–12427. [CrossRef]
18. Verma, M.; Kedia, A.; Newmai, M.B.; Kumar, P.S. Differential role of PVP on the synthesis of plasmonic gold nanostructures and their catalytic and SERS properties. *RSC Adv.* **2016**, *6*, 80342–80353. [CrossRef]
19. Pelipenko, J.; Kristl, J.; Janković, B.; Baumgartner, S.; Kocbek, P. The impact of relative humidity during electrospinning on the morphology and mechanical properties of nanofibers. *Int. J. Pharm.* **2013**, *456*, 125–134. [CrossRef]
20. Haider, A.; Haider, S.; Kang, I.-K. A comprehensive review summarizing the effect of electrospinning parameters and potential applications of nanofibers in biomedical and biotechnology. *Arab. J. Chem.* **2018**, *11*, 1165–1188. [CrossRef]
21. Hekmati, A.H.; Rashidi, A.; Ghazisaeidi, R.; Drean, J.-Y. Effect of needle length, electrospinning distance, and solution concentration on morphological properties of polyamide-6 electrospun nanowebs. *Text. Res. J.* **2013**, *83*, 1452–1466. [CrossRef]
22. Hu, X.; Kaplan, A.D.; Cebe, P. Determining Beta-Sheet Crystallinity in Fibrous Proteins by Thermal Analysis and Infrared Spectroscopy. *Macromolecules* **2006**, *39*, 6161–6170. [CrossRef]
23. Kim, S.H.; Nam, Y.S.; Lee, T.S.; Park, W.H. Silk Fibroin Nanofiber. Electrospinning, Properties, and Structure. *Polym. J.* **2003**, *35*, 185–190. [CrossRef]
24. Jin, H.-J.; Fridrikh, S.V.; Rutledge, A.G.C.; Kaplan, D.L. Electrospinning *Bombyx mori* Silk with Poly(ethylene oxide). *Biomacromolecules* **2002**, *3*, 1233–1239. [CrossRef] [PubMed]
25. Chutipakdeevong, J.; Ruktanonchai, U.R.; Supaphol, P. Process optimization of electrospun silk fibroin fiber mat for accelerated wound healing. *J. Appl. Polym. Sci.* **2013**, *130*, 3634–3644. [CrossRef]
26. Akturk, O.; Kismet, K.; Yasti, A.C.; Kuru, S.; Duymus, M.E.; Kaya, F.; Caydere, M.; Hucumenoglu, S.; Keskin, D. Wet electrospun silk fibroin/gold nanoparticle 3D matrices for wound healing applications. *RSC Adv.* **2016**, *6*, 13234–13250. [CrossRef]
27. Yi, B.; Zhang, H.; Yu, Z.; Yuan, H.; Wang, X.; Zhang, Y. Fabrication of high performance silk fibroin fibers via stable jet electrospinning for potential use in anisotropic tissue regeneration. *J. Mater. Chem. B* **2018**, *6*, 3934–3945. [CrossRef]
28. Cao, H.; Chen, X.; Huang, L.; Shao, Z. Electrospinning of reconstituted silk fiber from aqueous silk fibroin solution. *Mater. Sci. Eng. C* **2009**, *29*, 2270–2274. [CrossRef]
29. Tan, H.-L.; Teow, S.-Y.; Pushpamalar, J. Application of Metal Nanoparticle–Hydrogel Composites in Tissue Regeneration. *Bioengineering* **2019**, *6*, 17. [CrossRef]
30. Elgrishi, N.; Rountree, K.; McCarthy, B.D.; Rountree, E.; Eisenhart, T.T.; Dempsey, J.L. A Practical Beginner's Guide to Cyclic Voltammetry. *J. Chem. Educ.* **2017**, *95*, 197–206. [CrossRef]
31. Walters, C.C. Oil–oil and oil–source rock correlation. In *Encyclopedia of Earth Science*; Kluwer Academic Publishers: Dordrecht, The Netherlands, 1998; pp. 442–444.
32. Tschulik, K.; Palgrave, R.G.; Batchelor-McAuley, C.; Compton, R.G. 'Sticky electrodes' for the detection of silver nanoparticles. *Nanotechnology* **2013**, *24*, 295502. [CrossRef]
33. Fox, C.M. Electrochemical Synthesis of Silver Nanoparticles for Applications in Nitrate Detection, Catalysis and Antibacterial Activity. Ph.D. Thesis, National University of Ireland, Maynooth, Ireland, 2014.
34. Kohl, P.A. Electrodeposition of Gold. In *Modern Electroplating*; John Wiley & Sons, Inc.: Hoboken, NJ, USA, 2011; pp. 115–130.
35. Plowman, B.J.; Sidhureddy, B.; Sokolov, S.V.; Young, N.P.; Chen, A.; Compton, R.G. Electrochemical Behavior of Gold-Silver Alloy Nanoparticles. *ChemElectroChem* **2016**, *3*, 1039–1043. [CrossRef]

36. Klabunde, R. *Cardiovascular Physiology Concepts*; Wolters Kluwer Health/Lippincott Williams & Wilkins: Philadelphia, PA, USA, 2011.
37. Lasia, A. *Electrochemical Impedance Spectroscopy and Its Applications*; Springer: New York, NY, USA, 2014.
38. Mehmood, S.; Ciancio, R.; Carlino, E.; Bhatti, A.S. Role of Au(NPs) in the enhanced response of Au(NPs)-decorated MWCNT electrochemical biosensor. *Int. J. Nanomed.* **2018**, *13*, 2093–2106. [CrossRef] [PubMed]
39. Bonanni, A.; Pumera, M.; Miyahara, Y. Influence of gold nanoparticle size (2–50 nm) upon its electrochemical behavior: An electrochemical impedance spectroscopic and voltammetric study. *Phys. Chem. Chem. Phys.* **2011**, *13*, 4980–4986. [CrossRef] [PubMed]
40. Kim, S.; Le, T.-H.; Park, C.S.; Park, G.; Kim, K.H.; Kim, S.; Kwon, O.S.; Lim, G.T.; Yoon, H. A Solution-Processable, Nanostructured, and Conductive Graphene/Polyaniline Hybrid Coating for Metal-Corrosion Protection and Monitoring. *Sci. Rep.* **2017**, *7*, 15184. [CrossRef] [PubMed]
41. Abouzari, M.S.; Berkemeier, F.; Schmitz, G.; Wilmer, D. On the physical interpretation of constant phase elements. *Solid State Ionics* **2009**, *180*, 922–927. [CrossRef]
42. Ramanavicius, A.; Genys, P.; Ramanaviciene, A. Electrochemical Impedance Spectroscopy Based Evaluation of 1,10-Phenanthroline-5,6-dione and Glucose Oxidase Modified Graphite Electrode. *Electrochim. Acta* **2014**, *146*, 659–665. [CrossRef]
43. Jorcin, J.-B.; Orazem, M.E.; Pébère, N.; Tribollet, B. CPE analysis by local electrochemical impedance spectroscopy. *Electrochim. Acta* **2006**, *51*, 1473–1479. [CrossRef]
44. Guler, Z.; Erkoc, P.; Sarac, A.S. Electrochemical impedance spectroscopic study of single-stranded DNA-immobilized electroactive polypyrrole-coated electrospun poly(I/I-caprolactone) nanofibers. *Mater. Express* **2015**, *5*, 269–279. [CrossRef]
45. Boukamp, B.A. *Equivalent Circuit: (equivcrt.pas): Users Manual*; Department of Chemical Technology, University of Twente: Enschede, The Netherlands, 1989.
46. Beltran-Vargas, N.E.; Peña-Mercado, E.; Sánchez-Gómez, C.; Garcia-Lorenzana, M.; Ruiz, J.-C.; Arroyo-Maya, I.; Huerta-Yepez, S.; Campos-Terán, J. Sodium Alginate/Chitosan Scaffolds for Cardiac Tissue Engineering: The Influence of Its Three-Dimensional Material Preparation and the Use of Gold Nanoparticles. *Polymers* **2022**, *14*, 3233. [CrossRef]
47. Löfdahl, A.; Jern, A.; Flyman, S.; Kåredal, M.; Karlsson, H.L.; Larsson-Callerfelt, A.-K. Silver Nanoparticles Alter Cell Viability Ex Vivo and in Vitro and Induce Proinflammatory Effects in Human Lung Fibroblasts. *Nanomaterials* **2020**, *10*, 1868. [CrossRef]

membranes

*Review*

# An Investigation of the Constructional Design Components Affecting the Mechanical Response and Cellular Activity of Electrospun Vascular Grafts

**Suzan Ozdemir [1], Ipek Yalcin-Enis [1,*], Baturalp Yalcinkaya [2] and Fatma Yalcinkaya [3,*]**

[1] Textile Engineering Department, Textile Technologies and Design Faculty, Istanbul Technical University, Beyoglu, 34467 Istanbul, Turkey
[2] Department of Material Science, Faculty of Mechanical Engineering, Technical University of Liberec, 461 17 Liberec, Czech Republic
[3] Department of Environmental Technology, Institute for Nanomaterials, Advanced Technologies and Innovations, Technical University of Liberec, 461 17 Liberec, Czech Republic
* Correspondence: ipekyalcin@itu.edu.tr (I.Y.-E.); fatma.yalcinkaya@tul.cz (F.Y.); Tel.: +90-212-2931-300 (I.Y.-E.); +420-48535-3482 (F.Y.)

**Abstract:** Cardiovascular disease is anticipated to remain the leading cause of death globally. Due to the current problems connected with using autologous arteries for bypass surgery, researchers are developing tissue-engineered vascular grafts (TEVGs). The major goal of vascular tissue engineering is to construct prostheses that closely resemble native blood vessels in terms of morphological, mechanical, and biological features so that these scaffolds can satisfy the functional requirements of the native tissue. In this setting, morphology and cellular investigation are usually prioritized, while mechanical qualities are generally addressed superficially. However, producing grafts with good mechanical properties similar to native vessels is crucial for enhancing the clinical performance of vascular grafts, exposing physiological forces, and preventing graft failure caused by intimal hyperplasia, thrombosis, aneurysm, blood leakage, and occlusion. The scaffold's design and composition play a significant role in determining its mechanical characteristics, including suturability, compliance, tensile strength, burst pressure, and blood permeability. Electrospun prostheses offer various models that can be customized to resemble the extracellular matrix. This review aims to provide a comprehensive and comparative review of recent studies on the mechanical properties of fibrous vascular grafts, emphasizing the influence of structural parameters on mechanical behavior. Additionally, this review provides an overview of permeability and cell growth in electrospun membranes for vascular grafts. This work intends to shed light on the design parameters required to maintain the mechanical stability of vascular grafts placed in the body to produce a temporary backbone and to be biodegraded when necessary, allowing an autologous vessel to take its place.

**Keywords:** vascular grafts; biopolymers; physiological forces; compliance; burst pressure; cellular activity; permeability; porosity; fiber orientation; wall thickness

**Citation:** Ozdemir, S.; Yalcin-Enis, I.; Yalcinkaya, B.; Yalcinkaya, F. An Investigation of the Constructional Design Components Affecting the Mechanical Response and Cellular Activity of Electrospun Vascular Grafts. *Membranes* **2022**, *12*, 929. https://doi.org/10.3390/membranes12100929

Academic Editor: Andrea Ehrmann

Received: 2 September 2022
Accepted: 22 September 2022
Published: 25 September 2022



## 1. Introduction

Cardiovascular diseases (CVDs) remain the major cause of death worldwide, and an estimated 17.9 million individuals died in 2019. Additionally, prior CVDs are a significant risk factor for coronavirus disease of 2019 (COVID-19)-related complications and fatalities [1,2]. The main risk factors for CVD include smoking, being overweight, having diabetes, high blood pressure or hypertension, dyslipidemia, not exercising enough, eating poorly, and experiencing a lot of stress, all of which are extremely prevalent problems in today's society [3,4]. The number of CVD cases increased from 271 million in 1990 to 523 million in 2019, while CVD fatalities increased from 12.1 million in 1990 to 18.6 million in 2019 [5]. The World Health Organization estimates that by 2030, there will be a 24.5%

increase in the number of fatalities [6]. Coronary artery disease, the most prevalent form of CVD, necessitates surgery based on arterial replacement, known as bypass grafting [7]. The blood vessel that is injured or obstructed is replaced during bypass surgeries with an autologous vein or synthetic graft. Autologous grafts have significant disadvantages because of their scarcity and difficulties with graft harvesting [8]. Despite being the most common autograft, the saphenous vein has low patency and a failure rate of about 50% after ten years of implantation [9]. Vascular grafts made of synthetic materials can be used in place of autologous vessels. Expanded polytetrafluoroethylene (ePTFE, Gore-Tex, California, USA) and polyethylene terephthalate (PET, Dacron, Invista, Kansas, USA) are the most widely used commercial synthetic materials because they are effective at replacing large-diameter arteries and have shown successful long-term results. However, they are ineffective when used as smaller diameter vascular grafts (<6 mm), such as coronary arteries, because of low patency rates, thrombogenicity, and compliance mismatch [10,11]. The compliance mismatch between the native artery and the inelastic synthetic graft at the anastomosis sites results in low blood flow rates and turbulent blood flow in small-diameter grafts.

Due to these mechanical issues, the thrombogenic nature of the scaffold material, poor endothelialization, luminal narrowing, and thrombosis are brought on by intimal hyperplasia, causing low patency rates [12]. Enhancing the mechanical performance and biocompatibility of small-caliber vascular prostheses is necessary to satisfy a clinical requirement and offer patients alternative scaffolds due to the current limitations of clinically approved grafts [13]. However, the clinical applicability of vascular grafts is still constrained by problems with the intrinsic thrombogenic character of synthetic polymers, inability to sustain somatic growth and repair, inappropriate mechanical qualities, and severe intimal hyperplasia [14]. Thus, novel approaches for fabricating TEVGs, including electrospinning, decellularization, lyophilization, and 3D printing by utilizing biopolymers, have been explored to eliminate these issues and provide the ideal small-caliber graft that may be used in the clinic and can imitate the native artery in all aspects [15]. The selection of the material and the production technique is based on the determination and optimization of the design parameters, which require a better understanding of the vascular environment, the properties and needs of native vessels, and the correlation between the constructional criteria and graft properties. In this regard, this review covers vascular grafts, scaffold fabrication methods, biopolymers utilized in these prostheses, and mechanical forces acting on vascular grafts in detail. In addition, the impact of constructional design parameters on mechanical as well as permeability properties is discussed, and recent studies have been reviewed in the literature to give a broad perspective for the researchers to discover the necessities and limitations in this field and help to find alternative ways to meet the requirements of vascular grafts improved in the future.

## 2. Anatomy of Blood Vessels

Arteries, capillaries, and veins are all linked in series to make up the pulmonary vasculature [16]. Arteries and veins accomplish effective blood circulation across the lumen to distant locations. Arteries transport oxygenated blood from the heart to the tissues, whereas veins transport waste, nutrients, and oxygen from the capillaries back to the heart while returning deoxygenated blood to it [17,18]. The extracellular matrix (ECM) comprises 70% water, and the remaining 30% comprises a vascular wall consisting of collagen, elastin, proteoglycans, and vascular cells [19]. The three layers that form the typical arterial wall are the *tunica intima*, which consists of a single layer of endothelial cells (ECs) that exists in the internal elastic *lamina*, which is a dense elastic membrane that divides the *intima* from the *media* and is oriented parallel to the blood flow; the *tunica media*, which is composed of concentric layers of smooth muscle cells (SMCs) between the elastic *lamina* layers; and the *tunica adventitia*, which is formed of myofibroblasts involving connective tissue with nerve fibers and the *vasa vasorum* that nurtures the blood vessel wall and is divided from the *media* by an external elastic *lamina* [20,21] (Figure 1). The *tunica intima*, also known as the endothelium layer, controls the tone of blood vessels, platelet activation, adhesion and aggregation, leukocyte adherence, SMC migration, and proliferation and serves as a thrombo-resistant, continuous selective permeable wall that permits laminar blood flow throughout the blood vessel [22]. ECs and SMCs play crucial roles in preserving the vessel's mechanical efficiency and structural integrity. The *tunica media*, in which collagen, elastin fibers, and SMCs are radially aligned, offers the vessel mechanical strength and regulates vessel diameter by contracting or relaxing [23]. High blood pressure causes the arteries to experience significant mechanical stress. In the physiological pressure range, the load on the vessel is distributed between the collagen and elastin fibers. At higher blood pressures, where a greater amount of force is needed for a change in diameter, the stiffer collagen fibers dominate the mechanical behavior and protect the blood vessel from failure. In contrast, the elastic components, which are less stiff and more elastic chains, dominate mechanical behavior at lower pressures [19]. Collagen fibers, elastic fibers, elastic lamellae, and proteoglycans, which provide vessel elasticity and radial compliance, are secreted by SMCs [22]. The flexibility and structural stability of the artery are supported by elastic *laminae*. Intimal hyperplasia is avoided thanks to elastic fibers and *lamina* that slow down SMC development [24]. Elastin, which relieves stress on the heart and permits vasodilation and vasoconstriction in arteries with pulsatile flow, is responsible for reversible elasticity [25]. The *vasa vasorum* and vascular innervation are supported by the *tunica adventitia*. This outermost layer comprises fibroblasts, extracellular matrix, and fibrillar types I and III collagen and is placed between the exterior elastic *lamina* on the *media* layer and interstitial matrix [23,24].

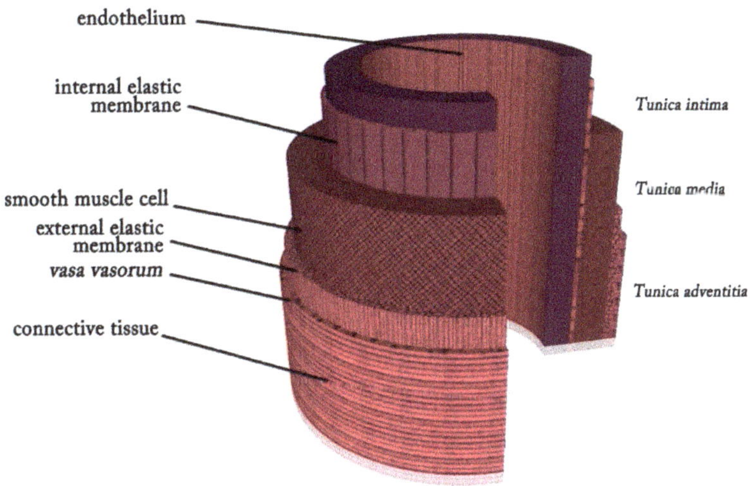


**Figure 1.** Layers of an artery.

## 3. Requirements for Vascular Grafts

The fundamental concern with vascular tissue engineering is still creating an ideal vascular graft that can replicate the structural, biological, and mechanical characteristics of the native blood vessels and be used as a replacement for the damaged blood vessel. When selecting a polymer and a method of fabrication for the construction of synthetic blood vessels, some fundamental properties to consider are processability, mechanical behavior, morphology and porosity, hydrophilicity, biodegradability, and biocompatibility [26]. The vascular scaffolds should offer an ideal topographic and structural framework for cell adhesion, proliferation, and diffusion [27]. Developing a suitable microstructure and functional material that encourages endothelialization requires proper processing and surface modification techniques. Additionally, the native tissues must be compatible with the mechanical characteristics of small-diameter vascular grafts, including modulus, nonlinear elasticity, compliance, burst pressure, and suture retention strength, as even a slight mechanical discrepancy between the prostheses and the native vessel can lead to graft failure [28]. A combination of dynamic mechanical forces, including hemodynamic forces originating from fluid flow, cyclic stretch, lateral pressure, and vessel wall forces created by the vasculature, creates the complex mechanical microenvironment [29]. The compliance mismatch, which is the incompatible dimensional change of the vascular graft and the native blood vessel in response to pressure variations inside the lumen, causes the hemodynamic flow imbalance in the vascular graft and stress concentration at the anastomosis. The pressure difference at the anastomosis can result in intimal hyperplasia and thrombosis [30]. Additionally, just like the native artery, the designed blood vessel's burst pressure must be higher than 1000 mmHg in order to resist blood pressure (200 mmHg) [31].

The pore size is a crucial structural element since it directly impacts cell migration, leakproofness, and mechanical qualities. Large pores obviously cause blood leakage, but small pores prevent cells from penetrating. SMCs must enter TEVGs through larger pores since they are bigger than ECs [32]. Additionally, in vitro research on the development of the macrovascular endothelium has shown that materials with smaller pore diameters and lower porosity promote better EC adherence. ECs range from 10 to 40 μm, and adhesion is necessary for their proliferation. As a result, the proliferation is limited to materials with pores larger than a cell, particularly those with diameters of more than 30 μm [33]. Moreover, a reduction in mechanical characteristics is observed with an increase in pore size and porosity [34]. In addition, vascular scaffolds should be suturable for convenience of use, and for usage in ophthalmic and microvascular procedures, a 0.6 N suture retention is adequate [35,36]. The choice of materials for vascular grafts is also heavily influenced by biodegradation. A fast degradation rate can limit neointimal hyperplasia by limiting the activation of inflammatory cells and reducing the probability and intensity of a foreign body response. However, a high rate of tissue degradation can compromise its performance. Thus, the vascular tissue regeneration rate should be harmonious with the biodegradation rate [37].

The topography and morphology of the lumen surface are other essential aspects affecting thrombosis and material selection. In addition, gradients strongly influence platelet adhesion and activation in the nanotopography of the material surface. Thus, a TEVG surface designed with the proper topography and roughness could significantly increase the hemocompatibility of the lumen surface of the scaffold [38].

Thus, for obtaining an applicable small-diameter vascular graft that satisfies all of the necessary features, this scaffold needs to be mechanically strong and compliant to withstand hemodynamic stress; suturable; available in various sizes in case of emergency; easy to use to reduce the time, cost, and risk; resistant to thrombus and infection; biocompatible to integrate with the body and allow the formation of neo-vessels similar to native arteries in characteristics and performance; low-cost; patent for long-term; able to show rapid endothelialization; and porous enough for easy cell diffusion; it is necessary to combine biomimetic design with improved cellular and molecular knowledge of the biology of the vessel wall [39].

## 4. Electrospinning Technique

The ability to have sufficient multicellular activities, nutrition delivery, and mechanical qualities is essential to successfully produce scaffolds at the macro- and microscales [40]. Solvent casting with particle leaching, thermally induced phase separation, freeze drying, electrospinning, 3D printing, and combination molding techniques are some methods for constructing tissue-engineered scaffolds that imitate the ECM [41]. Among these methods, electrospinning has emerged as a leading technology for creating synthetic polymer grafts because it is configurable, and the electrospun fibrous structure closely resembles the fibrous structure of native ECM in the vessel wall, enabling cell infiltration and cellularization of the grafts and having a high surface-to-volume ratio thanks to 3D fibrous matrices with varying fiber sizes [42].

Nanoscale (<1000 nm) and microscale (>1 μm) polymer fibers can be created using the electrospinning technique [43]. The electrospinning setup consists of three main parts: a high voltage supplier, a capillary tube with a tip, and a collector [27] (Figure 2). When a voltage is applied during electrospinning, the electrostatic force affecting the droplet overcomes the surface tension and forces a liquid jet to move out from the tip. This whipping action results in the drawing and thinning of the polymer and causes fiber production as the solvent evaporates simultaneously. The geometry of the resultant scaffold is influenced by the collector type and electrospinning arrangement [44]. Electrospinning can be used to create biomimetic degradable scaffolds for essential cellular and molecular activities using both natural and synthetic biopolymers [45].

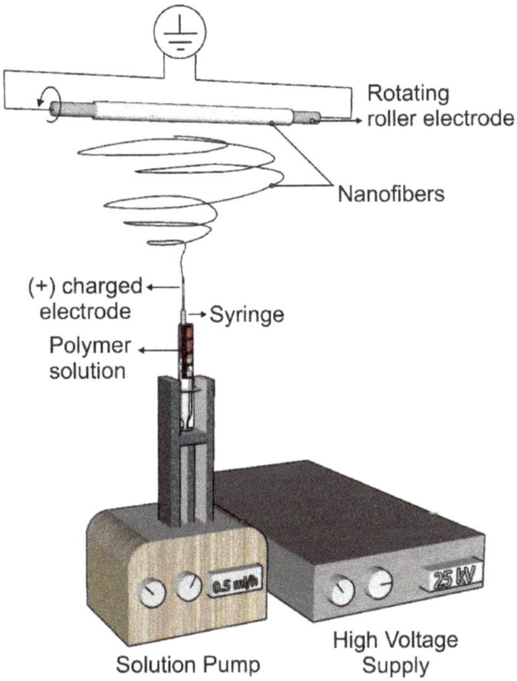


**Figure 2.** Electrospinning set-up.

To encourage capillary ingrowth and graft regeneration, vascular grafts must have an acceptable level of porosity, with an average pore diameter of 10 μm and a minimum pore area of 20–80 μm$^2$. Macrophages, cells, fibroblasts, and capillaries deposit on the wall with pore sizes of 25–40 μm and higher in the first few weeks, allowing cellular infiltration. Electrospinning may be a useful technology among other manufacturing processes for creating an appropriate microstructure for full neovessel production because of its ability to adjust fiber and pore diameters [20]. As a result, electrospun fibrous scaffolds have become much more promising candidates for vascular tissue engineering due to their highly porous structure and high surface-to-volume ratios, which encourage cell interactions, including EC adhesion and SMC diffusion into the porous outer layer [46,47].

The electrospinning parameters can be divided into three groups, which are solution parameters (viscosity, conductivity, molecular weight, and surface tension), process parameters (voltage, tip to collector distance, and flow rate), and ambient parameters (humidity and temperature) [48]. In order to produce scaffolds satisfying the requirements structurally, mechanically, and biologically, the optimum parameters must be chosen in accordance with the needs of the electrospinning method and the final prosthesis.

## 5. Design Components for Electrospun Vascular Prosthesis

The design parameters for electrospun vascular grafts can be divided into two categories: the constructional parameters, which involve fiber diameter, pore size, porosity, fiber orientation, wall thickness, the number of layers, and material selection. The scaffold's configuration and material choice are both essential because they have a significant impact on mechanical and biological characteristics, including compliance, tensile strength, burst pressure, blood permeability, and suturability, as well as biological processes such as cell phenotype, ECM formation, and cell diffusion.

### *5.1. Constructional Parameters*

It is challenging to design ideal 3D scaffolds that replicate the properties of ECM; thus, electrospinning is becoming more popular for making vascular grafts due to its potential to create scaffolds with micro/nano-scale topography, high surface area-to-volume proportions, and highly interconnected pores. Researchers can optimize the properties of prostheses and produce scaffolds with higher cell infiltration and proliferation and adequate mechanical properties by modifying the construction parameters of fibrous scaffolds by altering the electrospinning parameters [49].

#### 5.1.1. Fiber Diameter, Pore Size, Porosity, and Permeability

Ideal scaffolds are frequently fabricated to be very porous for cell diffusion, nutrient and oxygen delivery, and metabolic disposal of wastes to promote the development of targeted neotissues [50]. Small pore sizes are the major issue concerning electrospinning because they result in inadequate cell penetration and compliance mismatch [31,51]. This issue can be resolved by regulating porosity using various techniques, including salt/polymer leaching, collector modification, post-treatment with laser radiation, and adjusting the electrospinning conditions. It has been demonstrated that the pore size of electrospun webs is directly associated with the fiber diameter, suggesting that the pore size increases with an increase in fiber diameter. Thus, the diameter of the fiber can be easily modified by changing electrospinning variables such as the polymer concentration, voltage, and solvent type [51]. Even though electrospun prostheses made of nanofibers have a greater capacity for cell adhesion and proliferation than scaffolds made of microfibers, they frequently have lower cell infiltration levels. This is typically due to the small pore sizes, complex distribution, and lack of pore connectivity of scaffolds made of nanofibers, which have an impact on long-term matrix regeneration. Thus, using microfibers and nanofibers together encourages cell adhesion and proliferation with the help of nanofibers and gives more void areas for cell penetration through less dense microfibers [52].

The pore size of vascular grafts is recognized as an essential design parameter in the production of TEVGs because the vascular cells must be effectively settled with ECs on the lumen surface and SMCs in the outer layers. While ECs on the luminal surface prevent thrombosis, SMCs on the outer wall of vascular scaffolds support the scaffolds' activities such as vasoconstriction and vasodilatation [53]. It is claimed that electrospun scaffolds with fiber diameters of more than 1 μm allow larger pore diameters and encourage cell penetration, whereas smaller fiber diameters of less than 1 μm dramatically restrict diffusion for the majority of cell types, and so the ideal pore diameter necessary for sufficient cellular penetration is greater than 10 μm [53–55]. Small pore diameters are acceptable for the ECs to accumulate, proliferate, and infiltrate on the graft surface, which encourages ECM regeneration; however, they hinder SMCs' infiltration and colonization around the neo-vessel [53]. It has also been stated that the optimum scaffold pore diameter ranges from 5 to 500 μm since distinct cell types have unique dimensions and morphologies [56,57]. Large pores are ideal for better cell diffusion but can also promote blood leakage through the graft wall. With a homogeneous design, it is challenging to achieve a balance among enhanced tissue regeneration, decreased blood leakage, and sufficient mechanical characteristics; for this reason, multilayered vascular prostheses with different pore diameters have been thought to be useful [56]. Additionally, it is claimed that grafts with a porosity of 90% and pore sizes between 100 and 300 μm can effectively support cell adhesion and matrix development. When SMCs are cultured on these scaffolds, the mechanical behavior can be changed from elastic to viscoelastic, more closely approximating the mechanical characteristics of the native vessels [58].

The mechanical characteristics of the vascular scaffolds are also greatly influenced by the fiber diameter, pore size, and porosity, in addition to their biological impacts. Fluid permeability, thermal conductivity, diffusion coefficient, elastic modulus, yield, rupture, stiffness, fatigue resistance, and ductile strength are all significantly affected by porosity, a microstructural feature [59,60]. The superior mechanical properties are often seen in scaffolds with low porosity [61]. In nanofibrous scaffolds, mechanical characteristics are typically reported to decrease as porosity and pore diameter increase [34,62]. The stiff porous nanofibers located in the nanofibrous webs with strongly packed structures and enhanced molecular orientation have high tensile modulus and strength and low elongation at breakage. Reduced porosity and smaller pore sizes also improve the ductility of the material [63]. On the other hand, larger pore sizes lead to massive surrounding fibrous tissue accumulation post-implantation, which significantly reduces compliance, whereas low porosity limits endothelialization, negatively impacting antithrombogenicity [64]. The graft's flexibility is reduced due to the extensive fibrous accumulation caused by the large pore diameters, and high-porosity scaffolds are weaker than low-porosity ones. On the other hand, sufficient porosity (>80%) is usually necessary to simulate vascular distensibility. A detailed examination of burst strength is also essential to bring a promising graft through the stages of in vivo investigation and further clinical studies [65,66]. Interestingly, it has been demonstrated that the burst strength can decrease significantly after a certain porosity level because the low-porosity scaffolds are too fragile to withstand high pressures [67]. The more flexible high-porosity scaffold had a larger strain at rupture, burst pressure, and suture retention strength than the low-porosity scaffold with more closely packed fibers. When Young's modulus of the two grafts was compared, the low porosity graft had higher maximum stress and was stiffer than the high porosity grafts. On the other hand, bilayered grafts having layers of both high and low porosity exhibited performance outcomes that were comparable to those of monolayer grafts. Therefore, despite the use of the same polymer, different microarchitectures may provide mechanical properties that are noticeably different [68]. Additionally, the in vitro and in vivo mechanical performances of the grafts should be considered.

In addition to the porosity, pore size, and inner connectivity of pores, static permeability is another critical parameter that influences the penetration and proliferation of cells as vascular graft performance. The permeability affects the molecular exchange between the enclosed graft and the surrounding blood environment. Permeability depends on the electrospun scaffolds' packing density, porosity, and pore size. Densely packed fibers result in poor porosity and permeability, which hinder cellular infiltration inside the scaffolds, thus limiting the penetration distance of cells. In these circumstances, the oxygen and nutrient diffusion is limited, and cells can survive only on the surface. A perfectly permeable vascular graft should prevent immunogenic molecules from entering and permit the free transportation of oxygen, essential nutrients, and metabolic waste of cells [69]. To maintain a cell's expected growth, the permeability of TEVGs must be sufficient to transport oxygen and nutrients and export waste between the microenvironment of cells and the blood.

#### 5.1.2. Fiber Orientation

Recent studies on the electrospinning of aligned fibers mainly concentrate on the configuration of the collector system, such as parallel electrodes, metal rotating discs, and mandrels. The main concerns of researchers are the linear velocity of the collector surface and the effects of the collection settings on the electric field. It has been stated that the electrical properties of the solvent, along with the collector speed, have a significant impact on the level of fiber orientation [70]. It is challenging to achieve the high speeds greater than 10,000 $min^{-1}$ required to obtain fiber orientation by using rotating mandrels with diameters less than 6 mm. As a result, large-diameter rotating collectors (630 mm, 100 mm, 32 mm, and 640 mm) were used in many studies to achieve high rotational speeds and eliminate the resonance frequency concern [71]. It has also been shown in the literature that the polymer type is another factor that affects the aligned fiber morphology. Some polymers can align crimp-like, whereas others are oriented in the flat form [72].

Fiber orientation has been regarded as one of the most important characteristics of scaffolds since it affects both cellular orientation and the mechanical characteristics of prostheses used as vascular grafts [73,74]. The main factor influencing cell development behavior is fiber orientation, and cells on scaffolds typically create a phenotypic morphology and grow effectively based on fiber alignment [75,76]. Furthermore, it has been demonstrated in studies that radially oriented fibers encourage SMC penetration and alignment [77].

On the other hand, there is a significant correlation between the radial elastic modulus of the tubular scaffolds and the direction of fiber orientation. Circumferentially aligned fibers provide higher radial elastic modulus, and the Poisson effect confirms the distribution of fiber orientations in terms of mechanical characteristics [78]. In contrast to their orientated counterparts, randomly distributed fibers significantly improve the suture retention strength (SRS). This result is unexpected as efficient scaffold designs are usually approached with orientation to enhance mechanical properties. Thus, a multilayer strategy for vascular substitutes with carefully selected fiber orientations is necessary to provide the ideal balance of compliance, burst pressure strength, and SRS, particularly at the anastomotic site [79]. Modifying fiber orientation enables the control of graft compliance [80]. Oriented fibers display better modulus, tensile strength, and burst strength values, as well as reduced compliance when strained in the direction of orientation, which is related to the stiff structure of the material [81].

#### 5.1.3. Wall Thickness

Along with the previously mentioned factors, wall thickness is a crucial factor in designing vascular grafts since it affects the biomechanical characteristics, compliance, burst pressure resistance, and biological activities. Native vessels are reported to have walls with thicknesses ranging between 400 and 1000 μm [82]. Increasing the electrospinning duration will result in larger walls for the vascular scaffolds, significantly enhancing their circumferential tensile strength and suture retention strength [83]. Suture movement is more challenging in grafts with thicker walls, which provide increased fiber overlapping and enhanced binding force. However, the increased wall thickness is unfavorable for graft porosity and compliance [84]. Vascular grafts with a thinner wall thickness are more permeable and have greater mass transfer than the ones with a greater wall thickness in vivo. Hence, they have better cell proliferation and attachment performance than grafts constructed with thick layers [85]. Additionally, studies have demonstrated that as wall thickness is increased, vascular graft compliance decreases [86]. Compliance mismatch among the synthetic vascular scaffold and the native blood vessel also causes a change in hemodynamics, which then affects wall shear stress (WSS) and creates irregular flow patterns. Thus, undesirable biological responses are triggered by inconsistent mechanical signals that result in intimal hyperplasia [87]. In several investigations, the wall thickness has been decreased to produce compliant grafts similar to the native vessels; nevertheless, this can lead to poor bursting strength, which might not be adequate for implantation. The thickness of the graft wall also has an impact on blood permeability and graft handling during surgical procedures. Hence, it may be challenging to achieve a proper balance between all mechanical and biological properties and design parameters, especially blood leakage, cell permeability, burst strength, and compliance, when deciding on the wall thickness of the vascular grafts [88]. This makes determining the ideal wall thickness for vascular grafts extremely important.

#### 5.1.4. Number of Layers

As previously mentioned, vascular tissue engineering aims to imitate the construction and activities of native vessels that are composed of three layers known as the *tunica intima, tunica media,* and *tunica adventitia,* which provide high strength, elasticity, and compliance as well as outstanding hemodynamic function and anti-thrombogenicity [89]. Different roles are accomplished by each layer within the blood vessels. For instance, the endothelium layer of a native blood vessel is a well-organized monolayer, and the alignment of endothelial cells can regulate biological signaling such as intracellular protein expression, cytoskeleton development, and cellular interactions, whereas the middle layer involves spindle-shaped and circumferentially oriented SMCs that significantly affect the elasticity, mechanical strength, and vasoactive reactivity of blood vessels [90]. The reported mechanical and biological incompatibility of monolayered electrospun vascular scaffolds has led to the development of vascular prostheses with multilayers as an alternative technique for mimicking the characteristics of these layers [91]. In this regard, the middle and outermost layers should have a higher porosity to encourage SMC migration, whereas the inner layer should have a lower porosity to promote EC proliferation and limit blood permeability [92]. According to the researchers, creating multi-layered vascular scaffolds that imitate the mechanical and structural features of the native vessel walls is a useful way to mimic the functions of the *media* and *intima* layers [90,93]. Additionally, fabricating synthetic vascular scaffolds consisting of multiple layers with unique mechanical characteristics enables achieving a particular J-shaped stress–strain curve as in native blood vessels that show non-linear stress–strain behavior that provides the vessel's resilience and, as a result, helps prevent aneurysms [94]. Due to the integrated mechanical features of the layers, the composite effect has been observed in stress–strain graphs in the work by Yalcin Enis et al. [71] by creating bilayered scaffolds with layers of random and orientated fibers of PCL and PLC polymers with various molecular weights. Therefore, by optimizing the fiber diameter, fiber alignment, pore size, wall thickness, material type, or their combinations, multilayer designs should be created to meet the requirements of vascular grafts in separate layers.

### *5.2. Material Selection*

Synthetic vascular grafts made of non-biodegradable materials, including ePTFE, Dacron, and PU, which are commercially utilized, are not suitable for manufacturing grafts with diameters smaller than 6 mm, which are required to replace the saphenous vein, internal mammary artery, or radial artery as a vascular substitute because of poor patency, compliance mismatch, thrombosis, and ineffective neo-vessel development [12,77,95]. The drawbacks of currently available materials have prompted scientists to design biodegradable synthetic vascular grafts to enhance native vessel regeneration and reconstitute a functional arterial composition. However, when employed in animal experiments, these grafts have shown severe failure because of aneurysms, intimal hyperplasia, and thrombosis. These outcomes are probably brought on by the regenerated grafts with an insufficient amount of elastin [96].

Multiple biopolymers, including synthetic and natural ones, can be used to construct vascular grafts from electrospun fibers [97]. These materials are utilized to create a vascular scaffold that is physiologically suitable, and they should be chosen based on the graft structure, desirable biodegradation rate, and capacity for cell adhesion [38]. The material and architecture of small-caliber TEVGs have a significant impact on their biocompatibility, non-toxicity, non-immunogenicity, mechanical properties, ease of handling, and storability [98]. Furthermore, the scaffolds need to support host tissue remodelling and tissue regeneration during biodegradation and withstand inherent biological stresses as in biological systems to resist long-term issues including infection, intimal hyperplasia, stenosis, calcification, and aneurysmal dilatation [99]. While a rapid degradation rate may improve regeneration efficiency, it may also diminish tissue performance and damage mechanical qualities. In contrast, a slow degradation rate may hinder the development of neo-tissues. Therefore, it is essential to maintain a balance between the rates of vascular regeneration and biodegradation [37]. Additionally, the significance of in vivo foreign body responses of monocytes and macrophages to biomaterials is crucial for developing neo-vessels and thrombosis; as a result, strategies for material choice and manufacturing that control macrophage phenotype have drawn considerable attention [100]. Since it is normally impossible for one material to satisfy all of these qualities, mixing several polymers to form a hybrid graft seems to be an effective way to fabricate TEVGs [8]. These composite scaffolds can be thought of as innovative smart biomaterials that have the potential to produce TEVGs since they blend the advantages of natural polymers, including biocompatibility and biochemical capabilities, with the benefits of synthetic polymers, consisting of high strength, modifiability, and processability [101]. Some of the most commonly used biopolymers in vascular tissue engineering applications are given in Table 1, with their advantages and disadvantages.

**Table 1.** Some of the advantages and disadvantages of natural and synthetic biopolymers used in tissue engineering.

| Type | Biopolymers | Advantages | Disadvantages | References |
| --- | --- | --- | --- | --- |
| Natural | Collagen | • Supports EC and SMC attachment<br>• Biocompatible | • Shows thrombogenicity<br>• Lack of mechanical strength | [102–104] |
| | Alginate | • Biocompatible<br>• Gel forming ability<br>• Non-toxic<br>• Biodegradable<br>• Easy to process | • Low stability<br>• Poor mechanical and barrier characteristics<br>• Requirement of combination with other biopolymers | [105–107] |
| | Chitosan | • Formability<br>• Fabricability with other biopolymers<br>• Anticoagulant ability<br>• Biocompatible<br>• Biodegradable | • Poor mechanical strength<br>• Low stability<br>• Low spinnability in electrospinning<br>• Toxicity | [108–110] |
| | Elastin | • Mechanically flexible<br>• Resisting physiological pressures | • Strong tendency to calcify<br>• Hard purification process | [111–113] |
| | Fibrin | • Biocompatible<br>• Simple extraction procedure from patient's blood<br>• Encourages cell adhesion and collagen synthesis<br>• Nonlinear elasticity<br>• Contributing compliance<br>• Resisting intense deformations | • Low mechanical strength<br>• Fast degradation rate | [114–117] |
| | Gelatin | • Low cost<br>• Biocompatible<br>• Biodegradable<br>• Low antigenicity<br>• No denaturation during electrospinning | • Dissolution and loss of gelatin matrices under physiological conditions | [118,119] |

Table 1. *Cont.*

| Type | Biopolymers | Advantages | Disadvantages | References |
| --- | --- | --- | --- | --- |
| Synthetic | PCL | • Flexibility<br>• Structural stability<br>• Thermoplasticity<br>• Biocompatible<br>• Slow biodegradation | • Slow biodegradation<br>• High hydrophobicity resulting in low cell affinity<br>• Low bioactivity<br>• Low oxygen permeability<br>• Low surface energy | [120–123] |
| | PLA | • High tensile strength<br>• Non-toxicity<br>• Opportunity to control the polymer crystallization, shape, and hydrolysis<br>• Biodegradable<br>• Biocompatible | • Acidic degradation byproducts<br>• Brittleness<br>• Poor wettability | [124–127] |
| | PGA | • Fast degradation rate<br>• Good mechanical characteristics<br>• Biodegradable<br>• Biocompatible | • Triggering inflammatory reaction<br>• Fast biodegradation rate resulting in a loss of mechanical performance | [128–130] |

An ideal vascular graft should possess mechanical strength, compliance, suture retention strength, and a J-shaped mechanical response close to physiological levels. The mechanical responses of both passive (elastin and collagen fibers) and active components (SMCs) affect the mechanical reaction of the artery wall. When arteries are subjected to blood pressure, non-linear elastic behavior is seen as a J-shaped curve [131]. Many biomaterials and biological tissues have what is known as a J-shaped strain–stress curve, which illustrates how small increases in stress initially lead to enormous elongation, but as the material stretches further, it stiffens and becomes more difficult to stretch [132]. Elastin fibers are mainly responsible for the compliance of vessel walls at low pressures, whereas high stiffness is mainly caused by the mechanical reaction of collagen at high pressures. As pressure increases, collagen fibers start aligning and orienting, lowering arterial compliance. Therefore, utilizing collagen and elastin and mimicking their crimpy architecture appears to be a suitable method for creating a small-caliber vascular graft that simulates this mechanical reaction [131]. In order to provide sufficient compliance and structural integrity of TEVGs, the co-spinning of both elastin and collagen has been used as a strategy to imitate the artery's three-layered architecture [133].

A natural contractile-like SMC phenotype can be differentiated to resemble the composition of native blood vessels by using ECM proteins such as collagen type I and insoluble elastin, which have superior viscoelastic capabilities. In addition to their outstanding biological features, natural polymers can be easily modified in terms of their mechanical characteristics and biodegradation rates through the change in their degree of crosslinking [134]. The polymers derived from ECM components collagen, elastin, fibrin, and gelatin are utilized to create TEVGs. There are many studies targeting the improvement of the integrity of collagen-based grafts to overcome the inadequate mechanical qualities, including anastomosis strength, burst pressure, and tensile strength. Elastin and gelatin grafts have similar behavior. In addition, dynamic culture has enhanced the mechanical behavior of fibrin grafts developed from in vitro-produced fibroblasts, providing compliance close to natural blood vessels [8]. The J-shaped mechanical behavior is generally achieved when natural biopolymers such as collagen, elastin, or fibrin are used.

However, natural polymers' properties differ from sample to sample, and it might be challenging to find consistently suitable production conditions. They are also weak, which makes it tricky to withstand intense physiological forces. Although synthetic biopolymers with mechanical properties similar to collagen and elastin appear to be promising materials for replacement, using them in a neat form causes a decrease in compliance as stress increases. This makes achieving J-shaped reactions in vascular grafts difficult [131]. Some of the most studied synthetic biopolymers for vascular scaffolds are biodegradable polyesters, including PGA, PLA, PLLA, their copolymer PLGA, and PCL [10]. When compared to natural polymers, synthetic polymers have several advantages. First, they are simple to produce due to their physical and chemical characteristics. Despite their ability to help restore damaged tissue structure and activity, these biomaterials have limited cell attachment locations and thus need chemical modifications. The tensile strength, Young's modulus, and degradation rate can be predicted and repeated over a wide range. These polymers vary in their degree of biodegradability, biocompatibility, and mechanical characteristics, but no single polymer provides the ideal mix of all of these crucial characteristics [135]. Therefore, utilizing various polymers to construct hybrid grafts that may offer ideal features similar to native vessels is considered a promising technique.

Tissue engineers have been able to adjust TEVG features thanks to various polymers and production methods, but choosing the ideal mix of graft properties is still difficult to accomplish [136]. Therefore, the design and material selection should be considered together rather than individually, as the scaffold's characteristics depend on its morphology and material.

## 6. Mechanical Forces Acting on the Vascular Grafts

Sufficient mechanical strength and Young's modulus in the longitudinal and radial directions are essential for clinical trials of vascular grafts because these scaffolds must withstand repeated mechanical stresses, including expansion, shrinkage, bending, and stretching under the in vivo conditions affected by blood flow and body movements [137]. In creating vascular grafts, it has been challenging to balance adequate mechanical strength to endure physiological pressures and compliance similar to native vessels to avoid unfavorable hemodynamic fluctuations [31]. In order to imitate in vivo conditions, vascular tissue engineering needs a platform that mimics the hemodynamic shear and regular forces that vascular tissues experience in the body in the radial, circumferential, and longitudinal directions. Shear stresses are tangential frictional forces directly affecting ECs and are delivered to SMCs by interstitial flow and signaling [138]. Numerous hemodynamic forces, including flow shear stress, frictional forces parallel to the vascular wall produced by blood flow, and circumferential stress perpendicular to the vascular wall brought on by transmural pressure, affect all blood vessels. Vascular SMCs and ECs are likewise affected by these physical stresses; variations in stress can trigger intracellular signaling pathways that affect the cellular activity and blood vessel formation [139]. The endothelium performs the role of a mechanoreceptor, sensing variations in blood flow, shear stress, and pressure and then causing the secretion of signaling molecules that cause the SMCs to dilate or constrict. Pulsatile flow and mechanical stresses in the blood vessel lumen also have an impact on patency levels. Laminar shear stress is essential for maintaining the required endothelium morphology [38]. The pulsatile blood pressure periodically exposes the vascular wall to cyclic circumferential stresses of about 100–150 kPa, causing average strains of 10–15%. Further, on the vascular wall of humans, blood flow produces an oscillatory shear stress of 1–5 Pa, which differs based on body size and vessel type. Even though shear stresses are five times less intense than circumferential stresses, they still have a considerable effect on cellular activity [140].

Therefore, it can be noted that vascular grafts are exposed to four hemodynamic stresses: shear stress ($\tau$, tangential frictional forces acting on ECs attributable to blood flow); luminal pressure ($\sigma_{nor}$, a cyclic normal force due to blood pressure); cyclic circumferential stress ($\sigma_{cir}$, a circumferential mechanical stretch due to blood pressure); and longitudinal stress ($\sigma_{L}$,) [141], which are illustrated in Figure 3. Thus, blood pressure produced by pulsatile blood flow also moves perpendicular to the EC matrix [142].

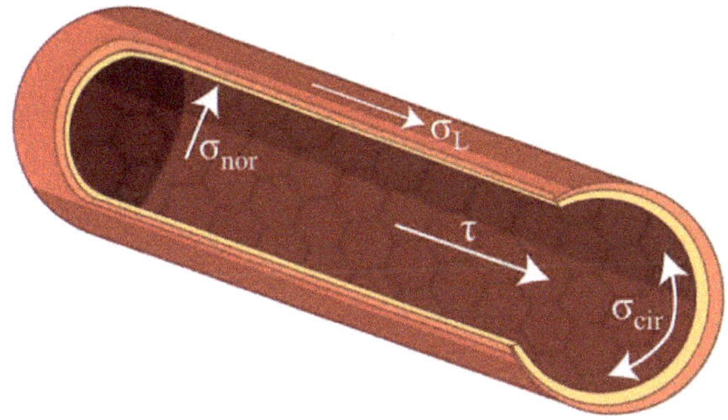


**Figure 3.** The mechanical stresses on blood vessels.

### *6.1. Shear Stress*

The tangential element of frictional forces brought on by the blood flow in a lumen is known as shear stress. The shear stress unit in the SI system is the Pascal (Pa). The cardiovascular system frequently makes use of dyn/cm$^2$ (1 Pa = 10 dyn/cm$^2$) [143]. Shear stress ($\tau$), which is parallel to the vessel wall, is used to describe the frictional force acting on the vascular endothelium. ECs face shear stress that ranges from 1 to 6 dynes/cm$^2$ in the circulatory system and from 10 to 70 dynes/cm$^2$ in the arteries, with an average of 20 dynes/cm$^2$ [144]. When blood shows laminar flow, shear stress is calculated as

$$\tau = 4\mu Q/(\pi r^3) \quad (1)$$

where $\mu$ is the viscosity, Q is the flow rate, and r is the radius of the vessel. The shear stress in arteries with large diameters typically ranges between 5 and 20 dyn/cm$^2$, although under conditions of high systolic pressures, significant instant values can reach 40 dyn/cm$^2$ [145].

As a result of blood flow, shear stress triggers ECs to convert mechanical sensations into intracellular signals that change cellular processes such as proliferation, apoptosis, infiltration, permeability, and regeneration as well as gene expression. These changes are crucial for maintaining the homeostasis of vascular system and the mechanisms underlying blood flow-induced circumstances such as angiogenesis, vascular regeneration, and atherogenesis [146,147]. Additionally, endothelial progenitor cells (EPCs) grown from human peripheral blood elongate and align longitudinally towards the flow direction when subjected to laminar shear stress [147]. Vasodilation and vasoconstriction triggered by flow are affected by WSS. Intimal hyperplasia is the outcome of alterations in the flow pattern that are brought on by both high and low shear stresses. Stress concentration in anastomotic regions and compliance mismatch between the native vessel and the scaffold are the two factors that contribute to abnormal WSS. Therefore, it is critical to avoid compliance mismatch in order to reduce these disruptive flow patterns [64]. Shear stress can also lead weakly adhered ECs to detach from the lumen, resulting in thrombosis in vascular grafts lacking endothelial lining when blood encounters a surface other than the endothelium. In numerous investigations, exposing vascular grafts to shear forces in pre-implantation improved EC adhesion [148].

### *6.2. Luminal Pressure*

Pulsatile blood flow creates a tensile stress as a result of the normal force that is perpendicular to the vessel wall and results in cyclic strain [149]. ECs can sense and respond to normal stresses. Cell adaptation systems can be inappropriate and cause disease and significant changes in the cell phenotype when they are subjected to extreme conditions such as mechanical stress from continuous and high-intensity stretching [150].

### 6.3. Cyclic Circumferential Stress

This is the periodic stretching of arterial wall components that is produced by periodic increments in transmural pressure difference and normal forces caused by blood flow [143]. Axial and circumferential strains have a significant impact on EC morphology, vascular cell proliferation, and matrix remodeling [151]. Endothelial cells adjust their phenotype and active signaling mechanisms in response to cyclic circumferential strain by orienting themselves perpendicular to the force vector [152]. Mechanical stretch, which is detected by mechanoreceptors, also controls how the SMC responds. Physiological pulsatile circumferential stress on the arterial wall enables SMCs to show contractile response. Circumferential stress consequently influences gene expression as well as SMC processes such as proliferation, survival/apoptosis, diffusion, and ECM remodeling. Circumferential stress varies from 1 to $2 \times 10^6$ dynes/cm$^2$, based on the anatomical region [151].

Circumferential stress is studied more extensively as it provides data on the dimensions of the vascular luminal diameter and wall thickness changes based on blood pressure variations. It is possible to determine circumferential stress ($\sigma_{cir}$), which is frequently represented in dynes/cm$^2$, as follows:

$$\sigma_{cir} = (P.r_i)/t \quad (2)$$

where P is the internal pressure, $r_i$ is the inner radius, and t is the wall thickness of the graft [153,154]. The mean circumferential stress is considered to better respond to slight variations in the *intima* thickness and increases with larger wall thicknesses [155,156]. The internal pressure also can be calculated as follows:

$$P = F/(2r_iL) \quad (3)$$

where F is the force when it is exposed to P, and L is the length in the z direction [157]. In addition, the circumferential strain measures the change in the internal diameter brought on by a change in the intraluminal pressure. The following formula can be used to determine strain ($\varepsilon_i$), which has no units:

$$\varepsilon i = (D_i - D_0)/D_0 \quad (4)$$

where $D_i$ is the diameter at a particular pressure, and $D_0$ is the reference diameter [153]. Under physiological blood pressure of 100 mmHg, the mean circumferential stress of the coronary arteries is around 150 kPa, whereas the strain is approximately 10–15% [158].

### 6.4. Longitudinal Stress

All arteries experience longitudinal (axial) stresses varying from 40% to 65% in vivo [159]. Physiological alterations can alter these longitudinal stresses because arterial tethering exposed by surrounding tissues maintains them. Key mechanical signals that encourage arterial remodeling are also provided by longitudinal stresses, such as changes in shear stress and circumferential strain [160]. A longitudinal strain also promotes cell proliferation in the artery wall while preserving arterial wall function [161]. The distending force in the longitudinal direction results in internal longitudinal stress ($\sigma_{L\text{-}P}$). On the other hand, a second tensile force creating longitudinal stress in the latter direction exists due to the arterial tethering ($\sigma_{L\text{-}T}$) caused by surrounding tissue along its length. The formulas given below can be used to calculate the internal longitudinal stress ($\sigma_{L\text{-}P}$), the longitudinal stress caused by arterial tethering ($\sigma_{L\text{-}T}$), and total longitudinal stress ($\sigma_L$):

$$\sigma_{L-P} = \frac{Pr_i}{2t} \quad (5)$$

$$\sigma_{L\text{-}T} = F_{L\text{-}T}/(\pi\,(r_e{}^2 - r_i{}^2) \quad (6)$$

$$\sigma_L = \sigma_{L\text{-}P} + \sigma_{L\text{-}T} \quad (7)$$

where the variables are the P, $r_i$, external radius ($r_e$), t, and the longitudinal forces comprising the forces resulting from the blood pressure ($F_{L\text{-}P}$) and tethering ($F_{L\text{-}T}$) in Equations (5) and (6). The total longitudinal stress ($\sigma_L$) equals the sum of stress due to pressure ($\sigma_{L\text{-}P}$) and stress due to tethering ($\sigma_{L\text{-}T}$) given in Equation (7) [19]. All dimensional variables and the force components used in these equations are represented based on the vascular graft in Figure 4.

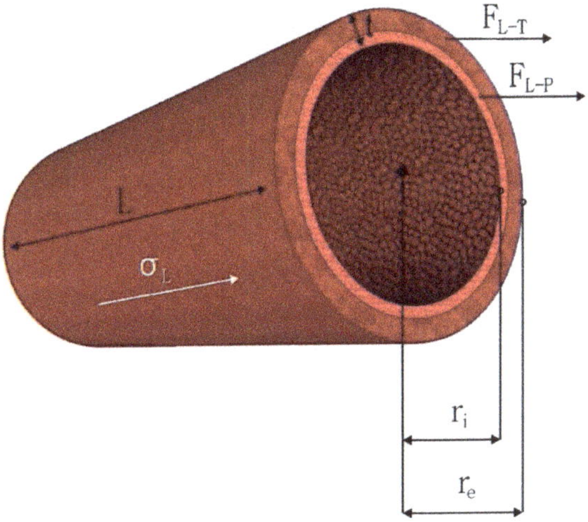


**Figure 4.** The dimensional variables of a vascular graft and longitudinal force components.

The human femoral artery has a longitudinal elastic modulus of 978 kPa and longitudinal maximum stress of 65 kPa, placing it at the maximum limit of small-diameter vascular grafts [162]. Additionally, it has been discovered that when the artery is stretched longitudinally by 48% of its load-free length in vivo, the longitudinal stresses are greater than the circumferential stresses [163].

Therefore, the physical and mechanical properties of vascular grafts, such as dimensions, compliance, bursting strength, elasticity, and Young's modulus, must be properly provided by considering the design criteria that have a major impact on their long-term performance in order for them to resist all of the aforementioned forces.

## 7. Mechanical Characteristics of Vascular Grafts

Recent methods for enhancing the clinical efficacy of vascular grafts and preventing graft failure brought on by intimal hyperplasia, thrombosis, aneurysm, blood leakage, and occlusion have been documented in the literature. These methods include creating grafts with adequate compliance, elasticity, ultimate tensile strength, ultimate strain, Young's modulus, burst pressure, and suture retention strength to withstand all of the physiological stresses that native vessels experience [81,164–167].

According to recent studies, it is essential to develop vascular prostheses that have sufficient tensile strength and strain and also exhibit a nonlinear, J-shaped stress–strain behavior similar to human coronary arteries [101]. The stress–strain curve shows that blood vessels have nonlinear J-shaped mechanical behavior, with an initial elastic region at low strain due to the presence of elastin, followed by an increase in stiffness with a curved transition at high strain due to the presence of collagen. Collagen and elastin, two structural proteins, work together synergistically to produce the J-shaped curve observed in strained materials. This mechanical response provides adaptation and protection to aneurysm generation, which is identified as a localized expansion or ballooning of a section of an artery higher than 50% of its normal diameter, primarily induced by a weakening of the vascular wall exposed to primary stress [168,169]. Unfortunately, the J-shaped mechanical behavior of native vessels has not been replicated in most trials fabricating vascular prostheses to substitute coronary arteries. Some techniques have been used to solve these issues, such as constructing the middle and outer layers to approximate the structural configurations of collagen and elastin fibers found in native blood vessels [170].

Native blood vessels are dynamic, flexible, and strong tissues that can withstand physiological stresses. The constructed vessel must resist pressure and pulsatile blood flow without rupturing or membrane leakage. Burst pressure is, therefore, among the most crucial elements in assessing whether a material is suitable for implantation and describes the amount of pressure the scaffolds can endure before failing [171]. Generally, TEVGs designed for replacement are expected to have burst pressures greater than 2000 mmHg [172]. In addition, TEVGs must be able to withstand distortion and compression and show adequate tensile and shear strength to guarantee that the scaffold can resist the tensile stress caused by suturing during implantation, to prevent rupture, scattering, wear of the edges, and tearing of the seams, and to sustain circumferential strength to withstand hemodynamic forces [84]. The suture retention strength analysis is employed to calculate the magnitude of force needed to rupture the scaffold wall or rip a suture from a structure. The suture retention strength of the human saphenous vein is 1.81 N [173].

Compliance, also known as the inverse of stiffness or the change in vessel diameter over time as a function of pressure, is a term used in vascular tissue engineering to describe the ability of the vascular scaffold to stretch circumferentially as a reaction to pulsatile pressure [87,88,174]. The performance of the prosthesis is affected by compliance mismatch between a native artery and a vascular replacement, which results from different mechanical behaviors that cause a mismatch in diameter change and can reduce patency due to intimal hyperplasia [87,175]. Low WSS is considered to be the outcome of the compliance mismatch interrupting the flow at the distal anastomosis. Because of the low WSS, the vessel wall tries to restore the flow disruption by thickening the *intima*, which finally causes the scaffold to occlude [174]. Additionally, a longer residence duration of atherogenic particles, associated with the low WSS, may trigger blood particle accumulation or adherence to the artery wall, encouraging plaque growth and intimal thickening. Thus, lower WSS may be negatively impacted by arterial compliance mismatch, which could lead to particle deposition in stiffer vascular wall regions and increase the risk of arterial disease [176]. In many studies, a compliance mismatch results from the stiffness of commercially available synthetic polymers, such as ePTFE and Dacron [177]. Moreover, it is well reported that intimal hyperplasia, anastomotic aneurysms, or pseudoaneurysms can result from excessive stress at the suture locations caused by a compliance mismatch between the native vessel and the Dacron scaffolds [178]. Therefore, avoiding these issues and obtaining effective graft performance depend on developing a scaffold with compliance properties close to native vessels. Figure 5 illustrates how the compliance level affects blood flow patterns and WSS.

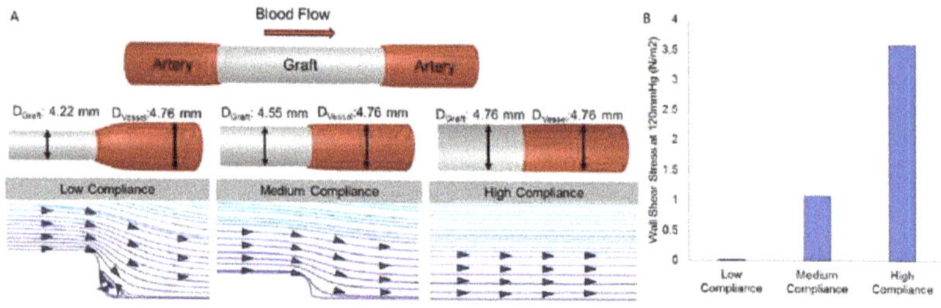


**Figure 5.** The effect of compliance on blood flow profiles and WSS [174].

## 8. Current Studies Guiding the Literature

Some studies focusing on the effect of design criteria on both cellular activities and mechanical properties and shaping the literature with seminal findings are discussed below.

Fiber diameter and pore geometry, which vary accordingly, are among the design parameters that should be prioritized in determining mechanical properties. Matsuzaki et al. [179] studied the influence of pore size on neoarterial tissue regeneration and graft stability over the regeneration phase using electrospun scaffolds. Scaffolds with an inner diameter of 5 mm were constructed using electrospun PCL fibers with four distinct porosity and pore sizes in the outermost layer (79% (4 μm), 82% (7 μm), 83% (10 μm), and 85% (15 μm) and heparin-conjugated PLCL sponge as the inner layer. All of the grafts were implanted in adult female sheep. Only scaffolds with pore sizes of 4 μm were demonstrated to resist dilatation for up to a year. Grafts with wider pore sizes improved cell infiltration, but neotissue could not regenerate quickly enough to provide the mechanical strength required to resist dilatation. Grafts with pore sizes of 10 and 15 μm had lower strength than that of the sheep carotid artery, which was approximately 3 MPa, but grafts with pore sizes of 4 and 7 μm had higher strength, around 4 MPa. Furthermore, the highest level of compliance was attained with the 10 μm pore-sized graft at roughly 1% $mmHg^{-1}$, which was still lower than that of the sheep carotid artery. It is also possible to examine the effect of pore size on mechanical properties through fiber diameter. Wang et al. [51] used electrospinning to manufacture PCL vascular grafts with an inner diameter of 2 mm, and thicker fibers were generated to achieve a structure with macroporosities to improve the scaffold's cell infiltration capabilities. While the fiber and pore diameters of these macroporous scaffolds were around 5–6 μm and 40 μm, respectively, these values were around 0.7 μm and 5 μm, respectively, in microporous scaffolds. The strain value of grafts with thicker fibers was approximately 639.20% higher than the strain value of grafts with thinner fibers, which was 168.40%, indicating that thicker-fiber grafts were substantially more rigid than thinner-fiber grafts. Grafts with thinner and thicker fibers had Young's moduli of 17.44 MPa and 21.00 MPa, respectively, indicating that an increase in fiber diameter resulted in a slight rise in Young's modulus. The ultimate tensile stress value of a thinner fiber graft was 13.35 MPa, but the ultimate tensile stress of thicker fiber grafts was 8.72 MPa, suggesting that increasing fiber diameter generated decreased tensile stress. Similarly, Valence et al. [56] used electrospinning to create PCL bilayered grafts with a 2 mm inner diameter, integrating a high-porosity layer with a low-porosity layer on either the luminal or adventitial side. There were four types of grafts produced: no barrier (high-porosity scaffold), inside barrier (low-porosity inner layer and high-porosity outer layer), outside barrier (high-porosity inner layer and low-porosity outer layer), and only a barrier (low-porosity scaffold). The results show that high-porosity grafts had a greater strain at break, burst pressure, and suture retention strength than low-porosity scaffolds. The maximum stress (6.09 MPa) and Young's modulus (12.0 MPa) of the low-porosity scaffold were greater than those of the high-porosity scaffold, at 4.98 MPa and 6.09 MPa, respectively. The only-barrier scaffold was more rigid than the no-barrier, inside-barrier, and outside-barrier grafts. On the other hand, the burst pressure and suture retention strength was higher for the more compliant no-barrier graft than for the only-barrier graft. Thus, it was emphasized that, despite using the same material, different microarchitectures could result in significantly different mechanical properties.

The wall thickness of vascular scaffolds, on the other hand, is another structural parameter that directly affects mechanical features, including compliance, burst pressure, suture retention, and tensile strength. Johnson et al. [31] investigated the effect of polymer choice and wall thickness on the mechanical properties of a vascular prosthesis. First, they evaluated the biomechanical properties of electrospun vascular grafts made of various biopolymers such as PCL, chitosan, PLCL, and PLLA, with a diameter of 6 mm and a wall thickness of 650 μm. The burst strength of the scaffold made of PLGA was 3.3 MPa, while the burst strength of the scaffold built of PCL was 0.8 MPa. The burst strengths of the human carotid artery, human saphenous vein, and ePTFE graft were all much lower than those of the electrospun scaffolds, ranging between 0.1 and 0.5 MPa. Among all, the PLLA graft exhibited the highest suture retention strength value of 1022 g. The best compliance value was 8.2% $mmHg^{-1}$ in PLCL grafts, while PLLA had a compliance value of approximately 3.8% $mmHg^{-1}$, close to that of the human coronary artery. The ePTFE graft had the lowest compliance value of 1.6% $mmHg^{-1}$. They also electrospun PCL grafts with varying wall thicknesses and studied their mechanical properties to see if sidewall thickness affects scaffold compliance and burst pressure. For wall thicknesses ranging from 400 to 1000 μm, compliance remained constant at 2–4% $mmHg^{-1}$. On the other hand, PCL grafts exhibited significant variance in compliance values, ranging from 2 to 11% $mmHg^{-1}$ at wall thicknesses of less than 400 μm. The burst pressure increased as the graft sidewall thickness increased. For PCL grafts with wall thicknesses ranging from 200 to 1000 μm, the burst pressure values increased linearly from 0.6 to 2.9 MPa. As a result, it was underlined that the wall thickness has a significant and more robust effect on the mechanical properties. The wall thickness can be adjusted by increasing or decreasing the electrospinning time. Therefore, optimizing this time to see the wall thickness effect is an important research topic. The goal of Bazgir et al. [180] was to evaluate the properties of PLGA and PCL-based nanofibrous scaffolds, as well as the effect of degradation on their structural properties. Six scaffolds were built by electrospinning, three with PCL and the others with PLGA, with processing times of 30, 60, and 90 min. It was observed that there was a clear relationship between the duration of the electrospinning and the tensile strength. In the case of PCL scaffolds, the ultimate stress and strain values were 0.99 MPa and 24.03% for the scaffolds fabricated in 30 min and 1.49 MPa and 28.15% for the scaffolds produced in 90 min, respectively, whereas 1.03 MPa and 34.36% for PLGA scaffolds manufactured in 30 min and 1.76 MPa and 36.33% for PLGA scaffolds produced in 90 min. Thus, both types of scaffolds demonstrated more robust mechanical properties with longer spinning periods, resulting in higher wall thicknesses. When the tensile strength and elongation values of PCL and PLGA scaffolds with a 90-min processing time were compared, the PLGA membrane was demonstrated to be more elastic and durable than the PCL membrane. The research reveals a relation between the electrospinning processing time and tensile strength. The scaffold became thicker and stronger when the electrospinning duration was increased.

Aside from the aforementioned criteria, several studies have focused on the effect of fiber orientation on the mechanical properties of vascular grafts. Many researchers claim that, based on the architecture of the native artery, fiber orientation can aid in improving mechanical properties, particularly tensile strength and burst pressure. Yalcin et al. [71] generated vascular scaffolds with a 6 mm diameter and a wall thickness ranging between 200 and 300 μm made of randomly distributed or radially oriented PCL and PLCL microfibers, and the effect of fiber orientation, polymer type, and the number of layers on the mechanical properties of the prosthesis was investigated. Bilayered grafts were created by combining layers with randomly distributed fibers in the inner layer and radially orientated fibers in the outer layer. In general, the fiber orientation in the radial direction utilized in the outer layers contributed to the burst and tensile strengths of the samples in the same direction. Among the single-layer grafts, the PCL grafts with oriented fibers had the highest ultimate tensile strength at 6.7 MPa in the radial direction and the lowest elongation at break, which was 80%. Oriented PLC samples, on the other hand, displayed elongation values greater than 900% due to their inherent high elasticity. In the case of bilayered samples, all of the samples had higher tensile strength and lower elongation values in the radial direction rather than the axial direction, owing to a greater number of fibers in that direction to bear stress but also fewer crossing points. Due to their extraordinarily elastic nature, both randomly distributed and radially orientated PLCL samples swelled at a lower pressure of around 562 mmHg but had the highest bursting resistance at 1500 mmHg. The findings supported the significant influence of polymer and graft construction on the mechanical behavior of scaffolds. Similarly, Grasl et al. [94] used an adjustable electrostatic field to alter the direction of the electrospinning jet, resulting in 2-mm-caliber vascular scaffolds containing circumferentially, axially, fenestrated, and randomly distributed PU and PLLA fibers. The influence of polymers and fiber orientations on the mechanical behavior of the prosthesis was evaluated and compared to that of the native rat aorta. The ultimate tensile force of PU and PLLA grafts composed of randomly oriented fibers was lower than that of the other fiber orientations. The grafts with circumferentially oriented fibers, on the other hand, had at least six times higher ultimate tensile forces than the grafts with randomly aligned fibers. Furthermore, the PU and PLLA prosthesis consisting of randomly oriented fibers showed a significantly lower maximum tensile force than the rat aorta (1.1 N) while the samples with circumferentially oriented fibers had much higher tensile forces than the rat aorta. Scaffolds built of both materials with randomly oriented fibers exhibited considerably lower burst pressures than other fiber orientations. Each fiber orientation contributed to bursting strength, close to or greater than the rat aorta's burst pressure (1043 mmHg). All directions of the alignment resulted in a decrease in compliance. The PU scaffolds with randomly oriented fibers displayed the maximum compliance of 29.7% 100 $mmHg^{-1}$, which is the closest value to the compliance of rat aortas (37.2% 100 $mmHg^{-1}$). Because of their rigidity, PLLA scaffolds have a lower compliance value. None of the grafts with alternative fiber alignments or polymers could replicate the J-shaped mechanical behavior of the rat aorta, supporting the researchers' view that multi-layered vascular scaffolds should be constructed to imitate the structure of the native blood vessel.

Polymers in mixed forms have been thought to be far more promising in tissue engineering applications than utilizing them alone. All polymers have unique properties that contribute to the various characteristics of the scaffolds. Finding the appropriate blend ratio is critical for attaining these scaffolds' required and optimum properties. Gao et al. [37] investigated the influence of polymer ratios on the mechanical properties and degradability of 2-mm-caliber electrospun vascular grafts made of PCL and PLGA in several blending ratios, including 95/5, 90/10, 80/20, and 60/40. When the findings were analyzed, it was revealed that the mechanical properties of PCL/PLGA (95/5) and PCL/PLGA (90/10) were adequate for use as vascular prostheses because they did not exhibit ultimate tensile stress, strain, or bursting strength values under the range of original grafts. The stress-strain graphs indicated that the strength and elongation of blended samples decreased as the proportion of PLGA increased. Furthermore, increasing the amount of PLGA reduced the suture strength and burst pressure of the PCL/PLGA blended scaffolds. The burst strength of the PCL/PLGA (95/5) and PCL/PLGA (90/10) scaffolds was found to be greater than 1500 mmHg, allowing them to be employed as a replacement for the native vessel. In another study examining the blend ratio effect on mechanical properties, Yang et al. [181] used electrospinning technology to create hybrid grafts of PCL and fibrin with blend ratios of 0/100, 10/90, 20/80, and 30/70. The results show that adding PCL to fibrin scaffolds significantly enhanced their mechanical properties. The burst pressure increased from 1347.43 to 1811.6 mmHg as the PCL ratio rose from 10% to 30%. On the other hand, the scaffold's Young's modulus dropped when the PCL content increased in all of the scaffolds. Finally, the PCL/fibrin (20/80) scaffold demonstrated balanced mechanical and degradability properties and high cell compatibility, indicating that it might be used as a tissue engineering platform for vascular grafts. However, the blend effect may not always combine the best result of both polymers as desired. Bolbasov et al. [182] used an electrospinning method to manufacture scaffolds comprised of PCL, PLLA, blended PCL/PLLA, and PLLA/PCL copolymer (PLC7015), emphasizing the relevance of material selection. Due to their highly crystalline structure, PLLA nanofibrous scaffolds showed the highest tensile strength (13.2 MPa) and lowest elongation (95%) values with uniaxial stretching. PCL and PLC7015 nanofibrous scaffolds' semicrystalline structures resulted in moderate strength values of 8.2 MPa and 9.1 MPa, respectively. PLC7015 has a lower crystalline structure than PCL, resulting in a maximum elongation value of 560%. According to the researchers, the PCL/PLLA exhibited the lowest strength at 4.7 MPa due to thermodynamic incompatibility at the molecular scale. The PCL/PLLA graft differed significantly from the PLC7015 graft in strength and elongation. Thus, when establishing design parameters, the effect of polymers and the method used to combine different types of polymers (either copolymerization or blending) on the properties of the scaffolds should be considered.

## 9. Conclusions

In this review, in-depth research has been conducted by firstly giving the fundamentals of the structure of the blood vessels, the requirements for the ideal vascular prosthesis, and the electrospinning method for fabricating vascular membranes. The significance of the correct determination of the design parameters to achieve the most ideal vascular grafts was then emphasized by giving their impact on material function, particularly from the standpoint of the mechanical and biological performance of the scaffolds. In order to have a better understanding of the importance of the mechanical properties of the vascular graft that can be used as a replacement to maintain the damaged tissue functions, the physiological forces acting on blood vessels and their consequences on physical and biological activities are also explained in detail. In the final stage, the theoretical part of the review is supported by the examination and comparison of the recent experimental findings in the literature. This study aims to guide researchers working in this field by providing the literature with a comprehensive and comparative summary of the most recent studies.

**Author Contributions:** Conceptualization, S.O. and I.Y.-E.; methodology, S.O. and I.Y.-E.; software, S.O., I.Y.-E., F.Y. and B.Y.; validation, S.O., I.Y.-E., F.Y. and B.Y.; formal analysis, S.O., I.Y.-E., F.Y. and B.Y.; investigation, S.O. and I.Y.-E.; resources, S.O., I.Y.-E., F.Y. and B.Y.; data curation, S.O. and I.Y.-E.; writing—original draft preparation, S.O. and I.Y.-E.; writing—review and editing, S.O., I.Y.-E., F.Y. and B.Y.; visualization, S.O., I.Y.-E., F.Y. and B.Y.; supervision, I.Y.-E.; project administration, I.Y.-E.; funding acquisition, I.Y.-E. All authors have read and agreed to the published version of the manuscript.

**Funding:** This research was funded by the Istanbul Technical University, Scientific Research Projects (grand no: 43368) and TUBITAK (grand no: 121M309).

**Institutional Review Board Statement:** Not applicable.

**Informed Consent Statement:** Not applicable.

**Data Availability Statement:** Not applicable.

**Acknowledgments:** The author F.Y. would like to acknowledge the funding from the Ministry of Education, Youth and Sports of the Czech Republic and the European Union - European Structural and Investment Funds in the frames of Operational Programme Research, Development and Education—project Hybrid Materials for Hierarchical Structures (HyHi, Reg. No. CZ.02.1.01/0.0/0.0/16_019/0000843).

**Conflicts of Interest:** The authors declare no conflict of interest.

## References

1. Ball, S.; Banerjee, A.; Berry, C.; Boyle, J.R.; Bray, B.; Bradlow, W.; Chaudhry, A.; Crawley, R.; Danesh, J.; Denniston, A.; et al. Monitoring Indirect Impact of COVID-19 Pandemic on Services for Cardiovascular Diseases in the UK. *Heart* **2020**, *106*, 1890–1897. [CrossRef]
2. Cardiovascular Diseases (CVDs). Available online: https://www.who.int/news-room/fact-sheets/detail/cardiovascular-diseases-(cvds) (accessed on 29 August 2022).
3. Kivimäki, M.; Steptoe, A. Effects of Stress on the Development and Progression of Cardiovascular Disease. *Nat. Rev. Cardiol.* **2018**, *15*, 215–229. [CrossRef]
4. US Preventive Services Task Force; Krist, A.H.; Davidson, K.W.; Mangione, C.M.; Barry, M.J.; Cabana, M.; Caughey, A.B.; Donahue, K.; Doubeni, C.A.; Epling, J.W., Jr.; et al. US Preventive Services Task Force Behavioral Counseling Interventions to Promote a Healthy Diet and Physical Activity for Cardiovascular Disease Prevention in Adults with Cardiovascular Risk Factors: US Preventive Services Task Force Recommendation Statement. *JAMA* **2020**, *324*, 2069–2075. [CrossRef]
5. Roth, G.A.; Mensah, G.A.; Johnson, C.O.; Addolorato, G.; Ammirati, E.; Baddour, L.M.; Barengo, N.C.; Beaton, A.Z.; Benjamin, E.J.; Benziger, C.P.; et al. Global Burden of Cardiovascular Diseases and Risk Factors, 1990–2019: Update From the GBD 2019 Study. *J. Am. Coll. Cardiol.* **2020**, *76*, 2982–3021. [CrossRef] [PubMed]
6. Louridi, N.; Amar, M.; Ouahidi, B.E. Identification of Cardiovascular Diseases Using Machine Learning. In Proceedings of the 2019 7th Mediterranean Congress of Telecommunications (CMT), Fez, Morocco, 24 October 2019; pp. 1–6.
7. Melly, L.; Torregrossa, G.; Lee, T.; Jansens, J.-L.; Puskas, J.D. Fifty Years of Coronary Artery Bypass Grafting. *J. Thorac. Dis.* **2018**, *10*, 1960–1967. [CrossRef]
8. Yuan, H.; Chen, C.; Liu, Y.; Lu, T.; Wu, Z. Strategies in Cell-Free Tissue-Engineered Vascular Grafts. *J. Biomed. Mater. Res. Part A* **2020**, *108*, 426–445. [CrossRef]
9. Kabirian, F.; Ditkowski, B.; Zamanian, A.; Heying, R.; Mozafari, M. An Innovative Approach towards 3D-Printed Scaffolds for the next Generation of Tissue-Engineered Vascular Grafts. *Mater. Today Proc.* **2018**, *5*, 15586–15594. [CrossRef]

10. Carrabba, M.; Madeddu, P. Current Strategies for the Manufacture of Small Size Tissue Engineering Vascular Grafts. *Front. Bioeng. Biotechnol.* **2018**, *6*, 41. [CrossRef]
11. Teebken, O.E.; Haverich, A. Tissue Engineering of Small Diameter Vascular Grafts. *Eur. J. Vasc. Endovasc. Surg.* **2002**, *23*, 475–485. [CrossRef]
12. Hiob, M.A.; She, S.; Muiznieks, L.D.; Weiss, A.S. Biomaterials and Modifications in the Development of Small-Diameter Vascular Grafts. *ACS Biomater. Sci. Eng.* **2017**, *3*, 712–723. [CrossRef]
13. Jouda, H.; Larrea Murillo, L.L.; Wang, T. Current Progress in Vascular Engineering and Its Clinical Applications. *Cells* **2022**, *11*, 493. [CrossRef] [PubMed]
14. Shakeel, A.; Corridon, P.R. Mitigating Challenges and Expanding the Future of Vascular Tissue Engineering—Are We There Yet? SSRN: Rochester, NY, USA, 2022.
15. Leal, B.B.J.; Wakabayashi, N.; Oyama, K.; Kamiya, H.; Braghirolli, D.I.; Pranke, P. Vascular Tissue Engineering: Polymers and Methodologies for Small Caliber Vascular Grafts. *Front. Cardiovasc. Med.* **2021**, *7*, 592361. [CrossRef] [PubMed]
16. Townsley, M.I. Structure and Composition of Pulmonary Arteries, Capillaries, and Veins. *Compr. Physiol.* **2012**, *2*, 675–709. [CrossRef] [PubMed]
17. Shrestha, B.; Prasai, P.K.; Kaskas, A.M.; Khanna, A.; Letchuman, V.; Letchuman, S.; Alexander, J.S.; Orr, A.W.; Woolard, M.D.; Pattillo, C.B. Differential Arterial and Venous Endothelial Redox Responses to Oxidative Stress. *Microcirculation* **2018**, *25*, e12486. [CrossRef]
18. Song, H.-H.G.; Rumma, R.T.; Ozaki, C.K.; Edelman, E.R.; Chen, C.S. Vascular Tissue Engineering: Progress, Challenges, and Clinical Promise. *Cell Stem Cell* **2018**, *22*, 340–354. [CrossRef] [PubMed]
19. Camasão, D.B.; Mantovani, D. The Mechanical Characterization of Blood Vessels and Their Substitutes in the Continuous Quest for Physiological-Relevant Performances. A Critical Review. *Mater. Today Bio* **2021**, *10*, 100106. [CrossRef]
20. Ercolani, E.; Del Gaudio, C.; Bianco, A. Vascular Tissue Engineering of Small-Diameter Blood Vessels: Reviewing the Electrospinning Approach. *J. Tissue Eng. Regen. Med.* **2015**, *9*, 861–888. [CrossRef]
21. Mitchell, R.N.; Schoen, F.J. Blood Vessels. In *Robbins and Cotran: Pathologic Basis of Disease*, 8th ed.; Saunders Elsevier: Philadelphia, PA, USA, 2010; pp. 516–517.
22. Zhang, W.J.; Liu, W.; Cui, L.; Cao, Y. Tissue Engineering of Blood Vessel. *J. Cell. Mol. Med.* **2007**, *11*, 945–957. [CrossRef]
23. MacNeill, B.D.; Pomerantseva, I.; Lowe, H.C.; Oesterle, S.N.; Vacanti, J.P. Toward a New Blood Vessel. *Vasc. Med.* **2002**, *7*, 241–246. [CrossRef]
24. Xu, J.; Shi, G.-P. Vascular Wall Extracellular Matrix Proteins and Vascular Diseases. *Biochim. Biophys. Acta* **2014**, *1842*, 2106–2119. [CrossRef]
25. Cocciolone, A.J.; Hawes, J.Z.; Staiculescu, M.C.; Johnson, E.O.; Murshed, M.; Wagenseil, J.E. Elastin, Arterial Mechanics, and Cardiovascular Disease. *Am. J. Physiol. Heart Circ. Physiol.* **2018**, *315*, H189–H205. [CrossRef] [PubMed]
26. Awad, N.K.; Niu, H.; Ali, U.; Morsi, Y.S.; Lin, T. Electrospun Fibrous Scaffolds for Small-Diameter Blood Vessels: A Review. *Membranes* **2018**, *8*, 15. [CrossRef]
27. Enis, I.Y.; Sadikoglu, T.G. Design Parameters for Electrospun Biodegradable Vascular Grafts. *J. Ind. Text.* **2018**, *47*, 2205–2227. [CrossRef]
28. Wang, D.; Xu, Y.; Li, Q.; Turng, L.-S. Artificial Small-Diameter Blood Vessels: Materials, Fabrication, Surface Modification, Mechanical Properties, and Bioactive Functionalities. *J. Mater. Chem. B* **2020**, *8*, 1801–1822. [CrossRef]
29. James, B.D.; Allen, J.B. Vascular Endothelial Cell Behavior in Complex Mechanical Microenvironments. *ACS Biomater. Sci. Eng.* **2018**, *4*, 3818–3842. [CrossRef]
30. Zhang, Y.; Li, X.S.; Guex, A.G.; Liu, S.S.; Müller, E.; Malini, R.I.; Zhao, H.J.; Rottmar, M.; Maniura-Weber, K.; Rossi, R.M.; et al. A Compliant and Biomimetic Three-Layered Vascular Graft for Small Blood Vessels. *Biofabrication* **2017**, *9*, 025010. [CrossRef]
31. Johnson, R.; Ding, Y.; Nagiah, N.; Monnet, E.; Tan, W. Coaxially-Structured Fibres with Tailored Material Properties for Vascular Graft Implant. *Mater. Sci. Eng. C Mater. Biol. Appl.* **2019**, *97*, 1–11. [CrossRef]
32. Wu, J.; Hu, C.; Tang, Z.; Yu, Q.; Liu, X.; Chen, H. Tissue-Engineered Vascular Grafts: Balance of the Four Major Requirements. *Colloid Interface Sci. Commun.* **2018**, *23*, 34–44. [CrossRef]
33. Hernandez, J.L.; Woodrow, K.A. Medical Applications of Porous Biomaterials: Features of Porosity and Tissue-Specific Implications for Biocompatibility. *Adv. Healthc. Mater.* **2022**, *11*, 2102087. [CrossRef]
34. Kim, G.H. Electrospun PCL Nanofibers with Anisotropic Mechanical Properties as a Biomedical Scaffold. *Biomed. Mater.* **2008**, *3*, 025010. [CrossRef]
35. McFadden, B.R.; Smyth, S.J. Perceptions of Genetically Engineered Technology in Developed Areas. *Trends Biotechnol.* **2019**, *37*, 447–451. [CrossRef]
36. Shariatzadeh, S.; Shafiee, S.; Zafari, A.; Tayebi, T.; Yazdanpanah, G.; Majd, A.; Haj-Mirzaian, A.; Bahrami, S.; Niknejad, H. Developing a Pro-Angiogenic Placenta Derived Amniochorionic Scaffold with Two Exposed Basement Membranes as Substrates for Cultivating Endothelial Cells. *Sci. Rep.* **2021**, *11*, 22508. [CrossRef]
37. Gao, J.; Chen, S.; Tang, D.; Jiang, L.; Shi, J.; Wang, S. Mechanical Properties and Degradability of Electrospun PCL/PLGA Blended Scaffolds as Vascular Grafts. *Trans. Tianjin Univ.* **2019**, *25*, 152–160. [CrossRef]
38. Radke, D.; Jia, W.; Sharma, D.; Fena, K.; Wang, G.; Goldman, J.; Zhao, F. Tissue Engineering at the Blood-Contacting Surface: A Review of Challenges and Strategies in Vascular Graft Development. *Adv. Healthc. Mater.* **2018**, *7*, e1701461. [CrossRef]

39. Obiweluozor, F.O.; Emechebe, G.A.; Kim, D.-W.; Cho, H.-J.; Park, C.H.; Kim, C.S.; Jeong, I.S. Considerations in the Development of Small-Diameter Vascular Graft as an Alternative for Bypass and Reconstructive Surgeries: A Review. *Cardiovasc. Eng. Technol.* **2020**, *11*, 495–521. [CrossRef]
40. Eltom, A.; Zhong, G.; Muhammad, A. Scaffold Techniques and Designs in Tissue Engineering Functions and Purposes: A Review. *Adv. Mater. Sci. Eng.* **2019**, *2019*, e3429527. [CrossRef]
41. Zhao, P.; Gu, H.; Mi, H.; Rao, C.; Fu, J.; Turng, L. Fabrication of Scaffolds in Tissue Engineering: A Review. *Front. Mech. Eng.* **2018**, *13*, 107–119. [CrossRef]
42. Li, S.; Sengupta, D.; Chien, S. Vascular Tissue Engineering: From in Vitro to in Situ. *Wiley Interdiscip. Rev. Syst. Biol. Med.* **2014**, *6*, 61–76. [CrossRef]
43. Lu, T.; Li, Y.; Chen, T. Techniques for Fabrication and Construction of Three-Dimensional Scaffolds for Tissue Engineering. *Int. J. Nanomed.* **2013**, *8*, 337–350. [CrossRef] [PubMed]
44. Kishan, A.P.; Cosgriff-Hernandez, E.M. Recent Advancements in Electrospinning Design for Tissue Engineering Applications: A Review. *J. Biomed. Mater. Res. A* **2017**, *105*, 2892–2905. [CrossRef]
45. Cui, W.; Zhou, Y.; Chang, J. Electrospun Nanofibrous Materials for Tissue Engineering and Drug Delivery. *Sci. Technol. Adv. Mater.* **2010**, *11*, 014108. [CrossRef]
46. Ahn, H.; Ju, Y.M.; Takahashi, H.; Williams, D.F.; Yoo, J.J.; Lee, S.J.; Okano, T.; Atala, A. Engineered Small Diameter Vascular Grafts by Combining Cell Sheet Engineering and Electrospinning Technology. *Acta Biomater.* **2015**, *16*, 14–22. [CrossRef]
47. Raeisdasteh Hokmabad, V.; Davaran, S.; Ramazani, A.; Salehi, R. Design and Fabrication of Porous Biodegradable Scaffolds: A Strategy for Tissue Engineering. *J. Biomater. Sci. Polym. Ed.* **2017**, *28*, 1797–1825. [CrossRef]
48. Yalcinkaya, F. Experimental Study on Electrospun Polyvinyl Butyral Nanofibers Using a Non-Solvent System. *Fibers Polym.* **2015**, *16*, 2544–2551. [CrossRef]
49. Tan, Z.; Gao, X.; Liu, T.; Yang, Y.; Zhong, J.; Tong, C.; Tan, Y. Electrospun Vein Grafts with High Cell Infiltration for Vascular Tissue Engineering. *Mater. Sci. Eng. C Mater. Biol. Appl.* **2017**, *81*, 407–415. [CrossRef] [PubMed]
50. Wang, W.; Nie, W.; Liu, D.; Du, H.; Zhou, X.; Chen, L.; Wang, H.; Mo, X.; Li, L.; He, C. Macroporous Nanofibrous Vascular Scaffold with Improved Biodegradability and Smooth Muscle Cells Infiltration Prepared by Dual Phase Separation Technique. *Int. J. Nanomed.* **2018**, *13*, 7003–7018. [CrossRef] [PubMed]
51. Wang, Z.; Cui, Y.; Wang, J.; Yang, X.; Wu, Y.; Wang, K.; Gao, X.; Li, D.; Li, Y.; Zheng, X.-L.; et al. The Effect of Thick Fibers and Large Pores of Electrospun Poly(ε-Caprolactone) Vascular Grafts on Macrophage Polarization and Arterial Regeneration. *Biomaterials* **2014**, *35*, 5700–5710. [CrossRef]
52. O'Connor, R.A.; Cahill, P.A.; McGuinness, G.B. Effect of Electrospinning Parameters on the Mechanical and Morphological Characteristics of Small Diameter PCL Tissue Engineered Blood Vessel Scaffolds Having Distinct Micro and Nano Fibre Populations—A DOE Approach. *Polym. Test.* **2021**, *96*, 107119. [CrossRef]
53. Ju, Y.M.; Choi, J.S.; Atala, A.; Yoo, J.J.; Lee, S.J. Bilayered Scaffold for Engineering Cellularized Blood Vessels. *Biomaterials* **2010**, *31*, 4313–4321. [CrossRef]
54. Huang, L.; Guo, S.; Jiang, Y.; Shen, Q.; Li, L.; Shi, Y.; Xie, H.; Tian, J. A Preliminary Study on Polycaprolactone and Gelatin-Based Bilayered Tubular Scaffolds with Hierarchical Pore Size Constructed from Nano and Microfibers for Vascular Tissue Engineering. *J. Biomater. Sci. Polym. Ed.* **2021**, *32*, 1791–1809. [CrossRef]
55. Woods, I.; Flanagan, T.C. Electrospinning of Biomimetic Scaffolds for Tissue-Engineered Vascular Grafts: Threading the Path. *Expert Rev. Cardiovasc. Ther.* **2014**, *12*, 815–832. [CrossRef]
56. de Valence, S.; Tille, J.-C.; Giliberto, J.-P.; Mrowczynski, W.; Gurny, R.; Walpoth, B.H.; Möller, M. Advantages of Bilayered Vascular Grafts for Surgical Applicability and Tissue Regeneration. *Acta Biomater.* **2012**, *8*, 3914–3920. [CrossRef]
57. Nottelet, B.; Pektok, E.; Mandracchia, D.; Tille, J.-C.; Walpoth, B.; Gurny, R.; Möller, M. Factorial Design Optimization and in Vivo Feasibility of Poly (Epsilon-Caprolactone)-Micro- and Nanofiber-Based Small Diameter Vascular Grafts. *J. Biomed. Mater. Res. A* **2009**, *89*, 865–875. [CrossRef] [PubMed]
58. Ratcliffe, A. Tissue Engineering of Vascular Grafts. *Matrix Biol.* **2000**, *19*, 353–357. [CrossRef]
59. Guarino, V.; Causa, F.; Ambrosio, L. Porosity and Mechanical Properties Relationship in PCL Porous Scaffolds. *J. Appl. Biomater. Biomech.* **2007**, *5*, 149–157. [CrossRef] [PubMed]
60. Ji, S.; Gu, Q.; Xia, B. Porosity Dependence of Mechanical Properties of Solid Materials. *J. Mater. Sci.* **2006**, *41*, 1757–1768. [CrossRef]
61. Ang, K.C.; Leong, K.F.; Chua, C.K.; Chandrasekaran, M. Investigation of the Mechanical Properties and Porosity Relationships in Fused Deposition Modelling-fabricated Porous Structures. *Rapid Prototyp. J.* **2006**, *12*, 100–105. [CrossRef]
62. Le, Q.P.; Uspenskaya, M.V.; Olekhnovich, R.O.; Baranov, M.A. The Mechanical Properties of PVC Nanofiber Mats Obtained by Electrospinning. *Fibers* **2021**, *9*, 2. [CrossRef]
63. Li, Y.; Lim, C.T.; Kotaki, M. Study on Structural and Mechanical Properties of Porous PLA Nanofibers Electrospun by Channel-Based Electrospinning System. *Polymer* **2015**, *56*, 572–580. [CrossRef]
64. Sarkar, S.; Salacinski, H.J.; Hamilton, G.; Seifalian, A.M. The Mechanical Properties of Infrainguinal Vascular Bypass Grafts: Their Role in Influencing Patency. *Eur. J. Vasc. Endovasc. Surg.* **2006**, *31*, 627–636. [CrossRef]
65. Sarkar, S.; Hillery, C.; Seifalian, A.; Hamilton, G. Critical Parameter of Burst Pressure Measurement in Development of Bypass Grafts Is Highly Dependent on Methodology Used. *J. Vasc. Surg.* **2006**, *44*, 846–852. [CrossRef] [PubMed]

66. Venugopal, J.; Vadgama, P.; Kumar, T.S.S.; Ramakrishna, S. Biocomposite Nanofibres and Osteoblasts for Bone Tissue Engineering. *Nanotechnology* **2007**, *18*, 055101. [CrossRef]
67. Li, L.; Hashaikeh, R.; Arafat, H.A. Development of Eco-Efficient Micro-Porous Membranes via Electrospinning and Annealing of Poly (Lactic Acid). *J. Membr. Sci.* **2013**, *436*, 57–67. [CrossRef]
68. de Valence, S.; Tille, J.-C.; Mugnai, D.; Mrowczynski, W.; Gurny, R.; Möller, M.; Walpoth, B.H. Long Term Performance of Polycaprolactone Vascular Grafts in a Rat Abdominal Aorta Replacement Model. *Biomaterials* **2012**, *33*, 38–47. [CrossRef]
69. Lovett, M.; Lee, K.; Edwards, A.; Kaplan, D.L. Vascularization Strategies for Tissue Engineering. *Tissue Eng Part B Rev* **2009**, *15*, 353–370. [CrossRef]
70. Putti, M.; Simonet, M.; Solberg, R.; Peters, G.W.M. Electrospinning Poly (ε-Caprolactone) under Controlled Environmental Conditions: Influence on Fiber Morphology and Orientation. *Polymer* **2015**, *63*, 189–195. [CrossRef]
71. Enis, I.Y.; Horakova, J.; Sadikoglu, T.G.; Novak, O.; Lukas, D. Mechanical Investigation of Bilayer Vascular Grafts Electrospun from Aliphatic Polyesters. *Polym. Adv. Technol.* **2017**, *28*, 201–213. [CrossRef]
72. Rowland, D.C.L.; Aquilina, T.; Klein, A.; Hakimi, O.; Alexis-Mouthuy, P.; Carr, A.J.; Snelling, S.J.B. A Comparative Evaluation of the Effect of Polymer Chemistry and Fiber Orientation on Mesenchymal Stem Cell Differentiation. *J. Biomed. Mater. Res. A* **2016**, *104*, 2843–2853. [CrossRef]
73. Hasan, A.; Memic, A.; Annabi, N.; Hossain, M.; Paul, A.; Dokmeci, M.R.; Dehghani, F.; Khademhosseini, A. Electrospun Scaffolds for Tissue Engineering of Vascular Grafts. *Acta Biomater.* **2014**, *10*, 11–25. [CrossRef] [PubMed]
74. Milleret, V.; Hefti, T.; Hall, H.; Vogel, V.; Eberli, D. Influence of the Fiber Diameter and Surface Roughness of Electrospun Vascular Grafts on Blood Activation. *Acta Biomater.* **2012**, *8*, 4349–4356. [CrossRef]
75. Malik, S.; Sundarrajan, S.; Hussain, T.; Nazir, A.; Berto, F.; Ramakrishna, S. Electrospun Biomimetic Polymer Nanofibers as Vascular Grafts. *Mater. Des. Process. Commun.* **2021**, *3*, e203. [CrossRef]
76. Murugan, R.; Ramakrishna, S. Design Strategies of Tissue Engineering Scaffolds with Controlled Fiber Orientation. *Tissue Eng.* **2007**, *13*, 1845–1866. [CrossRef]
77. Wang, Y.; Wu, T.; Zhang, J.; Feng, Z.; Yin, M.; Mo, X. A Bilayer Vascular Scaffold with Spatially Controlled Release of Growth Factors to Enhance in Situ Rapid Endothelialization and Smooth Muscle Regeneration. *Mater. Des.* **2021**, *204*, 109649. [CrossRef]
78. Hu, J.-J.; Chao, W.-C.; Lee, P.-Y.; Huang, C.-H. Construction and Characterization of an Electrospun Tubular Scaffold for Small-Diameter Tissue-Engineered Vascular Grafts: A Scaffold Membrane Approach. *J. Mech. Behav. Biomed. Mater.* **2012**, *13*, 140–155. [CrossRef] [PubMed]
79. Chaparro, F.J.; Matusicky, M.E.; Allen, M.J.; Lannutti, J.J. Biomimetic Microstructural Reorganization during Suture Retention Strength Evaluation of Electrospun Vascular Scaffolds. *J. Biomed. Mater. Res. Part B Appl. Biomater.* **2016**, *104*, 1525–1534. [CrossRef]
80. Caves, J.M.; Kumar, V.A.; Martinez, A.W.; Kim, J.; Ripberger, C.M.; Haller, C.A.; Chaikof, E.L. The Use of Microfiber Composites of Elastin-like Protein Matrix Reinforced with Synthetic Collagen in the Design of Vascular Grafts. *Biomaterials* **2010**, *31*, 7175–7182. [CrossRef]
81. Nezarati, R.M.; Eifert, M.B.; Dempsey, D.K.; Cosgriff-Hernandez, E. Electrospun Vascular Grafts with Improved Compliance Matching to Native Vessels. *J. Biomed. Mater. Res. Part B Appl. Biomater.* **2015**, *103*, 313–323. [CrossRef]
82. Yalcin, I.; Horakova, J.; Mikes, P.; Sadikoglu, T.G.; Domin, R.; Lukas, D. Design of Polycaprolactone Vascular Grafts. *J. Ind. Text.* **2016**, *45*, 813–833. [CrossRef]
83. Gao, J.; Huang, Z.; Guo, H.; Tian, S.; Wang, L.; Li, Y. Effect of Wall Structures on Mechanical Properties of Small Caliber PHBHHx Vascular Grafts. *Fibers Polym.* **2019**, *20*, 2261–2267. [CrossRef]
84. Meng, X.; Wang, X.; Jiang, Y.; Zhang, B.; Li, K.; Li, Q. Suture Retention Strength of P (LLA-CL) Tissue-Engineered Vascular Grafts. *RSC Adv.* **2019**, *9*, 21258–21264. [CrossRef]
85. Jang, B.S.; Cheon, J.Y.; Kim, S.H.; Park, W.H. Small Diameter Vascular Graft with Fibroblast Cells and Electrospun Poly (L-Lactide-Co-ε-Caprolactone) Scaffolds: Cell Matrix Engineering. *J. Biomater. Sci. Polym. Ed.* **2018**, *29*, 942–959. [CrossRef] [PubMed]
86. Wang, C.; Li, Z.; Zhang, L.; Sun, W.; Zhou, J. Long-Term Results of Triple-Layered Small Diameter Vascular Grafts in Sheep Carotid Arteries. *Med. Eng. Phys.* **2020**, *85*, 1–6. [CrossRef] [PubMed]
87. Jeong, Y.; Yao, Y.; Yim, E.K. Current Understanding of Intimal Hyperplasia and Effect of Compliance in Synthetic Small Diameter Vascular Grafts. *Biomater. Sci.* **2020**, *8*, 4383–4395. [CrossRef] [PubMed]
88. Bouchet, M.; Gauthier, M.; Maire, M.; Ajji, A.; Lerouge, S. Towards Compliant Small-Diameter Vascular Grafts: Predictive Analytical Model and Experiments. *Mater. Sci. Eng. C* **2019**, *100*, 715–723. [CrossRef]
89. Liu, K.; Wang, N.; Wang, W.; Shi, L.; Li, H.; Guo, F.; Zhang, L.; Kong, L.; Wang, S.; Zhao, Y. A Bio-Inspired High Strength Three-Layer Nanofiber Vascular Graft with Structure Guided Cell Growth. *J. Mater. Chem. B* **2017**, *5*, 3758–3764. [CrossRef]
90. Wu, T.; Zhang, J.; Wang, Y.; Li, D.; Sun, B.; El-Hamshary, H.; Yin, M.; Mo, X. Fabrication and Preliminary Study of a Biomimetic Tri-Layer Tubular Graft Based on Fibers and Fiber Yarns for Vascular Tissue Engineering. *Mater. Sci. Eng. C* **2018**, *82*, 121–129. [CrossRef]
91. Tolba, E. Diversity of Electrospinning Approach for Vascular Implants: Multilayered Tubular Scaffolds. *Regen. Eng. Transl. Med.* **2020**, *6*, 383–397. [CrossRef]

92. Oztemur, J.; Yalcin Enis, I. The Role of Biopolymer Selection in the Design of Electrospun Small Caliber Vascular Grafts to Replace the Native Arterial Structure. *Chapter* **2020**, *9*, 27.
93. Huang, R.; Gao, X.; Wang, J.; Chen, H.; Tong, C.; Tan, Y.; Tan, Z. Triple-Layer Vascular Grafts Fabricated by Combined E-Jet 3D Printing and Electrospinning. *Ann. Biomed. Eng.* **2018**, *46*, 1254–1266. [CrossRef]
94. Grasl, C.; Stoiber, M.; Röhrich, M.; Moscato, F.; Bergmeister, H.; Schima, H. Electrospinning of Small Diameter Vascular Grafts with Preferential Fiber Directions and Comparison of Their Mechanical Behavior with Native Rat Aortas. *Mater. Sci. Eng. C* **2021**, *124*, 112085. [CrossRef]
95. Ravi, S.; Chaikof, E.L. Biomaterials for Vascular Tissue Engineering. *Regen. Med.* **2010**, *5*, 107–120. [CrossRef] [PubMed]
96. Wang, Z.; Liu, L.; Mithieux, S.M.; Weiss, A.S. Fabricating Organized Elastin in Vascular Grafts. *Trends Biotechnol.* **2021**, *39*, 505–518. [CrossRef] [PubMed]
97. Zhu, J.; Chen, D.; Du, J.; Chen, X.; Wang, J.; Zhang, H.; Chen, S.; Wu, J.; Zhu, T.; Mo, X. Mechanical Matching Nanofibrous Vascular Scaffold with Effective Anticoagulation for Vascular Tissue Engineering. *Compos. Part B Eng.* **2020**, *186*, 107788. [CrossRef]
98. Yu, E.; Mi, H.-Y.; Zhang, J.; Thomson, J.A.; Turng, L.-S. Development of Biomimetic Thermoplastic Polyurethane/Fibroin Small-Diameter Vascular Grafts via a Novel Electrospinning Approach. *J. Biomed. Mater. Res. A* **2018**, *106*, 985–996. [CrossRef]
99. Ong, C.S.; Zhou, X.; Huang, C.Y.; Fukunishi, T.; Zhang, H.; Hibino, N. Tissue Engineered Vascular Grafts: Current State of the Field. *Expert Rev. Med. Devices* **2017**, *14*, 383–392. [CrossRef]
100. Szafron, J.M.; Khosravi, R.; Reinhardt, J.; Best, C.A.; Bersi, M.R.; Yi, T.; Breuer, C.K.; Humphrey, J.D. Immuno-Driven and Mechano-Mediated Neotissue Formation in Tissue Engineered Vascular Grafts. *Ann. Biomed. Eng.* **2018**, *46*, 1938–1950. [CrossRef]
101. Pashneh-Tala, S.; MacNeil, S.; Claeyssens, F. The Tissue-Engineered Vascular Graft-Past, Present, and Future. *Tissue Eng. Part B Rev.* **2016**, *22*, 68–100. [CrossRef]
102. Browning, M.B.; Dempsey, D.; Guiza, V.; Becerra, S.; Rivera, J.; Russell, B.; Höök, M.; Clubb, F.; Miller, M.; Fossum, T.; et al. Multilayer Vascular Grafts Based on Collagen-Mimetic Proteins. *Acta Biomater.* **2012**, *8*, 1010–1021. [CrossRef]
103. Copes, F.; Pien, N.; Van Vlierberghe, S.; Boccafoschi, F.; Mantovani, D. Collagen-Based Tissue Engineering Strategies for Vascular Medicine. *Front. Bioeng. Biotechnol.* **2019**, *7*, 166. [CrossRef]
104. Senthil, R.; Kavukcu, S.B.; Lakshmi, T.; Gülşah, T.; Candaş, A.Z.A. Collagen/Physiologically Clotted Fibrin-Based Nanobioscaffold Supported with Silver Nanoparticles: A Novel Approach. *Int. J. Artif. Organs* 2022, *ahead of print*. [CrossRef]
105. Antunes, M.; Bonani, W.; Reis, R.L.; Migliaresi, C.; Ferreira, H.; Motta, A.; Neves, N.M. Development of Alginate-Based Hydrogels for Blood Vessel Engineering. *Biomater. Adv.* **2022**, *134*, 112588. [CrossRef]
106. Gheorghita Puscaselu, R.; Lobiuc, A.; Dimian, M.; Covasa, M. Alginate: From Food Industry to Biomedical Applications and Management of Metabolic Disorders. *Polymers* **2020**, *12*, 2417. [CrossRef]
107. Sahoo, D.R.; Biswal, T. Alginate and Its Application to Tissue Engineering. *SN Appl. Sci.* **2021**, *3*, 30. [CrossRef]
108. Ahsan, S.M.; Thomas, M.; Reddy, K.K.; Sooraparaju, S.G.; Asthana, A.; Bhatnagar, I. Chitosan as Biomaterial in Drug Delivery and Tissue Engineering. *Int. J. Biol. Macromol.* **2018**, *110*, 97–109. [CrossRef]
109. Croisier, F.; Jérôme, C. Chitosan-Based Biomaterials for Tissue Engineering. *Eur. Polym. J.* **2013**, *49*, 780–792. [CrossRef]
110. Islam, M.M.; Shahruzzaman, M.; Biswas, S.; Nurus Sakib, M.; Rashid, T.U. Chitosan Based Bioactive Materials in Tissue Engineering Applications-A Review. *Bioact. Mater.* **2020**, *5*, 164–183. [CrossRef] [PubMed]
111. Foster, J.A. Elastin. In *Encyclopedia of Biological Chemistry*, 2nd ed.; Lennarz, W.J., Lane, M.D., Eds.; Academic Press: Cambridge, MA, USA, 2013; pp. 192–193, ISBN 9780123786319.
112. Gomes, M.; Azevedo, H.; Malafaya, P.; Silva, S.; Oliveira, J.; Silva, G.; Sousa, R.; Mano, J.; Reis, R. Chapter 6—Natural Polymers in Tissue Engineering Applications. In *Tissue Engineering*; van Blitterswijk, C., Thomsen, P., Lindahl, A., Hubbell, J., Williams, D.F., Cancedda, R., de Bruijn, J.D., Sohier, J., Eds.; Academic Press: Cambridge, MA, USA, 2008; pp. 145–192, ISBN 9780123708694.
113. Nasrollahzadeh, M.; Maham, M.; Nezafat, Z.; Shafiei, N. Chapter 4—Protein and Polypeptide Biopolymer Chemistry. In *Biopolymer-Based Metal Nanoparticle Chemistry for Sustainable Applications*; Nasrollahzadeh, M., Ed.; Elsevier: Amsterdam, The Netherlands, 2021; pp. 107–144. ISBN 9780128221082.
114. Janmey, P.A.; Winer, J.P.; Weisel, J.W. Fibrin Gels and Their Clinical and Bioengineering Applications. *J. R. Soc. Interface* **2009**, *6*, 1–10. [CrossRef]
115. Liu, R.H.; Ong, C.S.; Fukunishi, T.; Ong, K.; Hibino, N. Review of Vascular Graft Studies in Large Animal Models. *Tissue Eng. Part B Rev.* **2018**, *24*, 133–143. [CrossRef]
116. Shaikh, F.M.; Callanan, A.; Kavanagh, E.G.; Burke, P.E.; Grace, P.A.; McGloughlin, T.M. Fibrin: A Natural Biodegradable Scaffold in Vascular Tissue Engineering. *Cells Tissues Organs* **2008**, *188*, 333–346. [CrossRef]
117. Sundararaghavan, H.G.; Burdick, J.A. 5.509—Cell Encapsulation. In *Comprehensive Biomaterials*; Ducheyne, P., Ed.; Elsevier: Oxford, UK, 2011; pp. 115–13, ISBN 9780080552941.
118. Aldana, A.A.; Abraham, G.A. Current Advances in Electrospun Gelatin-Based Scaffolds for Tissue Engineering Applications. *Int. J. Pharm.* **2017**, *523*, 441–453. [CrossRef]
119. Asadpour, S.; Kargozar, S.; Moradi, L.; Ai, A.; Nosrati, H.; Ai, J. Natural Biomacromolecule Based Composite Scaffolds from Silk Fibroin, Gelatin and Chitosan toward Tissue Engineering Applications. *Int. J. Biol. Macromol.* **2020**, *154*, 1285–1294. Available online: https://www.sciencedirect.com/science/article/pii/S0141813019338577 (accessed on 29 August 2022). [CrossRef]

120. Deshmukh, K.; Basheer Ahamed, M.; Deshmukh, R.R.; Khadheer Pasha, S.K.; Bhagat, P.R.; Chidambaram, K. 3—Biopolymer Composites with High Dielectric Performance: Interface Engineering. In *Biopolymer Composites in Electronics*; Sadasivuni, K.K., Ponnamma, D., Kim, J., Cabibihan, J.-J., AlMaadeed, M.A., Eds.; Elsevier: Amsterdam, The Netherlands, 2017; pp. 27–128, ISBN 9780128092613.
121. McKeen, L. Chapter11—The Effect of Heat Aging on the Properties of Sustainable Polymers. In *The Effect of Long Term Thermal Exposure on Plastics and Elastomers*, 2nd ed.; McKeen, L., Ed.; Plastics Design Library; William Andrew Publishing: Norwich, NY, USA, 2021; pp. 313–332. ISBN 9780323854368.
122. Mohamed, R.M.; Yusoh, K. A Review on the Recent Research of Polycaprolactone (PCL). *Adv. Mater. Res.* **2016**, *1134*, 249–255. [CrossRef]
123. Patrício, T.; Domingos, M.; Gloria, A.; Bártolo, P. Characterisation of PCL and PCL/PLA Scaffolds for Tissue Engineering. *Procedia CIRP* **2013**, *5*, 110–114. [CrossRef]
124. Pavia, F.C.; Rigogliuso, S.; Carrubba, V.L.; Mannella, G.L.; Ghersi, G.; Brucato, V. Poly Lactic Acid Based Scaffolds for Vascular Tissue Engineering. *Chem. Eng. Trans.* **2012**, *27*, 409–414. [CrossRef]
125. Donate, R.; Monzón, M.; Alemán-Domínguez, M.E. Additive Manufacturing of PLA-Based Scaffolds Intended for Bone Regeneration and Strategies to Improve Their Biological Properties. *e-Polymers* **2020**, *20*, 571–599. [CrossRef]
126. Santoro, M.; Shah, S.R.; Walker, J.L.; Mikos, A.G. Poly(Lactic Acid) Nanofibrous Scaffolds for Tissue Engineering. *Adv. Drug Deliv. Rev.* **2016**, *107*, 206–212. [CrossRef]
127. Yazdanpanah, A.; Amoabediny, G.; Shariatpanahi, P.; Nourmohammadi, J.; Tahmasbi, M.; Mozafari, M. Synthesis and Characterization of Polylactic Acid Tubular Scaffolds with Improved Mechanical Properties for Vascular Tissue Engineering. *Trends Biomater. Artif. Organs* **2014**, *28*, 99–105.
128. Budak, K.; Sogut, O.; Sezer, U.A. A Review on Synthesis and Biomedical Applications of Polyglycolic Acid. *J. Polym. Res.* **2020**, *27*, 208. [CrossRef]
129. Hajiali, H.; Shahgasempour, S.; Naimi-Jamal, M.R.; Peirovi, H. Electrospun PGA/Gelatin Nanofibrous Scaffolds and Their Potential Application in Vascular Tissue Engineering. *Int. J. Nanomed.* **2011**, *6*, 2133–2141. [CrossRef]
130. Thomas, L.V.; Lekshmi, V.; Nair, P.D. Tissue Engineered Vascular Grafts—Preclinical Aspects. *Int. J. Cardiol.* **2013**, *167*, 1091–1100. [CrossRef]
131. Montini-Ballarin, F.; Calvo, D.; Caracciolo, P.C.; Rojo, F.; Frontini, P.M.; Abraham, G.A.; Guinea, G.V. Mechanical Behavior of Bilayered Small-Diameter Nanofibrous Structures as Biomimetic Vascular Grafts. *J. Mech. Behav. Biomed. Mater.* **2016**, *60*, 220–233. [CrossRef]
132. J-Shaped Curves. Available online: https://www.doitpoms.ac.uk/tlplib/bioelasticity/j-shaped-curves.php (accessed on 29 August 2022).
133. Benrashid, E.; McCoy, C.C.; Youngwirth, L.M.; Kim, J.; Manson, R.J.; Otto, J.C.; Lawson, J.H. Tissue Engineered Vascular Grafts: Origins, Development, and Current Strategies for Clinical Application. *Methods* **2016**, *99*, 13–19. [CrossRef]
134. Ryan, A.J.; Ryan, E.J.; Cameron, A.R.; O'Brien, F.J. Hierarchical Biofabrication of Biomimetic Collagen-Elastin Vascular Grafts with Controllable Properties via Lyophilisation. *Acta Biomater.* **2020**, *112*, 52–61. [CrossRef]
135. Reddy, M.S.B.; Ponnamma, D.; Choudhary, R.; Sadasivuni, K.K. A Comparative Review of Natural and Synthetic Biopolymer Composite Scaffolds. *Polymers* **2021**, *13*, 1105. [CrossRef]
136. Gupta, P.; Mandal, B.B. Tissue-Engineered Vascular Grafts: Emerging Trends and Technologies. *Adv. Funct. Mater.* **2021**, *31*, 2100027. [CrossRef]
137. Park, S.; Kim, J.; Lee, M.-K.; Park, C.; Jung, H.-D.; Kim, H.-E.; Jang, T.-S. Fabrication of Strong, Bioactive Vascular Grafts with PCL/Collagen and PCL/Silica Bilayers for Small-Diameter Vascular Applications. *Mater. Des.* **2019**, *181*, 108079. [CrossRef]
138. Seifu, D.G.; Purnama, A.; Mequanint, K.; Mantovani, D. Small-Diameter Vascular Tissue Engineering. *Nat. Rev. Cardiol.* **2013**, *10*, 410–421. [CrossRef]
139. Qiu, Y.; Myers, D.R.; Lam, W.A. The Biophysics and Mechanics of Blood from a Materials Perspective. *Nat. Rev. Mater.* **2019**, *4*, 294–311. [CrossRef]
140. van Haaften, E.E.; Bouten, C.V.C.; Kurniawan, N.A. Vascular Mechanobiology: Towards Control of In Situ Regeneration. *Cells* **2017**, *6*, 19. [CrossRef]
141. Isenberg, B.C.; Williams, C.; Tranquillo, R.T. Small-Diameter Artificial Arteries Engineered in Vitro. *Circ. Res.* **2006**, *98*, 25–35. [CrossRef]
142. Serbo, J.V.; Gerecht, S. Vascular Tissue Engineering: Biodegradable Scaffold Platforms to Promote Angiogenesis. *Stem Cell Res. Ther.* **2013**, *4*, 8. [CrossRef]
143. Chiu, J.-J.; Chien, S. Effects of Disturbed Flow on Vascular Endothelium: Pathophysiological Basis and Clinical Perspectives. *Physiol. Rev.* **2011**, *91*, 327–387. [CrossRef] [PubMed]
144. Zhou, T.; Zheng, Y.; Qiu, J.; Hu, J.; Sun, D.; Tang, C.; Wang, G. Endothelial Mechanotransduction Mechanisms for Vascular Physiology and Atherosclerosis. *J. Mech. Med. Biol.* **2014**, *14*, 1430006. [CrossRef]
145. de Mel, A.; Murad, F.; Seifalian, A.M. Nitric Oxide: A Guardian for Vascular Grafts? *Chem. Rev.* **2011**, *111*, 5742–5767. [CrossRef] [PubMed]
146. Li, Y.-S.J.; Haga, J.H.; Chien, S. Molecular Basis of the Effects of Shear Stress on Vascular Endothelial Cells. *J. Biomech.* **2005**, *38*, 1949–1971. [CrossRef]

147. Yamamoto, K.; Takahashi, T.; Asahara, T.; Ohura, N.; Sokabe, T.; Kamiya, A.; Ando, J. Proliferation, Differentiation, and Tube Formation by Endothelial Progenitor Cells in Response to Shear Stress. *J. Appl. Physiol.* **2003**, *95*, 2081–2088. [CrossRef]
148. Mitchell, S.L.; Niklason, L.E. Requirements for Growing Tissue-Engineered Vascular Grafts. *Cardiovasc. Pathol.* **2003**, *12*, 59–64. [CrossRef]
149. Dan, P.; Velot, É.; Decot, V.; Menu, P. The Role of Mechanical Stimuli in the Vascular Differentiation of Mesenchymal Stem Cells. *J. Cell Sci.* **2015**, *128*, 2415–2422. [CrossRef]
150. Jufri, N.F.; Mohamedali, A.; Avolio, A.; Baker, M.S. Mechanical Stretch: Physiological and Pathological Implications for Human Vascular Endothelial Cells. *Vasc. Cell* **2015**, *7*, 8. [CrossRef]
151. Kwak, B.R.; Bäck, M.; Bochaton-Piallat, M.-L.; Caligiuri, G.; Daemen, M.J.A.P.; Davies, P.F.; Hoefer, I.E.; Holvoet, P.; Jo, H.; Krams, R.; et al. Biomechanical Factors in Atherosclerosis: Mechanisms and Clinical Implications. *Eur. Heart J.* **2014**, *35*, 3013–3020. [CrossRef]
152. Green, D.J.; Hopman, M.T.E.; Padilla, J.; Laughlin, M.H.; Thijssen, D.H.J. Vascular Adaptation to Exercise in Humans: Role of Hemodynamic Stimuli. *Physiol. Rev.* **2017**, *97*, 495–528. [CrossRef]
153. Castillo-Cruz, O.; Pérez-Aranda, C.; Gamboa, F.; Cauich-Rodríguez, J.V.; Mantovani, D.; Avilés, F. Prediction of Circumferential Compliance and Burst Strength of Polymeric Vascular Grafts. *J. Mech. Behav. Biomed. Mater.* **2018**, *79*, 332–340. [CrossRef] [PubMed]
154. Castorena-Gonzalez, J.A.; Staiculescu, M.C.; Foote, C.; Martinez-Lemus, L.A. Mechanisms of the Inward Remodeling Process in Resistance Vessels: Is the Actin Cytoskeleton Involved? *Microcirculation* **2014**, *21*, 219–229. [CrossRef] [PubMed]
155. Bersi, M.R.; Bellini, C.; Humphrey, J.D.; Avril, S. Local Variations in Material and Structural Properties Characterize Murine Thoracic Aortic Aneurysm Mechanics. *Biomech. Model. Mechanobiol.* **2019**, *18*, 203–218. [CrossRef]
156. Sanft, R.; Power, A.; Nicholson, C. Modeling the Effects of Muscle Contraction on the Mechanical Response and Circumferential Stability of Coronary Arteries. *Math. Biosci.* **2019**, *315*, 108223. [CrossRef]
157. Thubrikar, M.J. Pressure Vessel Principles. In *Vascular Mechanics and Pathology*; Springer US: Boston, MA, USA, 2007; pp. 82–106, ISBN 9780387682341.
158. Ferreira, H.P.; Moura, D.; Pereira, A.T.; Henriques, P.C.; Barrias, C.C.; Magalhães, F.D.; Gonçalves, I.C. Using Graphene-Based Materials for Stiff and Strong Poly(Ethylene Glycol) Hydrogels. *Int. J. Mol. Sci.* **2022**, *23*, 2312. [CrossRef]
159. Olsen, T.R.; Casco, M.; Herbst, A.; Evans, G.; Rothermel, T.; Pruett, L.; Reid, J.; Barry, K.; Jaeggli, M.P.; Simionescu, D.T.; et al. Longitudinal Stretching for Maturation of Vascular Tissues Using Magnetic Forces. *Bioengineering* **2016**, *3*, 29. [CrossRef] [PubMed]
160. Jackson, Z.S.; Gotlieb, A.I.; Langille, B.L. Wall Tissue Remodeling Regulates Longitudinal Tension in Arteries. *Circ. Res.* **2002**, *90*, 918–925. [CrossRef] [PubMed]
161. Han, H.-C.; Ku, D.N.; Vito, R.P. Arterial Wall Adaptation under Elevated Longitudinal Stretch in Organ Culture. *Ann. Biomed. Eng.* **2003**, *31*, 403–411. [CrossRef] [PubMed]
162. Elliott, M.B.; Gerecht, S. Three-Dimensional Culture of Small-Diameter Vascular Grafts. *J. Mater. Chem. B* **2016**, *4*, 3443–3453. [CrossRef]
163. Das, A.; Paul, A.; Taylor, M.D.; Banerjee, R.K. Pulsatile Arterial Wall-Blood Flow Interaction with Wall Pre-Stress Computed Using an Inverse Algorithm. *BioMed. Eng. OnLine* **2015**, *14*, S18. [CrossRef]
164. Chaouat, M.; Le Visage, C.; Baille, W.E.; Escoubet, B.; Chaubet, F.; Mateescu, M.A.; Letourneur, D. A Novel Cross-linked Poly (Vinyl Alcohol)(PVA) for Vascular Grafts. *Adv. Funct. Mater.* **2008**, *18*, 2855–2861. [CrossRef]
165. Johnson, J.; Ohst, D.; Groehl, T.; Hetterscheidt, S.; Jones, M. Development of Novel, Bioresorbable, Small-Diameter Electrospun Vascular Grafts. *J. Tissue Sci. Eng.* **2015**, *6*, 1.
166. Greenwald, S.; Berry, C. Improving Vascular Grafts: The Importance of Mechanical and Haemodynamic Properties. *J. Pathol.* **2000**, *190*, 292–299. [CrossRef]
167. Inoue, T.; Kanda, K.; Yamanami, M.; Kami, D.; Gojo, S.; Yaku, H. Modifications of the Mechanical Properties of in Vivo Tissue-Engineered Vascular Grafts by Chemical Treatments for a Short Duration. *PLoS ONE* **2021**, *16*, e0248346. [CrossRef] [PubMed]
168. Marinov, G.; Guidoin, R.; Tse, L.W.; Ruthrauff, A.A.; Yao, T.; King, M.W. 21—Endovascular Prostheses for Aortic Aneurysms: A New Era for Vascular Surgery. In *Biotextiles as Medical Implants*; Woodhead Publishing Series in Textiles; King, M.W., Gupta, B.S., Guidoin, R., Eds.; Woodhead Publishing: Cambridge, UK, 2013; pp. 640–675. ISBN 9781845694395.
169. Rapoport, H.S.; Fish, J.; Basu, J.; Campbell, J.; Genheimer, C.; Payne, R.; Jain, D. Construction of a Tubular Scaffold That Mimics J-Shaped Stress/Strain Mechanics Using an Innovative Electrospinning Technique. *Tissue Eng. Part C Methods* **2012**, *18*, 567–574. [CrossRef] [PubMed]
170. Akentjew, T.L.; Terraza, C.; Suazo, C.; Maksimcuka, J.; Wilkens, C.A.; Vargas, F.; Zavala, G.; Ocaña, M.; Enrione, J.; García-Herrera, C.M.; et al. Rapid Fabrication of Reinforced and Cell-Laden Vascular Grafts Structurally Inspired by Human Coronary Arteries. *Nat. Commun.* **2019**, *10*, 3098. [CrossRef] [PubMed]
171. Kim, S.-H.; Mun, C.H.; Jung, Y.; Kim, S.-H.; Kim, D.-I.; Kim, S.H. Mechanical Properties of Compliant Double Layered Poly (L-Lactide-Co-ϵ-Caprolactone) Vascular Graft. *Macromol. Res.* **2013**, *21*, 886–891. [CrossRef]
172. Drilling, S.; Gaumer, J.; Lannutti, J. Fabrication of Burst Pressure Competent Vascular Grafts via Electrospinning: Effects of Microstructure. *J. Biomed. Mater. Res. Part A* **2009**, *88*, 923–934. [CrossRef]

173. Montini-Ballarin, F.; Abraham, G.; Caracciolo, P. Mechanical Behavior of Polyurethane-Based Small-Diameter Vascular Grafts. In *Advances in Polyurethane Biomaterials*; Elsevier: Amsterdam, The Netherlands, 2016; pp. 451–477.
174. Post, A.; Diaz-Rodriguez, P.; Balouch, B.; Paulsen, S.; Wu, S.; Miller, J.; Hahn, M.; Cosgriff-Hernandez, E. Elucidating the Role of Graft Compliance Mismatch on Intimal Hyperplasia Using an Ex Vivo Organ Culture Model. *Acta Biomater.* **2019**, *89*, 84–94. [CrossRef]
175. Goonoo, N.; Bhaw-Luximon, A.; Bowlin, G.L.; Jhurry, D. An Assessment of Biopolymer-and Synthetic Polymer-based Scaffolds for Bone and Vascular Tissue Engineering. *Polym. Int.* **2013**, *62*, 523–533. [CrossRef]
176. He, F.; Hua, L.; Gao, L. A Computational Model for Biomechanical Effects of Arterial Compliance Mismatch. *Appl. Bionics Biomech.* **2015**, *2015*, 213236. [CrossRef] [PubMed]
177. Wise, S.G.; Byrom, M.J.; Waterhouse, A.; Bannon, P.G.; Ng, M.K.; Weiss, A.S. A Multilayered Synthetic Human Elastin/Polycaprolactone Hybrid Vascular Graft with Tailored Mechanical Properties. *Acta Biomater.* **2011**, *7*, 295–303. [CrossRef] [PubMed]
178. Spadaccio, C.; Nappi, F.; Al-Attar, N.; Sutherland, F.W.; Acar, C.; Nenna, A.; Trombetta, M.; Chello, M.; Rainer, A. Old Myths, New Concerns: The Long-Term Effects of Ascending Aorta Replacement with Dacron Grafts. Not All That Glitters Is Gold. *J. Cardiovasc. Trans. Res.* **2016**, *9*, 334–342. [CrossRef] [PubMed]
179. Matsuzaki, Y.; Iwaki, R.; Reinhardt, J.W.; Chang, Y.-C.; Miyamoto, S.; Kelly, J.; Zbinden, J.; Blum, K.; Mirhaidari, G.; Ulziibayar, A.; et al. The Effect of Pore Diameter on Neo-Tissue Formation in Electrospun Biodegradable Tissue-Engineered Arterial Grafts in a Large Animal Model. *Acta Biomater.* **2020**, *115*, 176–184. [CrossRef] [PubMed]
180. Bazgir, M.; Zhang, W.; Zhang, X.; Elies, J.; Saeinasab, M.; Coates, P.; Youseffi, M.; Sefat, F. Degradation and Characterisation of Electrospun Polycaprolactone (PCL) and Poly (Lactic-Co-Glycolic Acid)(PLGA) Scaffolds for Vascular Tissue Engineering. *Materials* **2021**, *14*, 4773. [CrossRef] [PubMed]
181. Yang, L.; Li, X.; Wang, D.; Mu, S.; Lv, W.; Hao, Y.; Lu, X.; Zhang, G.; Nan, W.; Chen, H.; et al. Improved Mechanical Properties by Modifying Fibrin Scaffold with PCL and Its Biocompatibility Evaluation. *J. Biomater. Sci. Polym. Ed.* **2020**, *31*, 658–678. [CrossRef]
182. Bolbasov, E.; Goreninskii, S.; Tverdokhlebov, S.; Mishanin, A.; Viknianshchuk, A.; Bezuidenhout, D.; Golovkin, A. Comparative Study of the Physical, Topographical and Biological Properties of Electrospinning PCL, PLLA, Their Blend and Copolymer Scaffolds. *IOP Conf. Ser. Mater. Sci. Eng.* **2018**, *350*, 012012. [CrossRef]

membranes

MDPI

Article

# Electrospun Hydrophobic Interaction Chromatography (HIC) Membranes for Protein Purification

**Shu-Ting Chen [1], Sumith Ranil Wickramasinghe [1] and Xianghong Qian [2,*]**

1 Department of Chemical Engineering, University of Arkansas, Fayetteville, AR 72701, USA; sc086@uark.edu (S.-T.C.); swickram@uark.edu (S.R.W.)
2 Department of Biomedical Engineering, University of Arkansas, Fayetteville, AR 72701, USA
* Correspondence: xqian@uark.edu

**Abstract:** Responsive membranes for hydrophobic interaction chromatography have been fabricated by functionalizing poly(N-vinylcaprolactam) (PVCL) ligands on the substrate of electrospun regenerated cellulose nanofibers. Both static and dynamic binding capacities and product recovery were investigated using bovine serum albumin (BSA) and Immunoglobulin G (IgG) as model proteins. The effects of ligand chain length and chain density on static binding capacity were also studied. A static binding capacity of ~25 mg/mL of membrane volume (MV) can be achieved in optimal ligand grafting conditions. For dynamic binding studies, protein binding capacity increased with protein concentration from 0.1 to 1.0 g/L. Dynamic binding capacity increased from ~8 mg/mL MV at 0.1 g/L BSA to over 30 mg/mL at 1.0 g/L BSA. However, BSA recovery decreased as protein concentration increased from ~98% at 0.1 g/L BSA to 51% at 1 g/L BSA loading concentration. There is a clear trade-off between binding capacity and recovery rate. The electrospun substrate with thicker fibers and more open pore structures is superior to thinner fibrous membrane substrates.

**Keywords:** electrospun membranes; protein purification; hydrophobic interaction chromatography

**Citation:** Chen, S.-T.; Wickramasinghe, S.R.; Qian, X. Electrospun Hydrophobic Interaction Chromatography (HIC) Membranes for Protein Purification. *Membranes* **2022**, *12*, 714. https://doi.org/10.3390/membranes12070714

Academic Editor: Andrea Ehrmann

Received: 24 April 2022
Accepted: 6 July 2022
Published: 18 July 2022



## 1. Introduction

The rapid development of biopharmaceutical products has led to an increased demand for purification of high product titer feed, and regulatory agencies have implemented strict requirements for product purity [1]. However, downstream purification accounts for 50–80% of the entire production cost [2]. There is a severe bottleneck for cost-effective purification of biopharmaceuticals due to the dramatic increase in the number of products and the increased product titer and purity requirements [3]. Significant efforts have been made in recent years to overcome these challenges in order to improve the efficacy of purification during downstream processing.

Hydrophobic interaction chromatography (HIC) is often used as a chromatographic polishing step to remove the remaining host cell proteins (HCP), product aggregates and other more hydrophobic impurities during the downstream purification of biologics such as monoclonal antibodies (mAbs), Fc-fusion proteins and other recombinant therapeutic proteins as well as many other biologics including hormones, vaccines, growth factors and interferons [4,5]. The mechanism for HIC purification comes from the different hydrophobic binding interaction strengths between molecules with different hydrophobicity in the feed and the hydrophobic ligand immobilized on the stationary phase [4]. As hydrophobic interaction is modulated by the ionic strength of the solution, high ionic strength buffer is often used to bind the hydrophobic species in the feed whereas low ionic strength buffer is often used to elute the bound species. Currently, HIC is often operated in the flow-through mode where the product of interest passes through the stationary phase and impurities, including aggregates and other more hydrophobic species, are captured by the immobilized ligands. In addition, our previous studies [2,6–8] show that salt type and salt concentration, as well as ligand chain length and chain density, affect overall chromatographic performance of HIC membranes. Both protein binding capacity and recovery are observed to depend on both mobile phase conditions and stationary phase properties as well as to be protein dependent [6,8–10].

Packed-bed column chromatography has been widely used in downstream purification of proteins, nuclei acids and other biologics [2]. However, one major drawback of packed-bed column chromatography is the slow pore-diffusion which severely restricts its separation efficiency. The diffusion of targeted products to the ligands on the chromatographic bed is a slow process leading to a dramatic drop in binding capacity as the feed flow rate increases. Besides diffusion limitation, packed-bed chromatography also suffers from large buffer consumption as well as extra costs for packing and testing.

An alternative approach is to use adsorptive membrane chromatography during downstream processing [2]. Adsorptive membranes, known as membrane adsorbers, are macroporous membranes functionalized by ligands attached on the membrane pore surface to remove containments, such as product aggregates, viruses and DNAs [11]. Compared to resin-based chromatography, the pore diffusion limitation is eliminated in membrane adsorbers where convection becomes the dominant transport mechanism [12]. Moreover, the operation can be performed at relatively low pressure, which reduces denaturation and aggregation of the sensitive biologics. Buffer usage of membranes is lower than resins due to reduced void volume.

Membranes have substantially lower material cost compared to packed beds. Single-use membrane processes greatly reduce the cost of revalidation. In addition, the membrane system is easier to scale up and the cost of packing and testing is subsequently reduced significantly. Traditionally, membrane-based purification is always limited by its low capacity. Recent advances in materials engineering have led to high-capacity membranes that can now compete with resins [13,14]. Currently, there is interest in developing HIC membrane chromatography in the bind-and-elute mode for protein fractionation [5,7,9,12,14,15]. However, application of HIC has been limited due to the overall low capacity of the ligands and the efficiency of eluting the bound proteins [15].

Thermo-responsive polymer, such as poly(N-isopropylacrylamide) (PNIPAM) and poly(N-vinylcaprolactam) (PVCL), has a low critical solution temperature (LCST), above which the polymer adopts a collapsed hydrophobic conformation and below which the polymer has a coil-like hydrophilic conformation [16,17]. The LCST of the polymer is also affected by the salt type and salt concentration in the solution [16,18–22]. The reduction in LCST follows the Hofmeister series and sometimes an inverse Hofmeister series [5,18,21]. The microscopic mechanisms of the Hofmeister effect are not completely understood yet; however, the impact of salt ions and salt concentrations on the properties of polymers and biological molecules is tremendously significant. PVCL has a LCST between 30 °C and 50 °C in water [16,23]. Transition temperature is also affected by its molecular weight and polymer concentration in aqueous solution due to the presence of the bulky seven-member ring on the polymer side chain. The polymer switching to a hydrophobic state will promote stronger hydrophobic interaction and thus binding of the protein. In contrast, the polymer transitioning to a hydrophilic state will promote protein desorption [18,24]. Both PVCL and PNIPAM have been investigated for application as HIC ligands for protein purifications [4–8,10,24–27]. However, the advantages of using PVCL are its biocompatibility and low toxicity. The hydrolysis of PNIPAM at acidic or basic conditions could generate low molecular weight amines which are toxic to biological systems. On the other hand, PVCL does not produce small amine molecules since the amide bond is located on its seven-member ring [6,16].

During the past few years, we have investigated the responsive HIC membranes by grafting PVCL chains to the surface of regenerated cellulose (RC) membranes [6,8,10,23]. The static binding capacity for BSA at around 10–15 mg/mL MV (membrane volume) and dynamic binding capacity (DBC) at around 8–12 mg/mL MV in 1.8 M $(NH_4)_2SO_4$ solution are already comparable, if not better, than the commercial HIC membranes available. The effects of PVCL chain length and chain density on protein binding and recovery are also investigated. Our previous results suggest that there is a trade-off between capacity and recovery for the PVCL ligands immobilized on the flat-sheet commercial RC membrane substrates. Longer and denser chains lead to higher binding capacity but reduced recovery, due largely to constriction of the pores during elution. In order to further improve the performance of these responsive HIC membranes, it seems that a more porous membrane substrate with a higher surface area or higher surface-to-volume ratio than that of commercial RC membranes is more desirable. Previously, PNIPAM and its thermo-responsive copolymers were investigated for antibody purifications [25–27]. The functionalized copolymers possess negatively-charged sulfonic acid groups at neutral pH. As a result, these polymers can selectively bind antibodies based on both charge and hydrophobic interactions.

Membranes made from electrospun fibers have high porosity with a high surface-to-volume ratio [28]. Electrospun membranes have demonstrated broad applications in water treatment, tissue engineering and protein purifications with ion-exchange membranes [14,29–36]. We previously fabricated weak electrospun anion-exchange membranes, for protein capture [14], as well as mixed matrix membranes, for ammonium removal [36]. Compared to flat sheet membrane substrates, significantly higher binding capacity was obtained for protein capture [14]. However, there are no previously reported studies on the fabrication of electrospun HIC membranes. In order to overcome the current limitations of HIC membranes and to achieve higher protein binding capacity and better product recovery, electrospun RC membrane substrates were fabricated and used for grafting PCVL ligands using controllable atom-transfer radical polymerization (ATRP) while varying polymer chain density and chain length. The electrospun RC membrane substrates were characterized using Fourier transform infrared spectroscopy (FTIR), X-ray photoelectron spectroscopy (XPS) and scanning electron microscopy (SEM). The performance of these responsive electrospun HIC membranes was investigated to determine both static and dynamic binding capacities of BSA under industrially applied $(NH_4)_2SO_4$ salt solutions. Finally, the membranes were used in the bind-and-elute mode chromatography of an industrial $IgG_4$. The main objective of this research is to investigate electrospun HIC membranes for bind-and-elute application with both enhanced protein binding capacity and improved product recovery to meet the high product titer demands of the biopharmaceutical industry and the high product purity requirements of regulatory agencies.

## 2. Materials and Methods

### *2.1. Materials*

N,N,N,N,N-pentamethyl diethylenetriamine (PMDETA, 99%), 2-bromo-2-methylpropionyl bromide (BIB, 98%), 2-hydroxyethyl methacrylate (98%), N-vinylcaprolactam (98%), 4-(dimethylamino) pyridine (DMAP, >99%), copper (I) chloride (Cu(I)Cl, >99.99%), copper (II) chloride ($CuCl_2$, >99.99%) and cellulose acetate (CA, Mn ~30 kD) were obtained from Sigma-Aldrich (St. Louis, MO, USA). Triethylamine (TEA, >99%) and N,N dimethylacetamide (DMAc, 99%) were sourced from Alfa Aesar (Ward Hill, MA, USA). Acetonitrile (>99.8%), methanol (99.8%), acetone (>99.5%) and ammonium sulfate (proteomics grade) were purchased from VWR (Radnor, PA, USA). Boric anhydride and BSA were acquired from Avantor Performance Materials (Center Valley, PA, USA). Purified human $IgG_4$ monoclonal antibody was donated by industry. Deionized (DI) water was generated using the Thermo Fisher Scientific (Waltham, MA, USA) DI water system.

### *2.2. Fabrication of Electrospun Membrane Substrate*

In a typical experiment, cellulose acetate (13.8 wt%) was dissolved in an acetone/DMAc (2:1, $w/w$) mixture. It was then stirred at room temperature to obtain a homogenous solution and used as the casting solution. The casting solution was subsequently filled into a syringe pump and electrospun onto a piece of aluminum foil using an Electrospinning Apparatus. The applied voltage was kept at 12.5 kV while collector–needle tip distance was maintained at 15–20 cm. The flow rate of the syringe pump was kept at 0.3 mL/h.

A nanofiber mat was obtained after collecting the nanofibers for 8 h. The thus electrospun cellulose acetate (CA) membrane was placed in a fume hood for one day to let the residual solvent evaporate. After annealing, the cellulose acetate membrane was peeled from aluminum foil and hydrolyzed in aqueous 0.1 M NaOH solution overnight without stirring to remove the acetyl groups from the acetate. Thereafter, the regenerated cellulose (RC) membrane samples were stored in a DI water bath until further use.

### 2.3. Ligand Grafting

The reaction scheme for grafting the PVCL ligand using ATRP is shown in Figure 1 and follows our previously published protocol [6]. Briefly, the electrospun RC membrane was dried in a vacuum oven for 8 h to remove residual moisture. The membrane was then immersed in the solution containing 40, 80 and 200 mM initiator BIB in acetonitrile for a predetermined period of time. In the meantime, a mixed solution was prepared by adding monomer VCL, catalyst CuCl and $CuCl_2$ and conjugating ligand PMDETA into the 50:50 methanol/water $v/v$% solution. The ratio of VCL monomer, CuCl, $CuCl_2$ and PMDETA was fixed at 200:1:0.2:2. Argon was used to degas the mixture solution for 15–20 min. The previously initiator-immobilized membrane was then placed into a clean round-bottom three-necked flask. Finally, the polymerization process was initiated by pouring the derived mixture solution into the flask under an inert gas environment. After a predetermined period of ATRP reaction, the modified membranes were rinsed three times using 50:50 $v/v$% methanol/water and then three times with DI water. The resultant membrane was kept in a DI water bath on the shaker to remove any residual solvent.

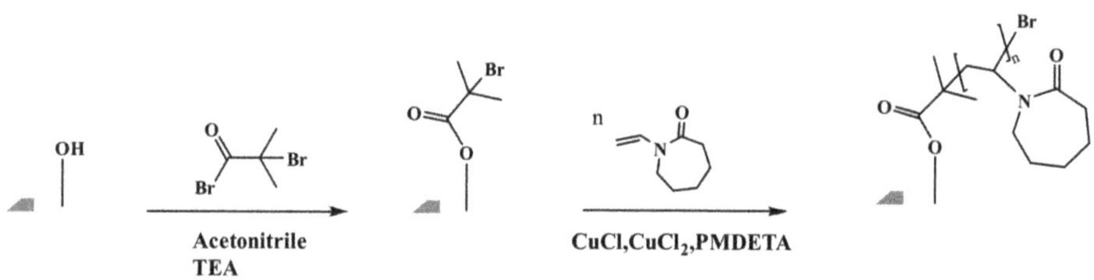


**Figure 1.** Reaction scheme for grafting poly(N-vinylcaprolactam) (PVCL) ligands on electrospun membranes with atom-transfer radical polymerization (ATRP).

### 2.4. Characterization of the Functionalized HIC Membranes

All membrane samples were cleaned with DI water and dried in a vacuum oven (12.5 L, VWR International, Radnor, PA, USA) overnight prior to characterization. Static contact angle measurements were carried out at different ionic strengths of solution. Quintuplicate measurements were made to determine the standard deviation.

FTIR (IRAffinity, Shimadzu, MD, USA) was performed to determine characteristic functional groups of the modified membranes. XPS (Thermo Fisher Scientific Inc., Waltham, MA, USA) was used for analyzing the chemical compositions of the membranes before and after modification. SEM (FESEM S-4800, Hitachi Co., Tokyo, Japan) was used to characterize the surface fiber structures of both modified and unmodified membranes.

Grafting degree (GD) was calculated to quantitatively determine the amount of PVCL grafted. The membranes before and after modification were dried in a vacuum oven at 40 °C and the weight of the samples was thereafter measured. GD was calculated based on the following equation:

$$GD\ (\%) = \frac{W_{modified} - W_{unmodified}}{W_{unmodified}} \times 100\% \quad (1)$$

*2.5. The Static and Dynamic Binding Capacities of HIC Membranes*

The fabricated HIC membranes were cut into 4.9 $cm^2$ disks for the static binding capacity tests. Each membrane was placed in a 60 mL glass bottle from VWR (Radnor, PA, USA) and equilibrated for 1 h with the adsorption buffer (20 mM phosphate buffer with 1.8 M $(NH_4)_2SO_4$ at pH 7.0, buffer A). Subsequently, the equilibrated membranes were challenged with model protein BSA at 5 different concentrations for 5 h at room temperature under gentle shaking. The final equilibrated BSA solutions were measured by UV absorbance at 280 nm with a UVVIS spectrophotometer (Thermo Scientific™ GENESYS 10S UV-Vis, Waltham, MA, USA) and their concentrations determined using a standard curve.

ÄKTA Pure (GE Healthcare Bio-Sciences Corp., Boston, MA, USA) was used to conduct the fast protein liquid chromatography (FPLC) experiments for dynamic binding capacity measurements. Unicorn software version 7.3 was used to automate experiments on BSA binding and elution. BSA solutions were prepared by dissolving 10 mg of BSA into 10 mL of buffer A, which contained 1.8 M $(NH_4)_2SO_4$. $IgG_4$ feed solutions were prepared by dissolving 1 mg of $IgG_4$ into 10 mL of buffer A. Prior to the binding tests, both protein and buffer solutions were filtered using Whatman 0.2-µm Polyethersulfone (PES) membrane filters. Subsequently, four fabricated HIC membranes (0.04 mL membrane volume) were loaded into a stainless-steel flow cell (Mustang Coin module, Pall Corporation, Port Washington, NY, USA) equipped with two flow distributors to establish a uniform flow across all membranes. The membranes were equilibrated in the forward flow configuration in buffer A (adsorption buffer) for 10 min at 1 mL/min. A protein solution (1 mg/mL) was loaded onto the membrane for 10 min at a flow rate of 1 mL/min. Unbound proteins were then washed from the membrane surface using adsorption buffer (buffer A) for 5 min at 1 mL/min, followed by a step change to running the elution buffer (20 mM phosphate buffer with 0 mM $(NH_4)_2SO_4$ at pH 7.0, buffer B) through the membranes at the same flow rate. The chromatographic run was stopped when the UV absorbance at 280 nm became constant. The elution and washing fractions, as well as the loading fraction, were collected and their protein concentrations were subsequently determined.

Both static and dynamic binding capacities were calculated based on the following equation:

$$\text{Binding capacity} = \frac{\text{Mass (bound protein, mg)}}{\text{Membrane Volume (mL)}} \tag{2}$$

Protein recovery from the FPLC chromatographic run was determined using the equation below:

$$\text{Recovery} = \frac{\text{Mass (eluted protein)}}{\text{Mass (bound protein)}} \times 100\% \tag{3}$$

## 3. Results and Discussion

*3.1. Physicochemical Properties of Fabricated HIC Membranes*

FTIR, XPS and SEM methods were used to characterize the chemical and structural properties of electrospun membranes before and after surface modification. The FTIR spectra of the originally fabricated cellulose acetate (CA) membrane, the subsequently hydrolyzed RC membrane and PVCL modified membranes are shown in Figure 2. As can be seen, the electrospun CA membrane exhibits a significant characteristic carbonyl (C=O) peak at around 1630 $cm^{-1}$ [8]. As the acetal group was hydrolyzed to the hydroxyl group in the RC membrane, a new broad peak was observed at around 3400 $cm^{-1}$, which is characteristic of the –OH group. In the meantime, the carbonyl peak completely disappeared after hydrolysis indicating the successful completion of the hydrolysis reaction. After grafting the PCVL ligand, a new weaker carbonyl group peak from the ligand reappears, indicating the successful grafting of PVCL on the membrane surface [8]. The stretching vibrations of the N–H and O–H groups of the PVCL chain on the RC membrane correspond to the broad bands at 3000–3700 $cm^{-1}$ [37]. Elemental composition was determined from XPS analysis as shown in Figure 3. The N peak was observed after surface modification with PVCL, indicating successful modification of the ligand.

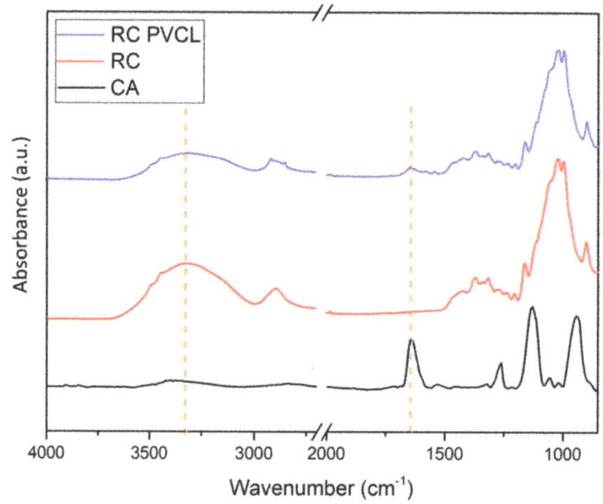


**Figure 2.** FTIR spectra of the cellulose acetate (CA), regenerated cellulose (RC) and surface–modified RC PVCL membranes.

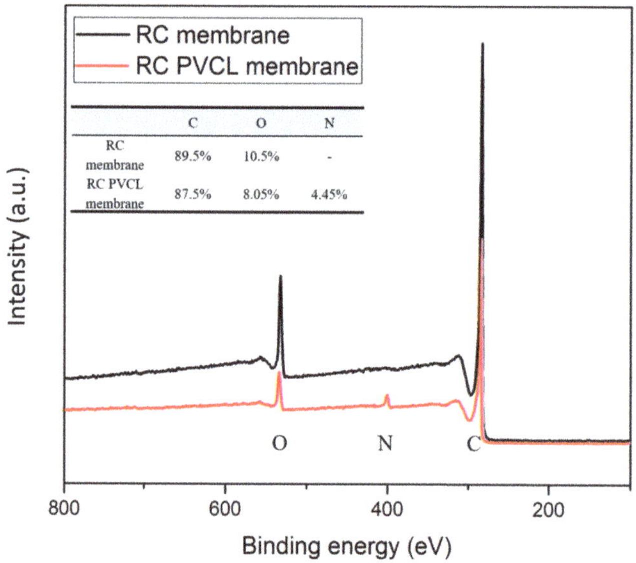


**Figure 3.** X-ray photoelectron (XPS) spectroscopy of the electrospun RC membranes before and after PVCL ligand grafting.

Figure 4 shows SEM images of the original electrospun CA membrane as well as the RC membranes after hydrolysis and after surface modification. Based on these SEM images, the average fiber diameter of the membranes was estimated. For the CA membrane, the diameters are estimated to range between 330 and 440 µm, similar to the previously reported results [29]. After hydrolysis converting CA to RC membranes, no significant change in fiber diameter or membrane morphology was observed except that the RC membrane became slightly rougher. However, after surface modification with PVCL, the membrane surface morphology altered significantly with much rougher surfaces. This is partly due to the organic solvent used, which could have affected the fiber morphology, and partly due to the presence of grafted PVCL ligands on the membrane substrate.

**Figure 4.** Scanning Electron Microscopy (SEM) images of cellulose acetate (CA) electrospun membrane (**a**), regenerated cellulose (RC) membrane (**b**) and PVCL modified RC membrane (**c**).

Contact angle measurements using the air bubble method were performed to determine the relative hydrophobicity of the fabricated membranes [38]. The converted water contact angles are shown in Figure 5. The electrospun CA membrane exhibited a contact angle of approximately 130°, whereas the flat-surface CA membrane had a contact angle of nearly 80° [39], which can be attributed to the presence of air gaps in its scaffold-like structure. After hydrolysis and conversion to the RC membrane, the contact angle of the electrospun membrane was reduced considerably to approximately 18°, mainly due to the presence of the hydroxyl groups. After PVCL ligand grafting, the water contact angle was increased to approximately 63°. The PVCL ligand grafted membrane is somewhat more hydrophobic compared to the RC membrane substrate due to the slightly more hydrophobic nature of the PVCL ligand compared to cellulose substrate. Since PVCL is temperature- and salt ion-responsive with its hydrophobicity affected by solution conditions, contact angles were also measured for the functionalized RC membrane by increasing the ionic strength of the ammonium sulfate solution from 0.4 to 1.8 M. The measurements are also shown in Figure 5. It can be seen that the membrane surface becomes more hydrophobic as the salt concentration increases with contact angle increasing from 63° in water to 120° in 1.8 M $(NH_4)_2SO_4$.

Grafting degree (GD) was determined based on Equation (1) by calculating the percentage of weight increase compared to the unmodified membrane. As shown in Figure 6, membrane GD increases as the ATRP time increases almost linearly, indicating the well-controlled nature of the polymerization reaction. However, it is still rather challenging to quantify the exact chain length and chain density of the PVCL grafted on the membrane substrate.

### 3.2. Static Binding Capacity

Static binding experiments were performed using BSA as the model protein to evaluate the performance of the fabricated HIC membranes. Initially, the static binding capacity of BSA was evaluated as a function of ATRP grafting time for 1 g/L BSA in 20 mM phosphate buffer with 1.8 M ammonium sulfate, as shown in Figure 7. The concentrations of BSA were measured after 5 h of equilibrium. The results show that static binding capacity increases as ATRP time increases, indicating that longer polymer chains have higher binding capacity. However, after more than 6 h of polymerization, further enhancement of binding capacity is not significant, increasing from ~22 mg/mL MV at 6 h to 25 mg/mL MV at 10 h. Further increases in ATRP time led to a slight decrease in BSA binding capacity. For the subsequent experiments, an ATRP time of 6 h was used for both static and dynamic binding studies.

The adsorption isotherm was investigated for 1 g/L BSA in 1.8 M ammonium sulfate buffer solution as shown in Figure 8. A Langmuir–Freundlich isotherm model [40,41] was able to describe the adsorption behavior using the following equation:

$$q = \frac{Q_{max}\left(a_{LF}C_{eq}\right)^{n_{LF}}}{1 + \left(a_{LF}C_{eq}\right)^{n_{LF}}} \quad (4)$$

where q represents the amount of proteins adsorbed at equilibrium (mg/mL MV), $Q_{max}$ represents the maximum adsorption capacity of a membrane (mg/mL MV), $C_{eq}$ represents the protein concentration at equilibrium (mg/mL), $a_{LF}$ represents the adsorption affinity constant (mL/mg) and $n_{LF}$ represents the heterogeneity index. In this study, a fitting $R^2$ value of 0.9988 was obtained indicating a high-quality fit. The fitting parameters are shown in Figure 8. It can be seen that $Q_{max}$ is about 24 mg/mL MV which is in agreement with earlier results. The heterogeneity index $n_{LF}$ is 0.99, indicating that BSA adsorption by the electrospun HIC membrane follows a Langmuir isotherm model with monolayer adsorption.

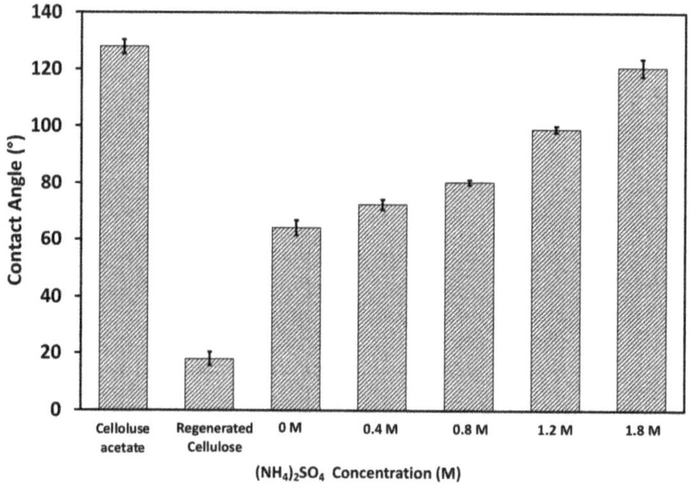


**Figure 5.** Contact angles of the electrospun CA, the subsequently hydrolyzed RC membrane substrates, as well as PVCL immobilized RC membranes under different concentrations of $(NH_4)_2SO_4$ salt solutions.

### *3.3. Protein Dynamic Binding Capacity and Recovery*

Experiments were conducted to evaluate the performance of the fabricated HIC membranes under actual chromatographic runs. The experiments were run with a range of BSA feed concentrations ranging from 0. 1 to 1 g/L BSA to determine protein dynamic binding capacity (DBC) and recovery. Initially, the electrospun RC membranes grafted with PVCL at two different chain lengths were tested for BSA binding and recovery, as shown in Table 1. At GD 6% (6 h ATRP) and 9% (10 h ATRP), DBC and recovery are quite different at the same BSA concentration. At 6% GD, a DBC of 8.0 mg/mL MV and a recovery of 98.2% can be achieved when the BSA concentration is 0.1 g/L. As the BSA concentration increased to 1.0 g/L, the DBC reached 30.7 mg/mL MV. However, recovery reduced to only 51.5%. At 9% GD, a DBC of 14.3 and 43.4 mg/mL MV was obtained at 0.1 and 1.0 g/L BSA feed solution, respectively, much higher than the corresponding values at 6% GD. However, recovery was only at 80.6% and 21.1%, respectively, significantly reduced compared to the values at 6% GD. Again, we can see that the longer polymer chain tends to have a higher DBC but also leads to reduced product recovery consistent with previous results [8]. Longer polymer chains lead to increased coverage of the ligand on the membrane surface and inside the membrane pores for protein binding which in turn leads to higher binding capacity. However, longer polymer chains with higher surface/pore coverage of the grafted polymer could also trap the proteins and prevent them from eluting, thereby leading to lower recovery. For subsequent studies, membranes with 6 h ATRP time were investigated.

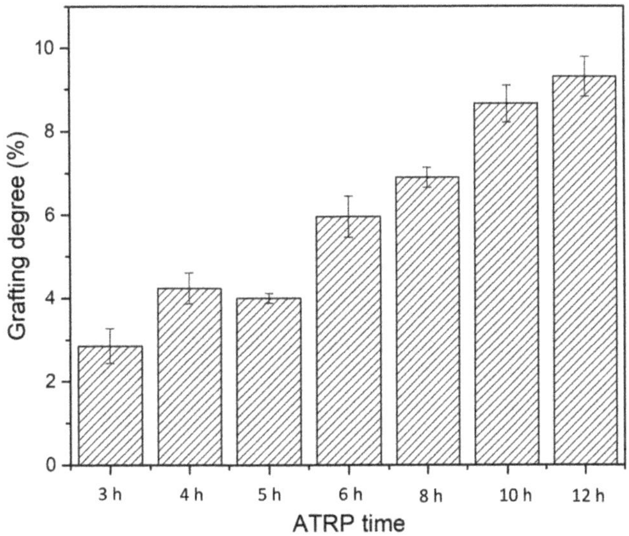


**Figure 6.** Grafting degree as a function of polymerization time for PVCL immobilized on the electrospun RC membrane substrate.

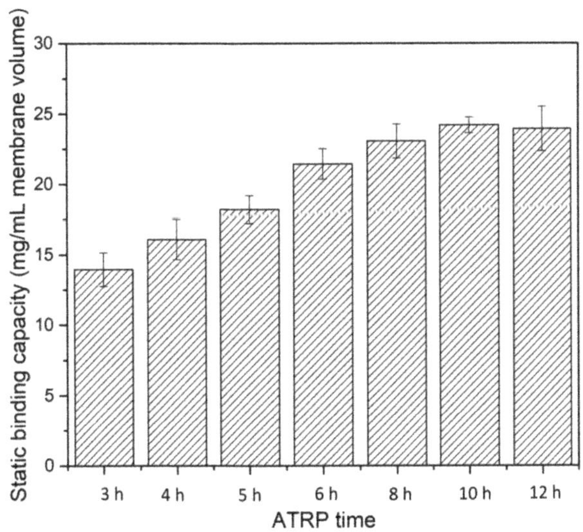


**Figure 7.** Static binding capacity as a function of polymerization time for PVCL grown on membranes.

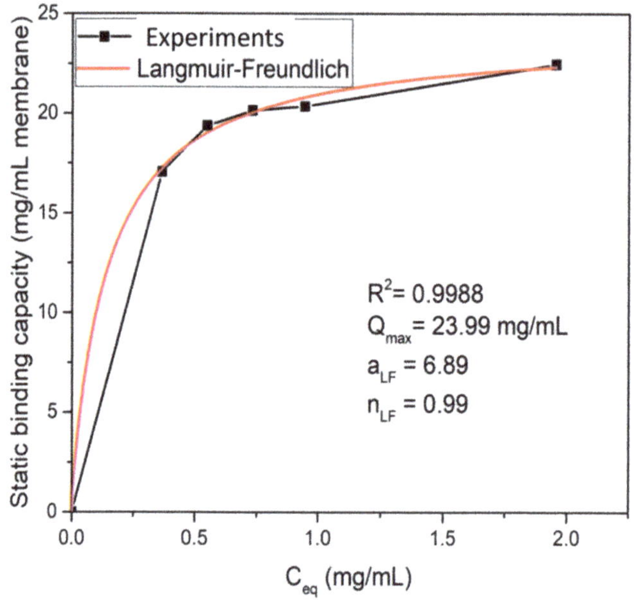


**Figure 8.** Langmuir–Freundlich curve for ATRP of a 6 h PVCL modified membrane.

**Table 1.** The effect of grafting degree on dynamic binding capacity (DBC) and recovery of bovine serum albumin (BSA) at two protein loading concentrations.

| Grafting Degree (DG) (%) | BSA Loading Concentration (g/L) | Protein Binding Capacity (mg/mL) | Recovery (%) |
|---|---|---|---|
| 6 | 0.10 | 7.95 ± 0.75 | 98.17 ± 0.02 |
| 6 | 1.0 | 30.69 ± 0.34 | 51.54 ± 1.96 |
| 9 | 0.10 | 14.29 | 80.64 |
| 9 | 1.0 | 43.43 | 21.09 |

In addition to the effect of polymer chain length on protein binding and elution, the effect of ligand density on DBC and product recovery was also investigated. Chromatographic runs for electrospun membranes with different ligand densities were performed. The variation of chain densities was achieved by immobilizing BIB initiator for 3 h at 40, 80 and 200 mM. Two sets of experiments were performed for each condition, with the final result as an average of the two. As shown in Table 2, the functionalized electrospun membrane using 80 mM initiator has the highest DBC of 8.0 and 30.7 mg/mL MV at 0.1 and 1.0 g/L BSA loading conditions, respectively. With 40 mM initiator and lower ligand density, DBCs are 4.8 and 25.0 mg/mL MV for the two protein feed conditions, respectively. Increasing initiator concentration to 200 mM and with higher ligand density, the DBC values are slightly reduced to 6.2 and 28.4 mg/mL MV, respectively. With regard to product recovery, at 40 and 80 mM initiator concentrations with relatively lower ligand densities, recovery at around 98% was achieved at 0.1 g/L BSA feed concentration. When feed concentration increases to 1.0 g/L, recovery is reduced to 57.2% and 51.5% for the two ligand densities. At the highest ligand density with 200 mM initiator concentration, the recovery is only 83.4% for 0.1 g/L BSA feed and 41.4% for 1 g/L feed. It can be seen that protein dynamic binding capacity of functionalized electrospun HIC membranes is significantly higher than the DBC of the functionalized flat sheet HIC membrane at similar conditions [5–8,10,23]. Previous studies [8] achieved an optimal dynamic binding capacity of 12.6 mg/mL MV and a recovery of 78% with 1.0 g/L BSA feed on HIC ligand functionalized on the flat sheet membrane. A binding capacity of 30.7 mg/mL MV can be achieved on electrospun membranes with a slightly lower recovery of 51.5% at the same feed condition. It seems that there is a trade-off between binding capacity and recovery [8]. High binding capacity leads to lower recovery which is in agreement with previous observations. Of the three ligand densities, 80 mM initiator concentration appears to provide the best compromise between binding capacity and recovery. Therefore, the ligand density with 80 mM initiator concentration was used.

**Table 2.** The effect of ligand density and BSA concentration on DBC and recovery.

| Initiator (mM) | BSA Loading Concentration (g/L) | Protein Binding Capacity (mg/mL) | Recovery (%) |
|---|---|---|---|
| 40 | 0.1 | 4.79 ± 0.11 | 97.54 ± 2.27 |
| 40 | 1.0 | 24.95 ± 0.56 | 57.24 ± 2.33 |
| 80 | 0.1 | 7.95 ± 0.75 | 98.17 ± 0.02 |
| 80 | 1.0 | 30.69 ± 0.34 | 51.54 ± 1.96 |
| 200 | 0.1 | 6.22 ± 0.11 | 83.39 ± 0.22 |
| 200 | 1.0 | 28.38 ± 1.75 | 41.44 ± 0.01 |

The effect of the amount of protein loaded on the functionalized membrane on DBC and recovery was also investigated systematically. BSA feed streams in 20 mM phosphate buffer with 1.8 M $(NH4)_2SO_4$ at 0.1, 0.3, 0.5 and 1.0 g/L concentrations were loaded on the functionalized HIC membranes. These membranes had a GD of 6% functionalized by a 3 h immobilization in 80 mM initiator followed by a 6 h ATRP reaction. Table 3 shows the DBC and recovery at different BSA loading concentrations. It can be seen that as BSA concentration increases DBC increases, whereas recovery decreases. This agrees with previous observations that higher binding capacity typically correlates with lower protein recovery for a given membrane. Nevertheless, a moderate DBC of 16.6 mg/mL MV and a recovery of 75.7% can be achieved when BSA loading concentration is 0.5 g/L.

**Table 3.** The effect of protein feed concentration on DBC and product recovery for PVCL grafted with 80 mM initiator and 6 h ATRP.

| BSA Loading Concentration (g/L) | Protein Binding Capacity (mg/mL) | Recovery (%) |
|---|---|---|
| 0.1 | 7.95 ± 0.75 | 98.17 ± 0.02 |
| 0.3 | 9.93 | 91.56 |
| 0.5 | 16.55 ± 2.09 | 75.67 ± 3.32 |
| 1.0 | 30.69 ± 0.34 | 51.54 ± 1.96 |

The effect of flow rate on dynamic protein binding capacity and recovery is shown in Table 4. In previous studies, flow rate was kept at 1.0 mL/min. Here, flow rates of 0.5 and 2 mL/min were investigated. It can be seen from Table 4 that as flow rate decreases from 1 to 0.5 mL/min, DBC decreases slightly from 30.7 to 28.9 mg/mL MV, whereas recovery increases from 51.5% to 57.5%. When flow rate increases from 1 to 2 mL/min, both DBC and recovery decrease to 27.4 mg/mL MV and 41.7%, respectively. It is clear that flow rate has a strong effect on product recovery since the elution and migration of the protein from the binding sites are affected by both hydrodynamic and diffusive forces. The slow flow rate will allow the protein molecules to migrate out of the membrane. On the other hand, the effect of flow rate on DBC only slightly indicates that it is possible to improve the performance of these functionalized responsive HIC membranes by reducing flow rate.

The effect of electrospun fiber diameter on protein binding capacity and recovery was also investigated, as shown in Table 5. The concentration of the polymer solution for electrospinning was kept the same at 13.8%. However, the voltage applied during electrospinning increased from 12.5 to 20 kV, which decreased fiber diameter from 330–440 to 220–350 μm as shown in Table 5. The pore size of the membrane formed also decreased from 3.24 to 2.88 μm. The modification conditions for these two membranes remained at 80 mM initiator concentration and 6 h ATRP time. The membrane with smaller fibers exhibited a smaller average pore size. The DBCs are 8.0 and 9.8 mg/mL MV for membranes with larger and smaller fiber diameters, respectively, when the BSA loading concentration is 0.10 g/L. Recovery was 98% for the larger fiber and was reduced to 79% for the smaller fiber. It seems that the small gain in binding capacity is compensated by the significant reduction in recovery. At a high BSA loading concentration of 1.0 g/L, the DBCs are 30.7 and 33.6 mg/mL MV, respectively, for the larger and smaller fiber membranes. However, recovery is reduced from 51.5% to only 37.9%. Clearly, larger fiber diameter membranes have better overall performance for bind-and-elute applications of these responsive HIC membranes.

**Table 4.** The effect of flow rate on DBC and recovery of BSA for PVCL grafted with 80 mM initiator and 6 h ATRP.

| BSA Loading Concentration (g/L) | Flowrate (mL/min) | Protein Binding Capacity (mg/mL) | Recovery (%) |
|---|---|---|---|
| 1 | 0.5 | 28.87 ± 0.79 | 57.51 ± 1.23 |
| 1 | 1 | 30.69 ± 0.34 | 51.54 ± 1.96 |
| 1 | 2 | 27.36 ± 0.93 | 41.72 ± 1.80 |

**Table 5.** The effect of fiber diameter on DBC and recovery.

| Voltage (kV) | Fiber Diameter (nm) | Pore Size (μm) | BSA (g/L) | DBC (mg/mL) | Recovery (%) |
|---|---|---|---|---|---|
| 12.5 | 330–440 | 3.24 ± 0.27 | 0.1 | 7.95 ± 0.75 | 98.17 ± 0.02 |
| | | | 1.0 | 30.69 ± 0.34 | 51.54 ± 1.96 |
| 20 | 220–350 | 2.88 ± 0.39 | 0.1 | 9.84 | 79.08 |
| | | | 1.0 | 33.58 | 37.86 |

Finally, binding capacity and recovery of $IgG_4$ were investigated using our responsive HIC membranes at three different loading concentrations of 0.1, 0.5 and 1 g/L. The membrane (80 mM initiator, 6 h ATRP) and feed buffer (20 mM phosphate buffer with 1.8 M $(NH_4)_2SO_4$) conditions remained the same as the previously performed experiments. The results are shown in Table 6. Average binding capacities of 16.3, 28.8 and 47.8 mg/mL MV were obtained at 0.1, 0.5 and 1.0 g/L IgG loading concentrations. However, recovery decreased from 85.3% to 24.0% and finally to 14.2% as IgG concentration increased. It is clear that IgG has an overall higher binding capacity compared to BSA. However, its recovery is much reduced compared to that of BSA, particularly at high protein concentrations. Himstedt et al. [23] observed an IgG capacity of 21 mg/mL, whereas Liu et al. [6] reported an IgG capacity of 12 mg/mL using flat sheet PVCL functionalized membranes with 1 g/L IgG loading concentration. These results highlight the potential benefits derived from using an open and more porous electrospun membrane as a substrate for ligand functionalization.

**Table 6.** Dynamic protein binding capacity and recovery results for $IgG_4$.

| $IgG_4$ Loading Concentration (g/L) | Protein Binding Capacity (mg/mL) | Recovery (%) |
|---|---|---|
| 0.1 | 16.31 ± 0.23 | 85.32 ± 1.38 |
| 0.5 | 28.78 ± 0.18 | 23.95 ± 1.17 |
| 1 | 47.68 ± 1.29 | 14.22 ± 1.93 |

## 4. Conclusions

Responsive HIC membranes were successfully fabricated by functionalizing electrospun RC membrane substrate. The PVCL functionalized electrospun membranes exhibit a much higher binding capacity compared to the corresponding flat sheet membranes. A maximum static BSA binding capacity of ~25 mg/mL MV can be achieved. The longer polymer chains lead to higher static binding capacity up to a grafting degree of 6%. Binding capacity also increases with salt concentration which is in agreement with previous studies. Static binding follows the monolayer Langmuir adsorption isotherm.

For the dynamic binding studies, protein binding capacity increased with protein concentration while protein recovery decreased as its concentration increased. There is a clear trade-off between binding capacity and recovery rate. There is an optimal chain density for the performance of these responsive HIC membranes. In addition, it also appears that larger fiber diameter leads to slightly reduced protein binding capacity but significantly higher recovery, indicating better performance for electrospun membranes with thicker fibers. This is due to the larger pores that resulted from the thicker fibers. There is some influence of flow rate on protein binding and elution. A slow flowrate (e.g., 0.5 mL/min) leads to slightly reduced binding capacity but slightly enhanced recovery compared to the corresponding values at 1 mL/min. On the other hand, a higher flowrate (e.g., 2 mL/min) leads to both slightly reduced binding capacity and much reduced recovery. The performance of these responsive HIC membranes is protein dependent due to their different hydrophobicity and other properties. The $IgG_4$ tested showed much higher binding capacity compared to BSA at the same conditions but highly reduced protein recovery.

**Author Contributions:** Conceptualization, S.R.W. and X.Q.; data curation, S.-T.C.; formal analysis, S.-T.C.; funding acquisition, X.Q. and S.R.W.; investigation, S.-T.C. and X.Q.; methodology, S.-T.C., X.Q. and S.R.W.; project administration, X.Q. and S.R.W.; software, S.-T.C.; supervision, X.Q. and S.R.W.; writing—original draft, X.Q.; writing—review and editing, X.Q. and S.R.W. All authors have read and agreed to the published version of the manuscript.

**Funding:** This research was funded by the Center for Advanced Surface Engineering at the University of Arkansas. The APC was funded by the University of Arkansas.

**Institutional Review Board Statement:** Not applicable.

**Informed Consent Statement:** Not applicable.

**Data Availability Statement:** The data presented in the study are already shown in the tables from this article.

**Conflicts of Interest:** The authors declare no conflict of interest.

## References

1. Cramer, S.M.; Holstein, M.A. Downstream bioprocessing: Recent advances and future promise. *Curr. Opin. Chem. Eng.* **2011**, *1*, 27–37. [CrossRef]
2. Liu, Z.; Wickramasinghe, S.R.; Qian, X. Membrane chromatography for protein purifications from ligand design to functionalization. *Sep. Sci. Technol.* **2017**, *52*, 299–319. [CrossRef]
3. Devi, N.; Patel, S.K.S.; Kumar, P.; Singh, A.; Thakur, N.; Lata, J.; Pandey, D.; Thakur, V.; Chand, D. Bioprocess Scale-up for Acetohydroxamic Acid Production by Hyperactive Acyltransferase of Immobilized Rhodococcus Pyridinivorans. *Catal. Lett.* **2022**, *152*, 944–953. [CrossRef]
4. Eriksson, K.O. *Hydrophobic Interaction Chromatography, Biopharmaceutical Processing*; Elsevier: Amsterdam, The Netherlands, 2018; pp. 401–408.
5. Ghosh, R. Separation of proteins using hydrophobic interaction membrane chromatography. *J. Chromatogr. A* **2001**, *923*, 59–64. [CrossRef]
6. Liu, Z.; Wickramasinghe, S.R.; Qian, X. Ion-specificity in protein binding and recovery for the responsive hydrophobic poly (vinylcaprolactam) ligand. *RSC Adv.* **2017**, *7*, 36351–36360. [CrossRef]
7. Lienqueo, M.E.; Mahn, A.; Salgado, J.C.; Asenjo, J.A. Current insights on protein behaviour in hydrophobic interaction chromatography. *J. Chromatogr. B* **2007**, *849*, 53–68. [CrossRef]
8. Liu, Z.; Wickramasinghe, S.R.; Qian, X. The architecture of responsive polymeric ligands on protein binding and recovery. *RSC Adv.* **2017**, *7*, 27823–27832. [CrossRef]
9. Baumann, P.; Baumgartner, K.; Hubbuch, J. Influence of binding pH and protein solubility on the dynamic binding capacity in hydrophobic interaction chromatography. *J. Chromatogr. A* **2015**, *1396*, 77–85. [CrossRef]
10. Vu, A.; Qian, X.; Wickramasinghe, S.R. Membrane-based hydrophobic interaction chromatography. *Sep. Sci. Technol.* **2017**, *52*, 287–298. [CrossRef]

11. Liu, Z.; Du, H.; Wickramasinghe, S.R.; Qian, X. Membrane surface engineering for protein separations: Experiments and simulations. *Langmuir* **2014**, *30*, 10651–10660. [CrossRef]
12. Boi, C.; Malavasi, A.; Carbonell, R.G.; Gilleskie, G. A direct comparison between membrane adsorber and packed column chromatography performance. *J. Chromatogr. A* **2020**, *1612*, 460629. [CrossRef] [PubMed]
13. Chenette, H.C.; Robinson, J.R.; Hobley, E.; Husson, S.M. Development of high-productivity, strong cation-exchange adsorbers for protein capture by graft polymerization from membranes with different pore sizes. *J. Membr. Sci.* **2012**, *423*, 43–52. [CrossRef] [PubMed]
14. Chen, S.-T.; Wickramasinghe, S.R.; Qian, X. Electrospun weak anion-exchange fibrous membranes for protein purification. *Membranes* **2020**, *10*, 39. [CrossRef] [PubMed]
15. Hall, T.; Kelly, G.M.; Emery, W.R. Use of mobile phase additives for the elution of bispecific and monoclonal antibodies from phenyl based hydrophobic interaction chromatography resins. *J. Chromatogr. B* **2018**, *1096*, 20–30. [CrossRef]
16. Sun, X.; Qian, X. Atomistic Molecular Dynamics Simulations of the Lower Critical Solution Temperature Transition of Poly (N-vinylcaprolactam) in Aqueous Solutions. *J. Phys. Chem. B* **2019**, *123*, 4986–4995. [CrossRef]
17. Zhang, Y.; Furyk, S.; Bergbreiter, D.E.; Cremer, P.S. Specific ion effects on the water solubility of macromolecules: PNIPAM and the Hofmeister series. *J. Am. Chem. Soc.* **2005**, *127*, 14505–14510. [CrossRef]
18. Du, H.; Wickramasinghe, R.; Qian, X. Effects of salt on the lower critical solution temperature of poly (N-isopropylacrylamide). *J. Phys. Chem. B* **2010**, *114*, 16594–16604. [CrossRef]
19. Du, H.; Qian, X. The Interactions between Salt Ions and Thermo-Responsive Poly (N-Isopropylacrylamide) from Molecular Dynamics Simulations. *Responsive Membr. Mater.* **2012**, *10*, 229–242. [CrossRef]
20. Du, H.; Wickramasinghe, S.R.; Qian, X. Specificity in cationic interaction with poly (N-isopropylacrylamide). *J. Phys. Chem. B* **2013**, *117*, 5090–5101. [CrossRef]
21. Zhang, Y.; Furyk, S.; Sagle, L.B.; Cho, Y.; Bergbreiter, D.E.; Cremer, P.S. Effects of Hofmeister anions on the LCST of PNIPAM as a function of molecular weight. *J. Phys. Chem. C* **2007**, *111*, 8916–8924. [CrossRef]
22. Hiruta, Y.; Nagumo, Y.; Suzuki, Y.; Funatsu, T.; Ishikawa, Y.; Kanazawa, H. The effects of anionic electrolytes and human serum albumin on the LCST of poly (N-isopropylacrylamide)-based temperature-responsive copolymers. *Colloids Surf. B Biointerfaces* **2015**, *132*, 299–304. [CrossRef] [PubMed]
23. Himstedt, H.H.; Qian, X.; Weaver, J.R.; Wickramasinghe, S.R. Responsive membranes for hydrophobic interaction chromatography. *J. Membr. Sci.* **2013**, *447*, 335–344. [CrossRef]
24. Darvishmanesh, S.; Qian, X.; Wickramasinghe, S.R. Responsive membranes for advanced separations. *Curr. Opin. Chem. Eng.* **2015**, *8*, 98–104. [CrossRef]
25. Okubo, K.; Ikeda, K.; Oaku, A.; Hiruta, Y.; Nagase, K.; Kanazawa, H. Protein purification using solid-phase extraction on temperature-responsive hydrogel-modified silica beads. *J. Chromatogr. A* **2018**, *1568*, 38–48. [CrossRef] [PubMed]
26. Nagase, K.; Ishii, S.; Ikeda, K.; Yamada, S.; Ichikawa, D.; Akimoto, A.M.; Hattori, Y.; Kanazawa, H. Antibody drug separation using thermoresponsive anionic polymer brush modified beads with optimised electrostatic and hydrophobic interactions. *Sci. Rep.* **2020**, *10*, 11896. [CrossRef]
27. Nomoto, D.; Nagase, K.; Nakamura, Y.; Kanazawa, H.; Citterio, D.; Hiruta, Y. Anion species-triggered antibody separation system utilizing a thermo-responsive polymer column under optimized constant temperature. *Colloids Surf. B Biointerfaces* **2021**, *205*, 111890. [CrossRef]
28. Huang, L.; Arena, J.T.; McCutcheon, J.R. Surface modified PVDF nanofiber supported thin film composite membranes for forward osmosis. *J. Membr. Sci.* **2016**, *499*, 352–360. [CrossRef]
29. Chitpong, N.; Husson, S.M. Polyacid functionalized cellulose nanofiber membranes for removal of heavy metals from impaired waters. *J. Membr. Sci.* **2017**, *523*, 418–429. [CrossRef]
30. Fu, Q.; Wang, X.; Si, Y.; Liu, L.; Yu, J.; Ding, B. Scalable fabrication of electrospun nanofibrous membranes functionalized with citric acid for high-performance protein adsorption. *ACS Appl. Mater. Interfaces* **2016**, *8*, 11819–11829. [CrossRef]
31. Dods, S.R.; Hardick, O.; Stevens, B.; Bracewell, D.G. Fabricating electrospun cellulose nanofibre adsorbents for ion-exchange chromatography. *J. Chromatogr. A* **2015**, *1376*, 74–83. [CrossRef]
32. Zhang, H.; Wu, C.; Zhang, Y.; White, C.J.B.; Xue, Y.; Nie, H.; Zhu, L. Elaboration, characterization and study of a novel affinity membrane made from electrospun hybrid chitosan/nylon-6 nanofibers for papain purification. *J. Mater. Sci.* **2010**, *45*, 2296–2304. [CrossRef]
33. Ma, H.; Hsiao, B.S.; Chu, B. Electrospun nanofibrous membrane for heavy metal ion adsorption. *Curr. Org. Chem.* **2013**, *17*, 1361–1370. [CrossRef]
34. Ma, Z.; Lan, Z.; Matsuura, T.; Ramakrishna, S. Electrospun polyethersulfone affinity membrane: Membrane preparation and performance evaluation. *J. Chromatogr. B* **2009**, *877*, 3686–3694. [CrossRef] [PubMed]
35. Schneiderman, S.; Zhang, L.; Fong, H.; Menkhaus, T.J. Surface-functionalized electrospun carbon nanofiber mats as an innovative type of protein adsorption/purification medium with high capacity and high throughput. *J. Chromatogr. A* **2011**, *1218*, 8989–8995. [CrossRef]
36. Chen, S.-T.; Wickramasinghe, S.R.; Qian, X. High Performance Mixed-Matrix Electrospun Membranes for Ammonium Removal from Wastewaters. *Membranes* **2021**, *11*, 440. [CrossRef]

37. Li, G.; Xiao, J.; Zhang, W. Knoevenagel condensation catalyzed by a tertiary-amine functionalized polyacrylonitrile fiber. *Green Chem.* **2011**, *13*, 1828–1836. [CrossRef]
38. Chiao, Y.-H.; Sengupta, A.; Chen, S.-T.; Huang, S.-H.; Hu, C.-C.; Hung, W.-S.; Chang, Y.; Qian, X.; Wickramasinghe, S.R.; Lee, K.-R. Zwitterion augmented polyamide membrane for improved forward osmosis performance with significant antifouling characteristics. *Sep. Purif. Technol.* **2019**, *212*, 316–325. [CrossRef]
39. Du, Y.; Li, Y.; Wu, T. A superhydrophilic and underwater superoleophobic chitosan–$TiO_2$ composite membrane for fast oil-in-water emulsion separation. *RSC Adv.* **2017**, *7*, 41838–41846. [CrossRef]
40. Umpleby, R.J.; Baxter, S.C.; Chen, Y.; Shah, R.N.; Shimizu, K.D. Characterization of molecularly imprinted polymers with the Langmuir–Freundlich isotherm. *Anal. Chem.* **2001**, *73*, 4584–4591. [CrossRef]
41. Jeppu, G.P.; Clement, T.P. A modified Langmuir-Freundlich isotherm model for simulating pH-dependent adsorption effects. *J. Contam. Hydrol.* **2012**, *129*, 46–53. [CrossRef]

*membranes*

*Article*

# Characterization of Electrospun Poly(ε-caprolactone) Nano/Micro Fibrous Membrane as Scaffolds in Tissue Engineering: Effects of the Type of Collector Used

**Dianney Clavijo-Grimaldo [1,2,*], Ciro Alfonso Casadiego-Torrado [2], Juan Villalobos-Elías [1], Adolfo Ocampo-Páramo [1] and Magreth Torres-Parada [1]**

1 School of Medicine, Universidad Nacional de Colombia, Bogotá 111321, Colombia; jvillalobos@unal.edu.co (J.V.-E.); admocampopa@unal.edu.co (A.O.-P.); mtorrespa@unal.edu.co (M.T.-P.)
2 School of Medicine, Fundación Universitaria Sanitas, Bogotá 111321, Colombia; ccasadie@gmail.com
* Correspondence: dclavijog@gmail.com

**Abstract:** Electrospinning is an electrohydrodynamic technique that transforms a polymer solution into nano/microscopic diameter fibers under the influence of a high-voltage electric field. Its use in the fabrication of nano/micro fibrous membranes as scaffolds for tissue engineering has increased rapidly in recent years due to its efficiency and reproducibility. The objective of this study is to show how the use of the same polymeric solution (polycaprolactone 9% *w/v* in chloroform: isopropanol 50:50) and identical electrohydrodynamic deposition parameters produces fibers with different characteristics using a flat collector platform with movements in the X and Y axes vs. a conventional rotary collector. The manufactured nano/microfibers show significant differences in most of their characteristics (morphology, roughness, hydrophilicity, and mechanical properties). Regarding the diameter and porosity of the fibers, the results were similar. Given that scaffolds must be designed to guarantee adequate survival and the proliferation and migration of a certain cell type, in this study we analyze how the variations in the characteristics of the fibers obtained are essential to defining their potential application.

**Keywords:** electrospinning; polycaprolactone; scaffold; tissue engineering

**Citation:** Clavijo-Grimaldo, D.; Casadiego-Torrado, C.A.; Villalobos-Elías, J.; Ocampo-Páramo, A.; Torres-Parada, M. Characterization of Electrospun Poly(ε-caprolactone) Nano/Micro Fibrous Membrane as Scaffolds in Tissue Engineering: Effects of the Type of Collector Used. *Membranes* **2022**, *12*, 563. https://doi.org/10.3390/membranes12060563

Academic Editor: Andrea Ehrmann

Received: 10 April 2022
Accepted: 26 April 2022
Published: 28 May 2022



## 1. Introduction

The low availability of donors and the morbidity associated with transplants mean that ET is considered a useful strategy to restore or reestablish the function of pathologically altered tissues and organs. The tissue engineering (TE) approach involves the regeneration of tissue on a suitable support and implanting it at the goal site. Tissue regeneration functionally requires a microenvironment that mimics the original site to obtain an adequate cellular response and to provide optimal conditions for regeneration. Traditionally, TE is defined as "the body persuasion of to re-pair itself, releasing, at the appropriate sites, molecular signals, cells and/or supporting structures". It seeks to manufacture tissues, directing molecular and mechanical signals to specific cells to restore or reestablish normal function [1]. TE uses three basic components, cells, scaffolds, and a chemical environment (hormones, growth factors, etc.). The scaffolds' role is to mimic the extracellular matrix (ECM), a multiphase nano/microstructure material that is essential for cell viability, and to maintain the morphological, mechanical, and functional characteristics of the tissue. Several strategies are used in TE. In one of them, the construct (cells + scaffold + growth factors) is placed in a bioreactor that reconstructs the designed tissue in vitro. Another strategy is the implantation of the scaffold in the patient so that it fulfills its regenerative role in vivo, The first strategy is known as TE in vitro (or ex vivo) and the second TE in vivo (or in situ). Bioreactors are used to mimic in vitro environmental conditions in vivo and/or to provide the chemical environment that regulates cell proliferation and differentiation and the production of ECM prior to implantation in vivo [2,3]. Nano/microfibrous polymeric membranes have been widely used as biomaterials for the fabrication of scaffolds in TE due to their properties (biocompatibility, high surface-to-volume ratio, high porosity, biodegradation, and mechanical properties). The most used natural polymers for the manufacture of scaffolds are silk, collagen, gelatin, fibrinogen, alginate, and chitosan. Among the most used synthetic polymers are polycaprolactone (PCL), poly(lactic acid), polyglycolide and poly(l-lactide-co-glycolide), among others [4].

Ideally, nano/microfibrous polymeric membrane used as scaffold should have the following characteristics: biocompatibility and nontoxicity properties, a three-dimensional structure, high porosity, it should be biodegradable or bioresorbable, have controllable degradation and reabsorption rates, a chemically appropriate surface to promote proliferation and cell differentiation, hierarchical organization, and mechanical properties similar to the tissue where it will be implanted. Additionally, the technique with which it is manufactured must be versatile and easily controllable and reproducible [5,6]. However, some of the scaffolds used have shown some limitations, such as inadequate physicochemical and mechanical properties, as well as inappropriate porosity, wettability, alignment, roughness, and surface-to-volume ratios, which leads to insufficient tissue regeneration. Therefore, there is a need to explore techniques and obtain materials for scaffold manufacturing that mimic and simulate the structural, topographic, mechanical, and conductive properties of a specific ECM to promote the regeneration of the tissue function [7–13].

Numerous techniques have been used to manufacture scaffolding: freeze-drying, thermal-induced phase separation, gas foaming, rapid prototyping, stereolithography, fused deposition modeling, selective laser sintering, three-dimensional printing, bioprinting, etc. [14]. However, these systems have drawbacks, such as their limited print resolution, in which cells cannot be formed and organized precisely. Another drawback is the solidification and gelation requirements during the printing process, which limit the materials (hydrogels) that can be used [15]. Indeed, it is necessary to rely on other systems and methods to complement and improve the efficiency, survival, and proliferation of cells. Electrohydrodynamic techniques improve the properties of materials and devices. Due to the simplicity and flexibility of the experimental setup of these techniques, they have been used successfully in the fabrication of particulate materials with controllable compositions, structures, sizes, morphologies, and shapes. These attributes, in addition to not having as many limitations in the working materials, make electrohydrodynamic techniques an extraordinary tool for preparing and assembling a wide range of micro- and nanostructured materials [16,17]. Electrospinning, in which a polymer in solution is subjected to a high voltage to produce nano/microfibers, is one of the most widely used techniques today due to its cost, ease of manufacture, low material requirements, and the evaporation of the solvents used, reducing the risks of toxicity inherent in some solvents and favoring greater biocompatibility. By adjusting variables such us the diameter of the fibers (and therefore the surface-to-volume ratio), the surface roughness of the fibers, and the porosity, the interconnectivity of the pores and mechanical properties of the scaffolds can be controlled to adjust them to specific conditions (bone regeneration, heart, skin, etc.). The main variables considered depend on the characteristics of the polymer solution (the molecular weight of the polymer, solvent, surface tension, concentration, and the viscosity of the solution, among others); the process (solution flow, the voltage applied between the needle through which the solution passes and the collector, the type of collector, needle–collector distance); and the environment (temperature and humidity) [11,13,14,18]. In addition, the fibers can be enriched with drugs and/or growth factors so that the scaffold works as a release system that favors or controls specific cellular functions, contributes to the management of pathologies, modulates the response of the immune system towards the biomaterial and, in general, increases its bioactivity [19–21].

The effect that the collector has on the evaporation of the solvent and on the orientation, diameter, density, fiber–fiber junction points, and mechanical properties of the fibers obtained is known. The rotary collector (RC) is the approach most frequently used in TE applications. There are numerous studies on how the diameter and rotation speed can determine the type of fibers obtained, and the changes that occur in them when compared to those obtained with flat collectors [22–25]. More recently, collectors have been adapted to combine electrospinning principles with 3D printing. In this sense, the 3D configuration depends on the deposition time and the displacement of the collector used, since it can be designed to move in several axes, with the aim of modifying certain characteristics (for example, tensile strength and elasticity) [26–29]. The effect on fiber characteristics of using an RC versus the configuration of the flat collector on a moving platform in the X–Y axes has not been compared and analyzed in detail.

In this study, PCL fiber scaffolds were constructed in a known standard solution [30]. The fibers were collected in two types of collectors: a conventional RC and a flat collector platform that moved in two axes (XYP) to compare the diameter and roughness of the individual fibers and the mechanical properties, porosity, and hydrophilicity of the fibers.

## 2. Materials and Methods

### 2.1. Materials

PCL (CAS # 134490-19-0 and average Mn = 80.000 g/mol), chloroform (99.5%, CAS # 67-66-3 and Mn = 119.38 g/mol), and isopropyl alcohol (99.7% CAS # 67-66-3 and Mn = 60 g/mol) were used. All the chemical agents used were supplied by Sigma-Aldrich (San Luis, MO, USA).

### 2.2. Preparation of the Solution

PCL was used in a 9% *w/v* solution in a 50:50 *v/v* mixture of chloroform and isopropyl alcohol. The resulting solution was stored at room temperature for 48 h and, prior to its use, it was subjected to homogenization using ultrasound (Model ATM40-2LCD, ATU UL-TRA-SON-IC) with a frequency of 50 Hz for 60 min at 17 °C.

### 2.3. Electrospinning Process

The scaffolds were developed using electrospinning equipment composed of a high-voltage source Model CZE1000R (Spellman high voltage corporation, Hauppauge, NY, USA), a dosing pump KDS 100 (KD Scientific Inc., Holliston, MA, USA), syringe and needle (Upchurch Scientific Inc., Oak Harbor, WA, USA), and two types of collectors: XYP (own manufacture) and RC ESD30s (Nanolab Instruments Sdn Bhd, Malaysia) (Figure 1). The definitive parameters were voltage (14 kV), distance between the needle tip and the collector (14 cm), solution flow (1 mL /h), and two deposition times (45 and 90 min). The collection of fibers was completed on aluminum sheets placed on the collector. A linear velocity was set in the XYP, in both axes, of 0.0025 m/s. To obtain equivalent parameters, the angular velocity of the RC (r: 0.04 m) was calculated based on the linear velocity of the platform, resulting in a speed of 6 rpm. The scaffolds produced were divided into four groups: scaffolds developed in the RC during 45 and 90 min of deposition (RC45 and RC90), and scaffolds developed in the XYP with equal deposition times (XYP45 and XYP90). The process was performed at 20 ± 2 °C room temperature and 60 ± 5% RH.

### 2.4. Roughness of the Individual Fibers

The surface roughness of the fibers was measured by means of atomic force microscopy AFM; (Asylum Research—MFP-3D-BIO (Oxford Instruments, Santa Barbara, CA, USA) in a sampling area of 1 × 1 μm$^2$. The roughness values (root mean square (Rms) and arithmetic mean (Ra)) were calculated in triplicate using the open-source software Gwyddion, version 2.56. Since the objective was to determine the roughness of individual fibers (and not of the scaffold produced) and taking in count compression forces with longer deposition times, the fibers used in this test were collected 30 s after their deposition.

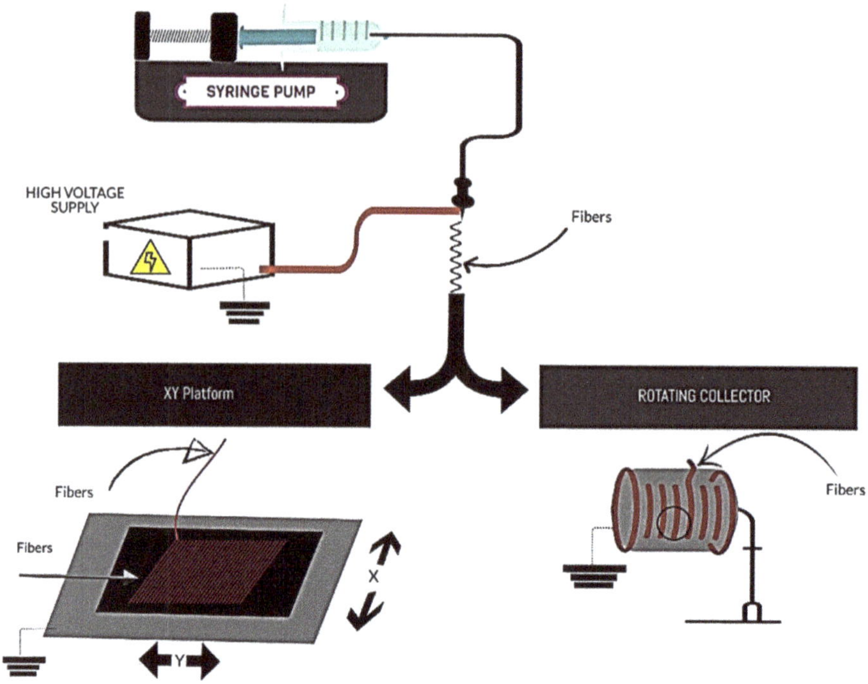


**Figure 1.** Schematic illustration of the electrospinning process performed: XY platform (flat collector platform with movements in the X and Y axes) vs rotary collector.

### 2.5. Mechanical Tests

Mechanical tests were performed on the four groups of scaffolds. The scaffolds were cut into rectangular samples of 10 mm × 110 mm with a thickness 0.0385 mm. The tensile strength testing of the samples was performed using a universal testing machine AG-IS 5KN, (Shimadzu corporation, Kyoto, Japan). The tests were carried out according to ASTM D882 with a preload of 0.003 N and at a speed of 20 mm/min at room temperature (20.8 °C) and 60% RH. The orientation of the samples for the mechanical tests was longitudinal. The tests were carried out in triplicate.

### 2.6. Morphology and Fibers Diameter

The fiber morphology of the four groups of the scaffolds was studied using high-vacuum scanning electron microscopy (SEM) Tescan Vega 3, (Tescan Analytics, Brno, Czech Republic) with an operating voltage of 10 kV and a magnification of 500× to find the areas of interest. The fiber diameter was analyzed using the public domain image analysis software Image J (National Institutes of Health) at a magnification of 5000×. Measurements were made before and after the mechanical tests.

### 2.7. Contact Angle of the Scaffolds

Contact angle tests, (ASTM D5725-99/2008, American Society for Testing and Materials, West Conshohocken, Pennsylvania, U.S) were performed on all four scaffold groups. Samples of 10 mm × 10 mm were placed on slides and 10 µL drops of deionized water were deposited on the surface. Images were obtained with a digital camera Canon eos rp, Lens: Canon rf 24–105 mm, f4 (Canon Inc., Melville, NY, USA) Contact angles were calculated using the public domain image analysis software Image J (National Institutes of Health).

### 2.8. Porosity Measurement of the Scaffolds

A methodology previously described in another study [31] was used due to its effectiveness in measuring various layers of nano/microfibers. Using the public domain image analysis software Image J, SEM images of the four groups of scaffolds were converted to binary images using three thresholds and the porosity of each scaffold was measured in three layers. Three thresholds were calculated to convert the original image to binary form based on the mean and standard deviation of the pixel values of the image across the equations: $T1 = (\mu + \sigma)/255$, $T2 = \mu/255$ and $T3 = (\mu - \sigma)/255$, where $\mu$ and $\sigma$ are the mean and the standard deviation of the image matrix, respectively. The percentage of porosity (P) of each binary image was obtained using the average intensity of the images, as indicated in the equation: $P = (1 - n/N) * 100$, where n is the number of white pixels and N is the total number of pixels in the total volume of the binary image [31]. Measurements were made before and after the mechanical tests.

### 2.9. Statistical Analysis

Variance analysis (ANOVA) and the F-test were performed to assess the amount of variability between group means in the context of within-group variation to determine if the mean differences were statistically significant. In this study, if the $p$-value was $\leq 0.05$ and the calculated F-value was greater than the critical value F (the ratio of two variances), the differences were considered significant.

## 3. Results

### 3.1. Roughness of the Individual Fibers

Figure 2 shows the appearance of a sector of fibers made on RC and XYP. Table 1 shows the Rm and Ra values obtained in the fibers, as well as their averages and standard deviation. The average Ra of the surface of the RC fibers was 42.72 ± 12.12 nm and the Rms was 51.47 ± 15.32 nm, whereas the surface roughness values for the XYP fibers decreased to 38.25 ± 23.7 nm and 36.94 ± 15.6 nm for Ra and Rms, respectively. Despite the differences in the averages of the RC fibers versus the XYP fibers, the analysis of variance did not show significant differences between the two groups either for Rm ($p = 0.31$, $F = 1.32$ and critical $F = 7.71$) or for Ra ($p = 0.79$, $F = 0.08$ and critical $F = 7.71$).

**Table 1.** Surface roughness of the samples collected using the RC and XYP.

| Collector Type | | Root Mean Square (Rms) (nm) | Arithmetical Average (Ra) (nm) |
|---|---|---|---|
| RC | Mean (nm) | 51.47 | 42.72 |
| | SD | 15.33 | 12.13 |
| XYP | Mean (nm) | 36.94 | 38.25 |
| | SD | 15.62 | 23.72 |

### 3.2. Mechanical Tests

Figure 3 shows the stress-strain curves for the scaffolds obtained on RC and XYP with deposition times of 45 min (Figure 3A) and 90 min (Figure 3B).

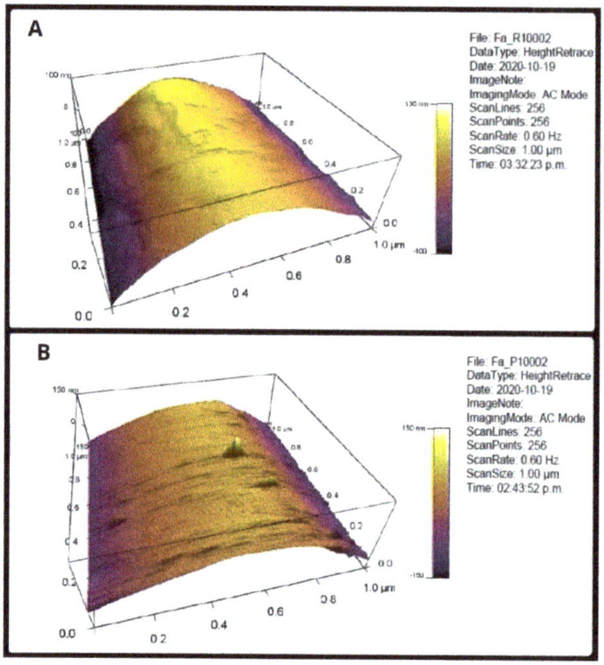


**Figure 2.** AFM of PCL fibers. (**A**) is for the RC scaffold, (**B**) for the XYP Scaffold. AFM: atomic force microscope, XYP: flat collector platform with movements in the X and Y axes, RC: rotary collector.

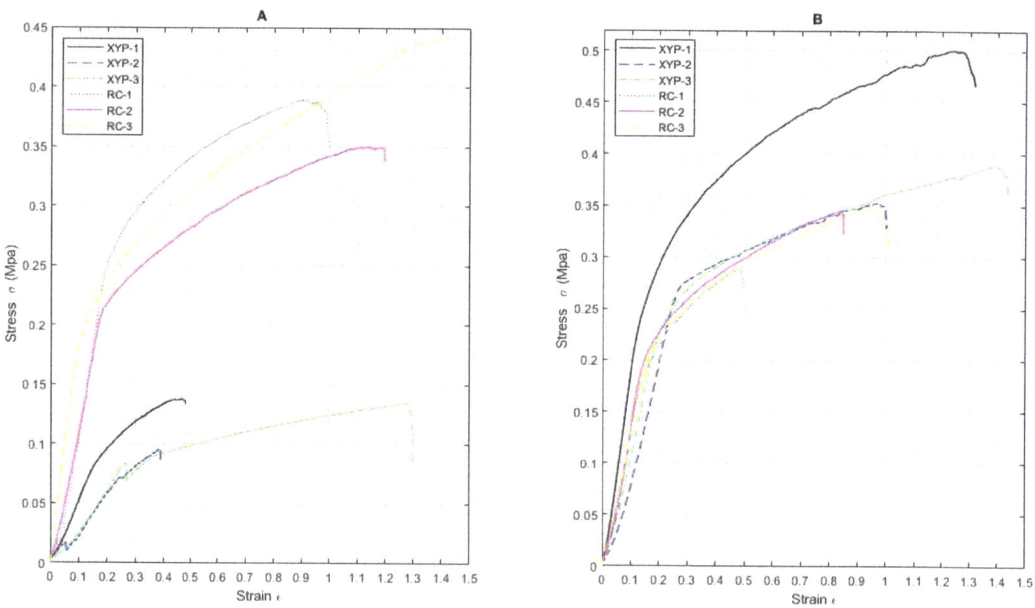


**Figure 3.** Stress-strain curves of PCL scaffolds obtained using RC and XYP. (**A**) Deposition time: 45 min, (**B**) deposition time: 90 min.

Table 2 shows the stress and strain values for the scaffolds obtained on RC and XYP at 45 min and 90 min of deposition, as well as their arithmetic averages and standard deviation. For a deposition time of 45 min, the scaffold deposited on RC showed a tensile stress value of 0.39 ± 0.05 MPa, Young's modulus of 1.11 MPa, and an elongation at break of 114.91% compared to the scaffold deposited in XYP, which showed lower average values, with a tensile stress value of 0.12 ± 0.02 MPa, Young's modulus of 2.49 MPa, and an elongation at break of 70.98%. For a deposition time of 90 min, the scaffold deposited on RC showed a tensile stress value of 0.41 ± 0.08 MPa, Young's modulus of 1.69 MPa, and an elongation at break of 119.88% compared to the scaffold deposited in XYP, which also showed lower average values: tensile stress of 0.33 ± 0.03 MPa, Young's modulus of 1.24 MPa, and an elongation at break of 75.87%. A highly significant difference was observed between the maximum stress value for the RC scaffolds versus XYP scaffolds when the deposition time was 45 min ($p = 0.0008$, F = 81.55, and critical F = 7.71); however, no significant difference was observed for the elongation at break ($p = 0.24$, F = 1.89, and critical F = 7.71). For the deposition time of 90 min, no significant differences were observed either in the maximum stress value ($p = 0.15$, F = 3.24, and critical F = 7.71) or in the elongation at break ($p = 0.07$, F = 5.45, and critical F = 7.71).

**Table 2.** Mechanical testing results of PCL membranes.

| Collector Type | Deposition Time | Sample | Max_Force N | Max_Disp mm | Max_Stress N/mm$^2$ MPa | Max_Strain % |
|---|---|---|---|---|---|---|
| RC | 45 min | Mean value | 0.18 | 105.06 | 0.39 | 114.91 |
| | | SD | 0.02 | 19.39 | 0.05 | 25.44 |
| | 90 min | Mean value | 0.19 | 118.87 | 0.41 | 119.88 |
| | | SD | 0.04 | 13.83 | 0.08 | 21.05 |
| XYP | 45 min | Mean value | 0.06 | 80.53 | 0.12 | 70.98 |
| | | SD | 0.01 | 32.99 | 0.02 | 49.14 |
| | 90 min | Mean value | 0.15 | 77.47 | 0.33 | 75.87 |
| | | SD | 0.02 | 24.51 | 0.03 | 24.95 |

Abbreviations: Max force, maximum force; Max Disp, maximum disposition; Max Stress, maximum Stress; Max Strain, maximum strain; min, minutes.

### 3.3. Morphology and Fiber Diameter

Table 3 shows the fiber diameters, their means and standard deviations for the four groups of scaffolds before and after the mechanical tests. No significant differences were found in the diameter of the fibers, according to the type of collector used, before the mechanical tests, or after 45 min of deposition ($p = 0.24$, F = 1.51, and critical F = 4.96) or 90 min of deposition ($p = 0.47$, F = 0.57, and critical F = 4.96). The differences are also not significant when comparing the type of collector at 45 min ($p = 0.41$, F = 0.74 and critical F = 4.96) and 90 min of deposition ($p = 0.55$, F = 0, 37 and critical F = 4.96) after mechanical tests. Figures 4 and 5 show the SEM morphology for deposition times of 45 min (Figure 4) and 90 min (Figure 5) in each collector used. The orientation of the fibers in both types of collectors was observed randomly, without any predominant pattern. The fibers obtained with RC were more uniform and did not present pearls or beads, whereas those obtained with XYP presented numerous pearls and beads, especially with 45 min of deposition.

**Table 3.** Diameter and statistics of different fibers of the samples collected using RC and XYP before and after mechanical testing.

| Collector Type | Deposition Time | Diameter (nm) | | | |
|---|---|---|---|---|---|
| | | Mean (nm) | | SD | |
| | | Before | After | Before | After |
| RC | RC45 | 1819 | 1147 | 549 | 424 |
| | RC90 | 1386 | 1267 | 674 | 404 |
| XYP | XYP45 | 548 | 1357 | 502 | 672 |
| | XYP90 | 1472 | 1005 | 846 | 820 |

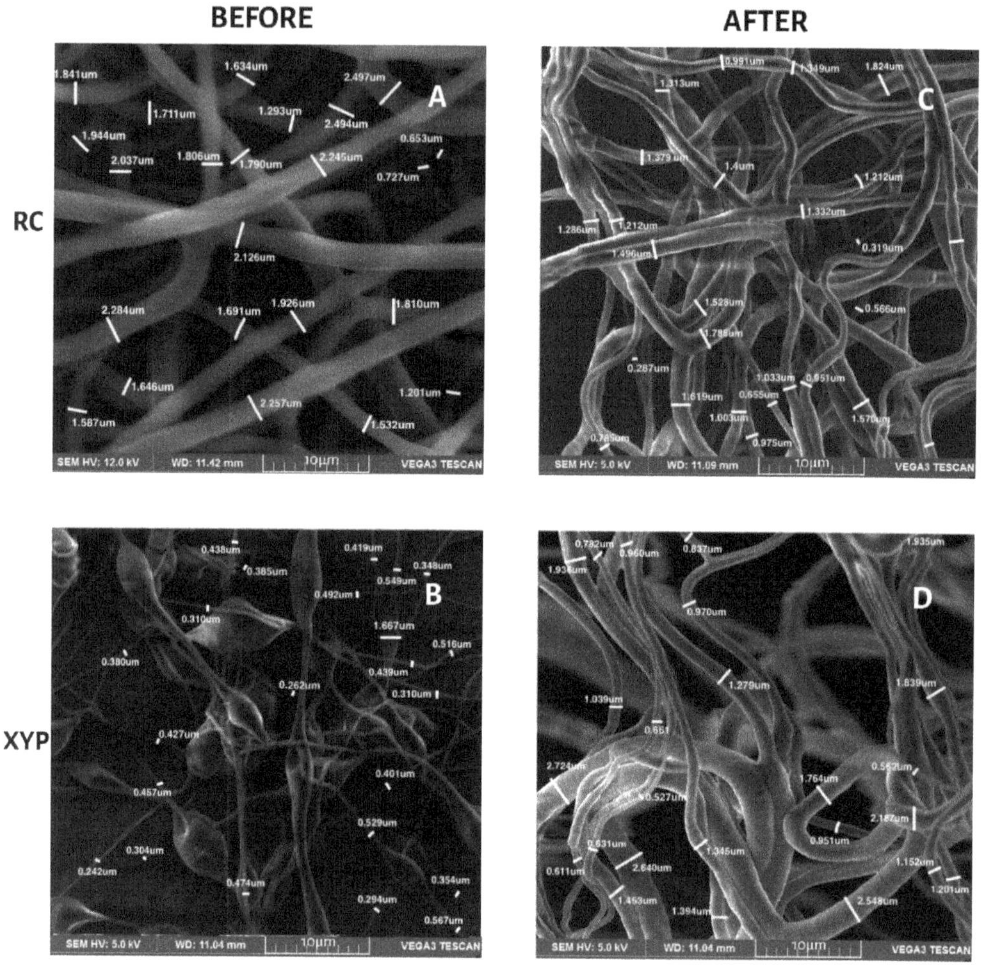


**Figure 4.** Fiber diameter in SEM micrographs, 45 min of deposition. (**A**) RC45 before mechanical tests, (**B**) XYP45 before mechanical tests, (**C**) RC45 after mechanical tests, (**D**) XYP45 after mechanical tests.

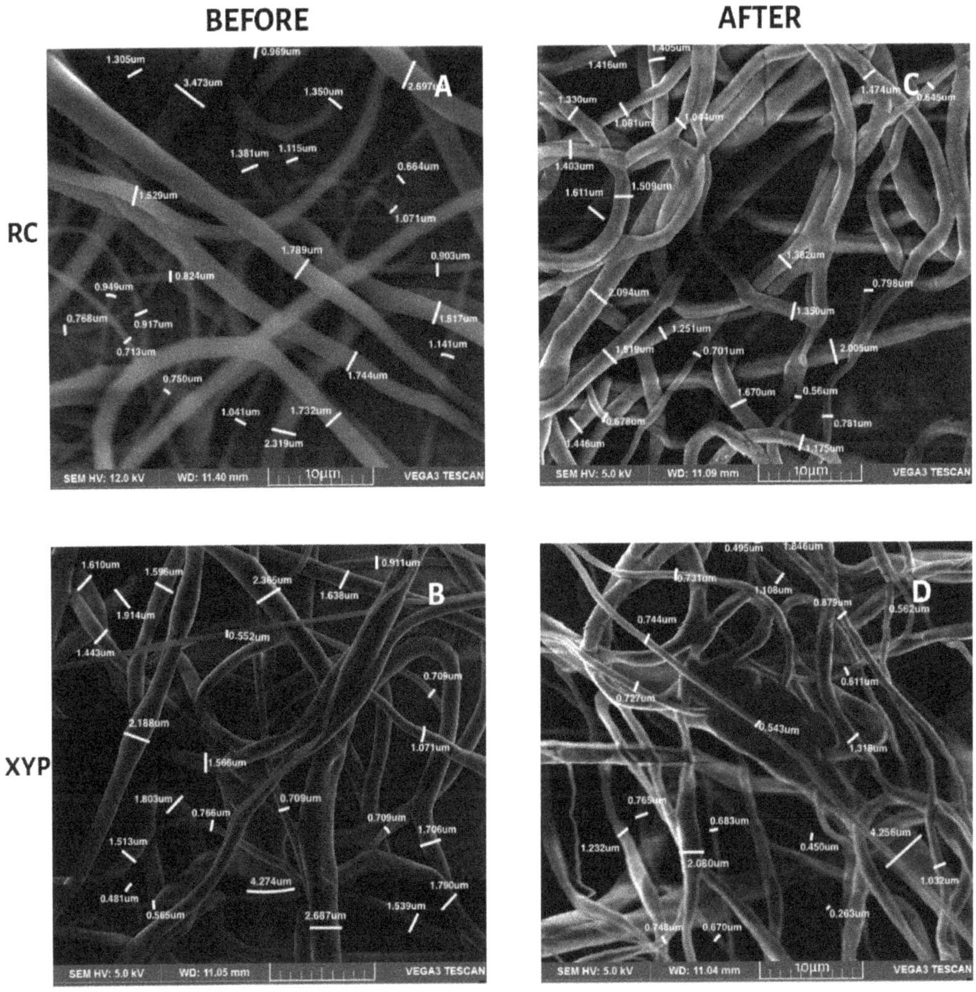


**Figure 5.** Fiber diameter in SEM micrographs, 90 min of deposition. (**A**) RC90 before mechanical tests, (**B**) XYP90 before mechanical tests, (**C**) RC90 after mechanical tests, (**D**) XYP90 after mechanical tests.

### *3.4. Contact Angle of the Scaffolds*

The contact angle values obtained are shown in Table 4. The RC45 scaffolds had an average contact angle of 51.95° ± 1.8°. The RC90 scaffolds achieved a 1.2% increase in their contact angles (58.19° ± 6.6°). For XYP45 scaffolds there was a significant 55% increase compared to RC45 contact angles (80.52° ± 3.8°) and a 7% decrease compared to XPY90 scaffold contact angles (74.77° ± 15.6°). An increase in hydrophilicity was observed in RC45 scaffolds, with significant differences in the contact angle ($p = 1.08^{-5}$, F = 178.77, and critical F = 5.98) when compared to XYP45 scaffolds. No significant differences were observed between RC90 versus XYP90 scaffolds.

**Table 4.** Contact angles according to collector type.

| Deposition Time | | Contact Angle |
|---|---|---|
| RC45 | Mean | 51.95° |
| | SD | 1.80 |
| RC90 | Mean | 58.19° |
| | SD | 6.66 |
| XYP45 | Mean | 80.52° |
| | SD | 3.87 |
| XYP90 | Mean | 74.77° |
| | SD | 15.63 |

### *3.5. Porosity Measurement*

Table 5 shows the values of the porosity measurements of binary images of different samples with various thresholds before and after mechanical tests. It was observed that the porosity was greater in the superficial layers of the scaffold and decreased with depth. On average, there was 11% less porosity per threshold. Additionally, the number of pores was reduced by approximately 10% when the XYP collector was used. The collection time showed no relationship with the porosity of the scaffolds. The binary SEM images of the three thresholds for the four samples produced using the RC and XYP as collectors with deposition times of 45 min and 90 min are shown in Figures 6–9.

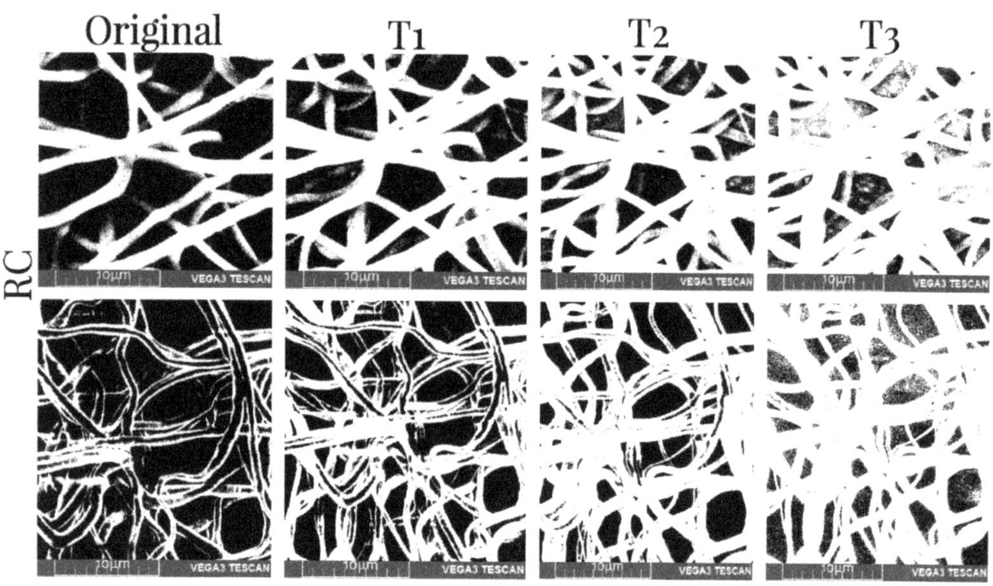


**Figure 6.** SEM images of the obtained RC45 scaffolds and binary images with three thresholds before (top row) and after mechanical tests (bottom row). T1–T3: threshold 1–3.

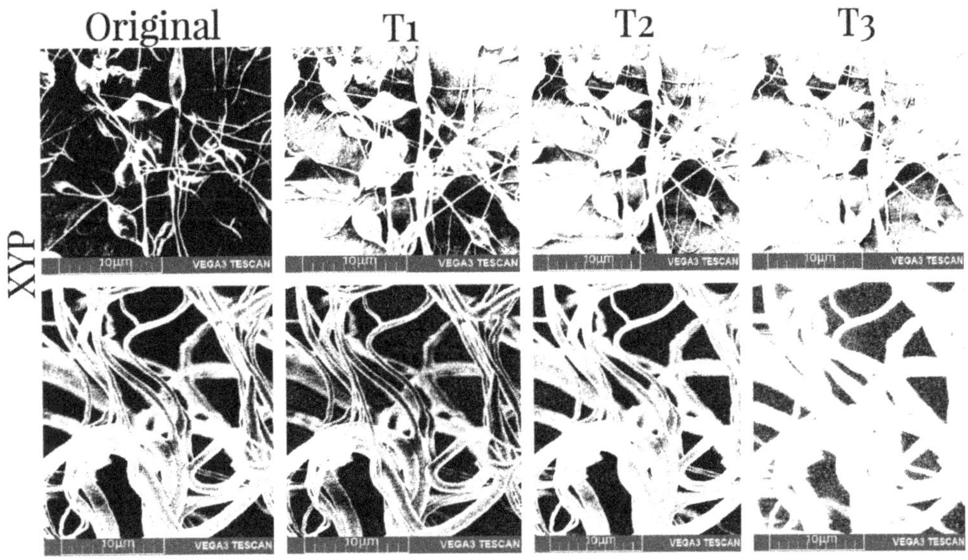


**Figure 7.** SEM images of the obtained XYP45 scaffolds and binary images with three thresholds before (top row) and after mechanical tests (bottom row). T1–T3: threshold 1–3.

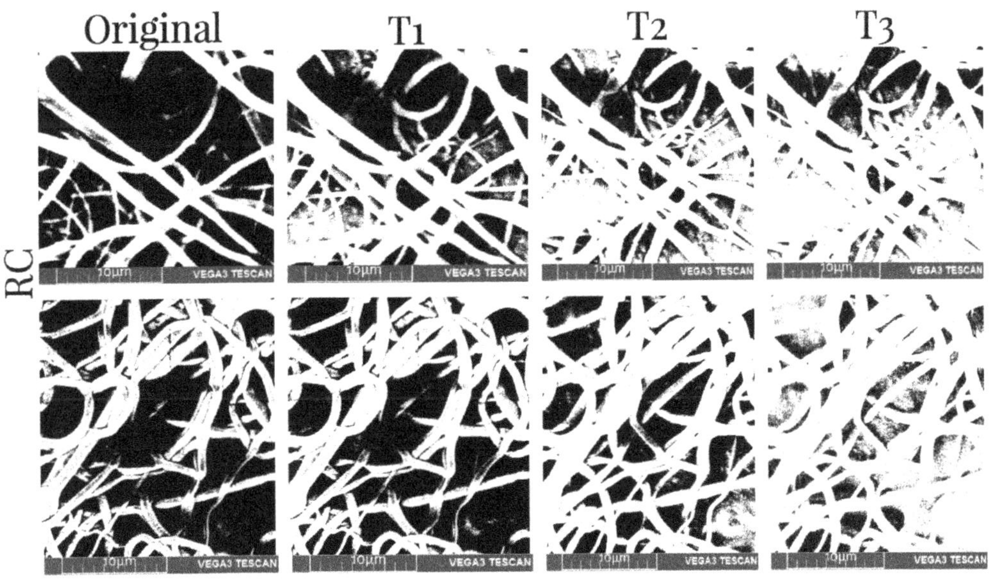


**Figure 8.** SEM images of the obtained RC90 scaffolds and binary images with three thresholds before mechanical tests (top row) and after mechanical tests (bottom row). T1–T3: threshold 1–3.

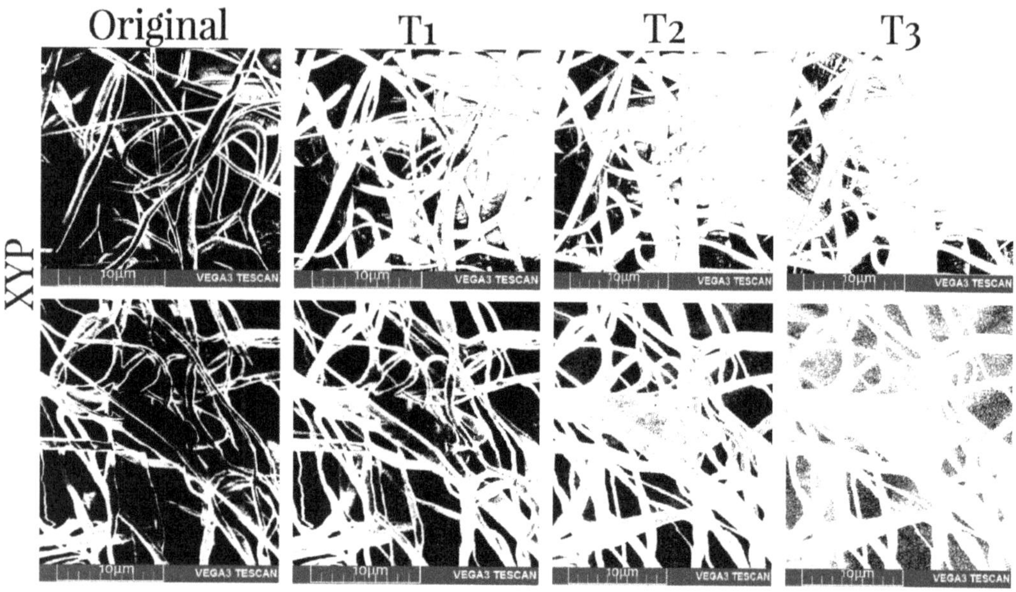


**Figure 9.** SEM images of the obtained XYP90 scaffolds and binary images with three thresholds before mechanical tests (top row) and after mechanical tests (bottom row). T1–T3: threshold 1–3.

**Table 5.** Porosity measurements of binary images of different samples with various thresholds before and after mechanical tests.

| Collector Type | Deposition Time | Sample Type | T1% | T2% | T3% |
|---|---|---|---|---|---|
| RC | 45 min | Before | 38.19 | 25.95 | 12.71 |
| | | After | 48.09 | 34.97 | 22.36 |
| | 90 min | Before | 39.32 | 26.39 | 17.11 |
| | | After | 29.29 | 19.88 | 15.68 |
| XYP | 45 min | Before | 28.06 | 18.89 | 9.26 |
| | | After | 47.80 | 31.42 | 15.45 |
| | 90 min | Before | 29.42 | 19.47 | 9.66 |
| | | After | 46.30 | 28.34 | 10.16 |

## 4. Discussion

The design and manufacture of nano/microfibrous membranes that function as scaffolds, mimic the ECM, and thus optimize tissue regeneration is one of the main challenges in TE and represents an alternative to the limited number of donors and the implications involved with the use of autologous grafts. Allografts may be capable of transmitting disease and eliciting host immune responses. A scaffold must be manufactured considering the porosity, the balance between hydrophilicity/hydrophobicity, the mechanical properties, the three-dimensional architecture, the biocompatibility, and the non-toxicity of its components. For its manufacturing, electrohydrodynamic techniques, such as electrospinning, are the most frequently used given their possibilities, versatility, low cost, and the simplicity and reproducibility of the process. The modification of parameters in the polymeric solution used and the process (voltage, solution flow, distance from the collector needle) and environmental conditions (temperature and RH), allows one to obtain scaffolds with particular characteristics in terms of their morphology, the dimensions and orientation of the fibers, their porosity, hydrophilicity, and their tensile strength for a specific use, especially in bone, musculoskeletal, cutaneous, cardiovascular, or neurological applications, among others [19–21,32]. Numerous studies have demonstrated the effect that the type of collector used has on the characteristics of the fibers and the scaffold itself. Multiple variants of RC and static flat collectors have traditionally been used [22–25]. More recently, the use of platforms with displacements in the X and Y axes has been introduced to facilitate the manufacturing of three-dimensional scaffolds with greater control over their shape and dimensions [33]. In this work, nano/microfibers were manufactured with PCL, a biodegradable polymer accepted by the FDA for surgical implants, drug delivery systems, and applications in TE and regenerative medicine [34]. To evaluate whether there were significant differences in the scaffolds produced on a RC versus XYP, all the parameters were standardized: the composition of the solution, the process variables, and the environmental conditions, including the displacement speed of the XYP in each axis with the angular velocity of the RC, although this implied an rpm much lower than the one normally used.

Determining whether there are differences in roughness is essential because it is considered a critical factor in cell adhesion, proliferation, and differentiation. To avoid compression deformations due to the deposition of several layers, the fibers used in the roughness tests were collected 30 s after their deposition. Although no significant differences were observed in the fibers obtained in the two types of collectors, the surface roughness of the XYP scaffolds showed a higher standard deviation than that of RC, especially the Ra value. This allows us to assume that the roughness of the scaffold deposited on the CR is more homogeneous due to the intrinsic rotation movement and the traction it exerts when picking up the fiber. High roughness favors osteogenic differentiation, neurite outgrowth, and Schwann cell proliferation, but impairs chondrogenic differentiation and endothelial function, to cite a few examples [33], hence the collecting type is an important consideration when scaffolds are to be used in a specific application.

Mechanical test results are affected by material composition, microscopic imperfections, the manufacturing process, the loading rate, and the temperature during testing [35]. The results showed that it is also necessary to consider the total deposition time when fabricating a scaffold by means of electrospinning. The results obtained using the two types of collectors showed that the collection time is an important parameter since, by increasing the deposition time, a greater number of fibers was generated on the surface, causing changes in the mechanical properties. That is, a greater presence of fibers produces a structure with greater mechanical resistance. When comparing the scaffolds, we determined that the increase in the elongation at break value when using RC was possibly due to the absence of pearls in the fibers obtained compared to those observed in XYP. The SEM images indicated that the RC fibers had a more homogeneous morphology in terms of the length of the fibers. The diameter of the fibers did not present great variation with respect to those observed in the XYP. Additionally, the area under the curve of the graphs (Figure 3) indicated that the fibers obtained via RC presented greater deformation due to the localized and directed shape as they fell into the collector. Scaffolds made in the RC system with a deposition time of 45 min had a significantly higher tensile strength than scaffolds made in XYP. This difference was not observed when the deposition time was 90 min. The higher tensile strength of the scaffolds obtained in RC correlated with their higher porosity. The larger spaces between the fibers allow the scaffolds to buffer stress effectively and slow down structural deterioration. This is essential in scaffolds designed for interfaces between soft and hard tissues (for example, in surgery to reconstruct ligaments or articular cartilage with their reattachment to bone tissue), in which the scaffold acts as a material developed to perform a gradual mechanical transfer between ECM of different properties [36]. RC usually produces aligned fibers, in the case of this work, due to the low angular velocity used, and scaffolds with randomly oriented fibers were obtained with both types of collectors, but with a higher number of pearls in the XYP. The presence of beads modifies the mechanical properties of the scaffolds and must be considered when using the mentioned interfaces [37].

In this study, our goal was to compare the type of collector, keeping the concentration of the polymer the same; therefore, it was not possible to analyze the influence of the amount of the polymer on the mechanical properties. On the other hand, in previous studies, they determined that, after increasing the concentration of PCL or the load of another component in the fibers, such as drugs and proteins, the tensile strength of the PCL nanofibrous membranes decreased notably [38,39]. It should be noted that the fibers produced by means of electrospinning exhibit a relatively smooth deposition process; therefore, collisions can occur with neighboring fibers that move in different directions. This can cause periodic blockages of larger-diameter fibers in which a microstructural failure appears to occur sequentially, involving a balance between localized strain in the direction of traction and anisotropic point junction that locally resists deformation [40]. This could explain why the scaffolds made in the RC90 system showed slightly higher Young's modulus values compared to the scaffolds in the XYP90 system.

Electrospinning allows one to obtain fibers with diameters between micrometers and nanometers. The critical factors that determine the diameter of the fibers are the flow, the concentration of the polymer solution, and the voltage used [41]. When the concentration of PCL solutions is increased, the diameter of the fibers generally increases, although with this technique it is necessary to consider the other variables involved [42]. When comparing fibers deposited on a fixed-plate collector versus RC, the latter have a smaller diameter that decreases as the rotation speed (rpm) increases [43]. Differences in mechanical properties were observed with speeds greater than 640 rpm. In this work, due to the low angular velocity used, there were no significant differences between the diameters of the fibers deposited on RC versus XPY, independently of the deposition time. The low speed can also be considered an important factor regarding why differences in fiber diameters were not observed before and after the mechanical tests [44].

The wettability of surfaces depends on the chemical composition and the microgeometry of their roughness. Models made to predict the contact angle of fibers manufactured via electrospinning show that the concentration of the solution is the determining parameter [45]. For other authors, surface roughness is the critical factor and, in some polymeric fibers, changes in wettability can be achieved by modifying roughness, without the need for chemical changes [46]. When other parameters were studied, it was observed that PCL fibers can behave in a hydrophilic or hydrophobic character, depending on the solvent used [47]. However, these studies did not consider the type of collector used. When the fibers deposited on copper collectors (sheet versus mesh) were compared, it was observed that the fibers collected in the copper sheet had a greater diameter, less roughness, and showed a more hydrophilic behavior than those collected in the copper sheet [48]. In this work, significant differences were observed in the fibers collected in XYP versus RC when the deposition time was 45 min. We observed that the fibers manufactured on RC had lower contact angle values and therefore a more hydrophilic behavior, which is a critical factor in the biocompatibility and bioactivity of a material, since wettability generally favors adhesion and cell proliferation and increases biomineralization in engineering for bone tissues [49]. Although the scaffolds manufactured in XYP cannot be considered hydrophobic, their values were close to 90° and they could be used in applications in which cell adhesion must be controlled or inhibited, for example, to reduce bacterial contamination or the formation of biofilms on medical devices, since decreasing their wettability results in a deterioration of the stability of the bacterial colonies and favors their detachment [50].

The effect of the ECM in cell proliferation, gene expression, differentiation, and migration highlight its importance as a design parameter for scaffolds with specific applications. Scaffold designs with controlled mechanical properties, roughness, porosity, and hydrophilicity, among others, can have a great impact on improving the success of TE applications in various medical treatments.

## 5. Conclusions

The scaffolds fabricated and observed in this study showed significant differences in most of their characteristics—morphology, roughness, hydrophilicity, and mechanical properties. Regarding the diameter and porosity of the fibers, the results were similar between the scaffolds. It has been validated that when scaffolds are manufactured using the electrospinning technique, differences in important properties can be obtained, even when using the same solution and the same manufacturing parameters, merely by changing the type of collector in the process. Given that scaffolds must be designed to guarantee the adequate survival, proliferation, and migration of a certain cell type, in this study we analyzed how the variations in the characteristics of the fibers obtained were essential to defining their potential application. For example, scaffolds with high roughness can be used in bone regeneration, whereas those with minimal roughness could be used to mimic the tunica intima of blood vessels. Therefore, the implementation of new technologies such as 2D moving platforms in this technique presents multiple applications in the design of scaffolds applied to TE. An important advantage of the XYP approach is the control of the shape of the scaffold produced. This is the first study to date, to our knowledge, that has compared the effects of electrospinning on the characteristics of the fibers and scaffolds while utilizing both a rotating collector and a displacement platform in the X-Y axes.

**Author Contributions:** Conceptualization, D.C.-G.; methodology, D.C.-G., C.A.C.-T., J.V.-E., A.O.-P. and M.T.-P.; software; validation, D.C.-G., C.A.C.-T., J.V.-E., A.O.-P. and M.T.-P.; formal analyses, D.C.-G., C.A.C.-T., J.V.-E., A.O.-P. and M.T.-P.; methodology, D.C.-G., C.A.C.-T., J.V.-E., A.O.-P. and M.T.-P.; investigation, D.C.-G., C.A.C.-T., J.V.-E., A.O.-P. and M.T.-P.; resources, D.C.-G.; data curation, D.C.-G., C.A.C.-T., J.V.-E., A.O.-P. and M.T.-P.; writing—original draft preparation, D.C.-G., C.A.C.-T., J.V.-E., A.O.-P. and M.T.-P.; writing—review and editing, D.C.-G., C.A.C.-T., J.V.-E., A.O.-P. and M.T.-P.; visualization, D.C.-G., C.A.C.-T., J.V.-E., A.O.-P. and M.T.-P.; supervision, D.C.-G.; project administration, D.C.-G.; funding acquisition, D.C.-G. All authors have read and agreed to the published version of the manuscript.

**Funding:** This research was funded by the Ministerio de Ciencia, Tecnología e Innovación de Colombia, Fundación Universitaria Sanitas and Universidad Nacional de Colombia, through project "Titanium biofunctionalization to improve osseointegration in osteoporotic patients who require orthopedic implants" Code:1101744555730—Contract: 630-2017.

**Institutional Review Board Statement:** Not applicable.

**Data Availability Statement:** Not applicable.

**Conflicts of Interest:** The authors declare no conflict of interest. The funders had no role in the design of the study; in the collection, analyses, or interpretation of data; in the writing of the manuscript, or in the decision to publish the results.

## References

1. Williams, D.F. On the mechanisms of biocompatibility. *Biomaterials* **2008**, *29*, 2941–2953. [CrossRef] [PubMed]
2. Ikada, Y. Challenges in tissue engineering. Journal of the Royal Society. *Interface* **2006**, *3*, 589–601. [PubMed]
3. Marijanovic, I.; Antunovic, M.; Matic, I.; Panek, M.; Ivkovic, A. Bioreactor-based bone tissue engineering. In *Advanced Techniques in Bone Regeneration*; IntechOpen: Rijeka, Croatia, 2016.
4. Dhandayuthapani, B.; Yoshida, Y.; Maekawa, T.; Kumar, D.S. Polymeric scaffolds in tissue engineering application: A review. *Int. J. Polym. Sci.* **2011**, *2011*, 290602. [CrossRef]
5. Lee, J.; Cuddihy, M.J.; Kotov, N.A. Three-dimensional cell culture matrices: State of the art. *Tissue Eng. Part B Rev.* **2008**, *14*, 61–86. [CrossRef] [PubMed]
6. Hutmacher, D.W. Scaffolds in tissue engineering bone and cartilage. *Biomaterials* **2000**, *21*, 2529–2543. [CrossRef]
7. Ratner, B.D.; Hoffman, A.S.; Schoen, F.J.; Lemons, J.E. *Biomaterials Science: An Introduction to Materials in Medicine*; Elsevier: Amsterdam, The Netherlands, 2004.
8. Tibbitt, M.W.; Rodell, C.B.; Burdick, J.A.; Anseth, K.S. Progress in material design for biomedical applications. *Proc. Natl. Acad. Sci. USA* **2015**, *112*, 14444–14451. [CrossRef]
9. Ratner, B.D. A pore way to heal and regenerate: 21st century thinking on biocompatibility. *Regen. Biomater.* **2016**, *3*, 107–110. [CrossRef]

10. Kumar, P.; Saini, M.; Dehiya, B.S.; Sindhu, A.; Kumar, V.; Kumar, R.; Lamberti, L.; Pruncu, C.I.; Thakur, R. Comprehensive survey on nanobiomaterials for bone tissue engineering applications. *Nanomaterials* **2020**, *10*, 2019. [CrossRef]
11. Doostmohammadi, M.; Forootanfar, H.; Ramakrishna, S. Regenerative medicine and drug delivery: Progress via electrospun biomaterials. *Mater. Sci. Eng. C* **2020**, *109*, 110521. [CrossRef]
12. Roseti, L.; Parisi, V.; Petretta, M.; Cavallo, C.; Desando, G.; Bartolotti, I.; Grigolo, B. Scaffolds for bone tissue engineering: State of the art and new perspectives. *Mater. Sci. Eng. C* **2017**, *78*, 1246–1262. [CrossRef]
13. Turnbull, G.; Clarke, J.; Picard, F.; Riches, P.; Ja, L.; Han, F.; Li, B.; Shu, W. 3D bioactive composite scaffolds for bone tissue engineering. *Bioact. Mater.* **2018**, *3*, 278–314. [CrossRef] [PubMed]
14. Eltom, A.; Zhong, G.; Muhammad, A. Scaffold techniques and designs in tissue engineering functions and purposes: A review. *Adv. Mater. Sci. Eng.* **2019**, *2019*, 3429527. [CrossRef]
15. Ozbolat, I.T.; Moncal, K.K.; Gudapati, H. Evaluation of bioprinter technologies. *Addit. Manuf.* **2017**, *13*, 179–200. [CrossRef]
16. Barrero, A.; González Loscertales, I.; Márquez, M. Microchorros y Nanochorros. *Investig. Cienc.* **2005**, *351*, 44–51.
17. Xie, J.; Jiang, J.; Davoodi, P.; Srinivasan, M.P.; Wang, C.-H. Electrohydrodynamic atomization: A two-decade effort to produce and process micro-/nanoparticulate materials. *Chem. Eng. Sci.* **2015**, *125*, 32–57. [CrossRef] [PubMed]
18. Rahmati, M.; Mills, D.K.; Urbanska, A.M.; Saeb, R.M.; Venugopal, J.R.; Ramakrishna, S.; Mozafari, M. Electrospinning for tissue engineering applications. *Prog. Mater. Sci.* **2020**, *117*, 100721. [CrossRef]
19. Wu, T.; Ding, M.; Shi, C.; Qiao, Y.; Wang, P.; Qiao, R.; Wang, X.; Zhong, J. Resorbable polymer electrospun nanofibers: History, shapes and application for tissue engineering. *Chin. Chem. Lett.* **2020**, *31*, 617–625. [CrossRef]
20. Soares, R.M.D.; Siqueira, N.M.; Prabhakaram, M.P.; Ramakrishna, S. Electrospinning and electrospray of bio-based and natural polymers for biomaterials development. *Mater. Sci. Eng. C* **2018**, *92*, 969–982. [CrossRef]
21. Calori, I.R.; Braga, G.; de Jesus, P.d.C.C.; Bi, H.; Tedesco, A.C. Polymer scaffolds as drug delivery systems. *Eur. Polym. J.* **2020**, *129*, 109621. [CrossRef]
22. Ojha, S. Structure—Property relationship of electrospun fibers. In *Electrospun Nanofibers*; Elsevier: Amsterdam, The Netherlands, 2017; pp. 239–253.
23. Munteanu, B.S.; Vasile, C. Electrospun Polymeric Nanostructures with Applications in Nanomedicine. In *Polymeric Nanomaterials in Nanotherapeutics*; Elsevier: Amsterdam, The Netherlands, 2019; pp. 261–297.
24. Neves, N.M.; Campos, R.; Pedro, A.; Cunha, J.; Macedo, F.; Reis, R.L. Patterning of polymer nanofiber meshes by electrospinning for biomedical applications. *Int. J. Nanomed.* **2007**, *2*, 433.
25. Kumar, P. Effect of Colletor on Electrospinning to Fabricate Aligned Nano Fiber. Ph.D. Thesis, National Institute of Technology, Rourkela, India, 2012.
26. Yeo, M.; Kim, G. Micro/nano-hierarchical scaffold fabricated using a cell electrospinning/3D printing process for co-culturing myoblasts and HUVECs to induce myoblast alignment and differentiation. *Acta Biomater.* **2020**, *107*, 102–114. [CrossRef] [PubMed]
27. Ambrus, R.; Alshweiat, A.; Csóka, I.; Ovari, G.; Esmail, A.; Radacsi, N. 3D-printed electrospinning setup for the preparation of loratadine nanofibers with enhanced physicochemical properties. *Int. J. Pharm.* **2019**, *567*, 118455. [CrossRef] [PubMed]
28. Kim, J.I.; Hwang, T.I.; Aguilar, L.E.; Park, C.H.; Kim, C.S. A controlled design of aligned and random nanofibers for 3D bi-functionalized nerve conduits fabricated via a novel electrospinning set-up. *Sci. Rep.* **2016**, *6*, 23761. [CrossRef] [PubMed]
29. Joshi, M.K.; Shrestha, R.M.; Pant, H.R. 3D Nonwoven Fabrics for Biomedical Applications. In *Generation, Development and Modifications of Natural Fibers*; IntechOpen: Rijeka, Croatia, 2020.
30. Clavijo-Grimaldo, D.; Ponce-Zapata, N.; Casadiego-Torrado, C. Control of Bacterial Proliferation and Formation of Biofilm in Membranes for Food Packaging Manufactured by Electrospinning. *Chem. Eng. Trans.* **2019**, *75*, 241–246.
31. Ghasemi-Mobarakeh, L.; Semnani, D.; Morshed, M. A novel method for porosity measurement of various surface layers of nanofibers mat using image analysis for tissue engineering applications. *J. Appl. Polym. Sci.* **2007**, *106*, 2536–2542. [CrossRef]
32. Vasita, R.; Katti, D.S. Nanofibers and their applications in tissue engineering. *Int. J. Nanomed.* **2006**, *1*, 15. [CrossRef] [PubMed]
33. Xue, J.; Wu, T.; Dai, Y.; Xia, Y. Electrospinning and electrospun nanofibers: Methods, materials and applications. *Chem. Rev.* **2019**, *119*, 5298–5415. [CrossRef]
34. Li, L.; LaBarbera, D.V. 3D High-content screening of organoids for drug discovery. *Compr. Med. Chem. III* **2017**, *119*, 388–415.
35. Sánchez Cepeda, Á.P.; Vera-Graziano, R.; Muñoz-Prieto, E.d.J.; Gómez-Pachón, E.Y.; Bernad-Bernad, M.J.; Maciel-Cerda, A. Preparación y caracterización de membranas poliméricas electrohiladas de policaprolactona y quitosano para la liberación controlada de clorhidrato de tiamina. *Cienc. Desarro.* **2016**, *7*, 133–151. [CrossRef]
36. Seidi, A.; Ramalingam, M.; Elloumi-Hannachi, I.; Ostrovidov, S.; Khademhosseini, A. Gradient biomaterials for soft-to-hard interface tissue engineering. *Acta Biomater.* **2011**, *7*, 1441–1451. [CrossRef]
37. Greenfeld, I.; Rodricks, C.W.; Sui, X.; Wagner, H.D. Beaded fiber composites—Stiffness and strength modeling. *J. Mech. Phys. Solids* **2019**, *125*, 384–400. [CrossRef]
38. Fujihara, K.; Kotaki, M.; Ramakrishna, S. Guided bone regeneration membrane made of polycaprolactone/calcium carbonate composite nano-fibers. *Biomaterials* **2005**, *26*, 4139–4147. [CrossRef] [PubMed]
39. Karuppuswamy, P.; Venugopal, J.R.; Navaneethan, B.; Laiva, A.L.; Ramakrishna, S. Polycaprolactone nanofibers for the controlled release of tetracycline hydrochloride. *Mater. Lett.* **2015**, *141*, 180–186. [CrossRef]
40. Gaumer, J.; Prasad, A.; Lee, D.; Lannutti, J. Structure—Function relationships and source-to-ground distance in electrospun polycaprolactone. *Acta Biomater.* **2009**, *5*, 1552–1561. [CrossRef] [PubMed]

41. Cramariuc, B.; Cramariuc, R.; Scarlet, R.; Manea, L.R.; Lupu, I.G.; Cramariuc, O. Fiber diameter in electrospinning process. *J. Electrost.* **2013**, *71*, 189–198. [CrossRef]
42. Fridrikh, S.V.; Jian, H.Y.; Brenner, M.P.; Rutledge, G.C. Controlling the fiber diameter during electrospinning. *Phys. Rev. Lett.* **2003**, *90*, 144502. [CrossRef]
43. He, H.; Wang, Y.; Farkas, B.; Nagy, Z.K.; Molnar, K. Analysis and prediction of the diameter and orientation of AC electrospun nanofibers by response surface methodology. *Mater. Des.* **2020**, *194*, 108902. [CrossRef]
44. El-Hadi, A.M.; Al-Jabri, F.Y. Influence of electrospinning parameters on fiber diameter and mechanical properties of poly (3-hydroxybutyrate) (PHB) and polyanilines (PANI) blends. *Polymers* **2016**, *8*, 97. [CrossRef]
45. Moghadam, B.H.; Hasanzadeh, M. Predicting contact angle of electrospun PAN nanofiber mat using artificial neural network and response surface methodology. *Adv. Polym. Technol.* **2013**, *32*, 21365. [CrossRef]
46. Szewczyk, P.K.; Ura, D.P.; Metwally, S.; Knapczyk-Korczak, J.; Gajek, M.; Marzec, M.M.; Bernasik, A.; Stachewicz, U. Roughness and fiber fraction dominated wetting of electrospun fiber-based porous meshes. *Polymers* **2019**, *11*, 34. [CrossRef]
47. Dias, J.; Bártolo, P. Morphological char influence. *Procedia CIRP* **2013**, *5*, 216–221. [CrossRef]
48. Alam, A.K.M.; Ewaldz, E.; Xiang, C.; Qu, W.; Bai, X. Tunable Wettability of Biodegradable Multilayer Sandwich-Structured Electrospun Nanofibrous Membranes. *Polymers* **2020**, *12*, 2092. [CrossRef] [PubMed]
49. Tiwari, A.P.; Joshi, M.K.; Lee, J.; Maharjan, B.; Ko, S.W.; Park, C.H.; Kim, C.S. Heterogeneous electrospun polycaprolactone/polyethylene glycol membranes with improved wettability, biocompatibility, and mineralization. *Colloids Surf. A Physicochem. Eng. Asp.* **2017**, *520*, 105–113. [CrossRef]
50. Al-Amshawee, S.; Yunus, M.Y.B.M.; Lynam, J.G.; Lee, W.H.; Dai, F.; Dakhil, I.H. Roughness and wettability of biofilm carriers: A systematic review. *Environ. Technol. Innov.* **2020**, *21*, 101233. [CrossRef]

*membranes*

*Article*

# Electrospun Biodegradable Nanofibers Coated Homogenously by Cu Magnetron Sputtering Exhibit Fast Ion Release. Computational and Experimental Study

**Anton M. Manakhov [1,*], Natalya A. Sitnikova [1], Alphiya R. Tsygankova [2], Alexander Yu. Alekseev [3,4], Lyubov S. Adamenko [3], Elizaveta Permyakova [1,5], Victor S. Baidyshev [6], Zakhar I. Popov [7], Lucie Blahová [8], Marek Eliáš [8], Lenka Zajíčková [8,9] and Anastasiya O. Solovieva [1,*]**

1 Research Institute of Clinical and Experimental Lymphology—Branch of the ICG SB RAS, 2 Timakova St., 630060 Novosibirsk, Russia; sitnikovanat9@gmail.com (N.A.S.); permyakova.elizaveta@gmail.com (E.P.)
2 Nikolaev Institute of Inorganic Chemistry SB RAS, 3 Acad. Lavrentiev Ave., 630090 Novosibirsk, Russia; alphiya@yandex.ru
3 Research Institute of Virology, The Federal Research Center of Fundamental and Translational Medicine, 2 Timakova St., 630060 Novosibirsk, Russia; al-alexok@ngs.ru (A.Y.A.); aminisib@yandex.ru (L.S.A.)
4 Research Institute of Applied Ecology, Dagestan State University, Dahadaeva 21, 367000 Makhachkala, Russia
5 Laboratory of Inorganic Nanomaterials, National University of Science and Technology "MISiS", Leninsky Prospekt 4, 119071 Moscow, Russia
6 Department of Computer Engineering and Automated Systems Software, Katanov Khakas State University, Pr. Lenin, 90, 655017 Abakan, Russia; bayd_vs@mail.ru
7 Laboratory of Acoustic Microscopy, Emanuel Institute of Biochemical Physics RAS, Kosygina 4, 119334 Moscow, Russia; zipcool@bk.ru
8 Central European Institute of Technology CEITEC-BUT, Purkyňova 123, 61200 Brno, Czech Republic; Lucie.Blahova@ceitec.vutbr.cz (L.B.); marek.elias@ceitec.vutbr.cz (M.E.); lenkaz@physics.muni.cz (L.Z.)
9 Department Condensed Matter Physics, Faculty of Science, Masaryk University, Kotlářská 2, 61137 Brno, Czech Republic
* Correspondence: ant-manahov@ya.ru (A.M.M.); solovey_ao@mail.ru (A.O.S.); Tel.: +7-915-8494059 (A.O.S.)

**Citation:** Manakhov, A.M.; Sitnikova, N.A.; Tsygankova, A.R.; Alekseev, A.Y.; Adamenko, L.S.; Permyakova, E.; Baidyshev, V.S.; Popov, Z.I.; Blahová, L.; Eliáš, M.; et al. Electrospun Biodegradable Nanofibers Coated Homogenously by Cu Magnetron Sputtering Exhibit Fast Ion Release. Computational and Experimental Study. *Membranes* **2021**, *11*, 965. https://doi.org/10.3390/membranes11120965

Academic Editor: Andrea Ehrmann

Received: 17 November 2021
Accepted: 4 December 2021
Published: 8 December 2021



**Abstract:** Copper-coated nanofibrous materials are desirable for catalysis, electrochemistry, sensing, and biomedical use. The preparation of copper or copper-coated nanofibers can be pretty challenging, requiring many chemical steps that we eliminated in our robust approach, where for the first time, Cu was deposited by magnetron sputtering onto temperature-sensitive polymer nanofibers. For the first time, the large-scale modeling of PCL films irradiation by molecular dynamics simulation was performed and allowed to predict the ions penetration depth and tune the deposition conditions. The Cu-coated polycaprolactone (PCL) nanofibers were thoroughly characterized and tested as antibacterial agents for various Gram-positive and Gram-negative bacteria. Fast release of $Cu^{2+}$ ions (concentration up to 3.4 μg/mL) led to significant suppression of *E. coli* and *S. aureus* colonies but was insufficient against *S. typhimurium* and *Ps. aeruginosa*. The effect of Cu layer oxidation upon contact with liquid media was investigated by X-ray photoelectron spectroscopy revealing that, after two hours, 55% of Cu atoms are in form of CuO or $Cu(OH)_2$. The Cu-coated nanofibers will be great candidates for wound dressings thanks to an interesting synergistic effect: on the one hand, the rapid release of copper ions kills bacteria, while on the other hand, it stimulates the regeneration with the activation of immune cells. Indeed, copper ions are necessary for the bacteriostatic action of cells of the immune system. The reactive $CO_2/C_2H_4$ plasma polymers deposited onto PCL-Cu nanofibers can be applied to grafting of viable proteins, peptides, or drugs, and it further explores the versatility of developed nanofibers for biomedical applications use.

**Keywords:** PCL nanofibers; XPS; copper; antibacterial coating; ion release; cytotoxicity

## 1. Introduction

The preparation of copper/copper oxide nanofibrous materials became a prevalent topic thanks to a vast range of applications of such nanomaterials, including catalysis [1,2], disinfection [3], antiviral nanocomposites [4], antibacterial wound dressings [5,6], sensors [7], $CO_2$ electrocatalytic reduction [8] and others. The advantage of using Cu nanofibers instead of other forms of nano or micromaterials is their high surface-to-volume ratio, the possibility to prepare sheets or foils of Cu nanofibers without any limitations in terms of their size, the opportunity to run the continuous roll-to-roll process, and its scalability.

Several approaches were employed to prepare Cu-containing nanofibers: electroless deposition on a nanofibrous foil as a template [8], self-assembly method using PANI nanofibers as a template [9], incorporation of Cu by admixture of Cu nanoparticles to the electrospinning solution [10], and decoration of nanofibers with Cu nanoparticles [5].

The area of applications of nanofibrous mats is vast. Mainly, the biomedical use of nanofibers attracts the attention of many researchers because they efficiently accelerate wound healing [11,12], may filter bacteria and viruses [13,14], and regenerate the bones [15–18]. The use of nanofibers with Cu/Cu-oxide coatings for biomedical applications was not fully covered in the literature compared to, e.g., Ag nanoparticles. In most instances, the material was prepared using methods with high consumption of chemicals, i.e., either by soaking the nanofibers in the $Cu^{2+}$ containing solution, the addition of Cu salts into the electrospinning solutions [19], grafting of Cu nanoparticles [20], or by admixture of Cu nanoparticles in the solution [6]. In general, authors witnessed both antibacterial and cytotoxic effects of their Cu-containing nanofibers and, interestingly, various forms of Cu ($Cu^0$, $Cu^+$, $Cu^{2+}$, CuO, and $Cu(OH)_2$) behave differently in different cultures.

Polyacrylonitrile (PAN) nanofibers with embedded CuO were tested for antimicrobial breathe masks with high antibacterial effects [21]. PVA nanofibers with Cu nanoparticles have shown significant inhibition zones against Gram-negative *Escherichia coli* and Gram-positive *Staphylococcus aureus* bacteria [10]. Haider et al. have shown that PLGA/CuO nanofiber scaffolds exhibited excellent antibacterial activity against *E. coli* and *S. aureus* bacterial strains [20]. The mechanism of the antibacterial action is based on the $Cu^{2+}$ ion release. Phan and co-authors made a similar conclusion for *E. coli* and *B. subtilis* [19]. They also concluded that CuO and $Cu(OH)_2$ embedded into polyacrylonitrile nanofibers would be less effective than $CuSO_4$ but more efficient than metallic Cu thanks to the faster release of $Cu^{2+}$ ions from oxidized surfaces. However, the authors did not reveal the cytotoxicity of such structures.

This work presents a facile, robust, and scalable method for preparing Cu-coated nanofibers based on magnetron sputtering of copper onto the FDA-approved biodegradable polycaprolactone (PCL) nanofibers. The temperature-sensitive polymer nanofibers have never been tested as a substrate for the Cu deposition by magnetron sputtering. The main challenge is the deposition of a well-adhered metallic coating with high Cu content PCL membrane without degradation of the nanofibrous structure. The antibacterial properties against Gram-negative and Gram-positive bacteria and cell viability of mesenchymal stromal cells were studied, and the mechanism of Cu-coated PCL nanofibers onto different strains was discussed.

## 2. Materials and Methods

### 2.1. Electrospinning of PCL Nanofibers

The electrospun nanofibers were prepared by the electrospinning of a 9 wt% solution of polycaprolactone PCL (80,000 g/mol). The processing of the sample can be found elsewhere [22]. Briefly, the granulated PCL was dissolved in a mixture of acetic acid (99%) and formic acid (98%). All compounds were purchased from Sigma Aldrich (Darmstadt, Germany). The weight ratio of acetic acid (AA) to formic acid (FA) was 2:1. The PCL solutions in AA and FA were stirred at 25 °C for 24 h. The PCL solution was electrospun with a 20 cm long wired electrode using a Nanospider™ NSLAB 500 machine (ELMARCO, Liberec, Czech Republic). The applied voltage was 50 kV. The distance between the electrodes was set to 100 mm. The as-prepared and non-treated PCL nanofibers are referred to as PCL-ref throughout the text.

### 2.2. Magnetron Sputtering

The Cu coatings were deposited by magnetron sputtering of a copper target in an ultra-high-vacuum deposition chamber (BESTEC, Germany). The input power to the magnetron was set to 37 W. Before the deposition, the chamber was evacuated down to $6.2 \cdot 10^{-8}$ mbar. A 30 sccm flow of a high purity Ar gas (99.99%) was introduced into the deposition chamber, setting the operation pressure to $1.5 \cdot 10^{-3}$ mbar. During the film deposition, the distance between the targets and the substrate was kept at 30 mm and the substrate holder was rotated at 10 rpm to obtain a homogenous film thickness. The deposition time was adjusted to deposit a 50 nm thick film (controlled by deposition onto Si wafer). The Cu-coated nanofibers are referred as PCL-Cu throughout the text.

### 2.3. Plasma COOH Coating

The COOH plasma polymer layers were deposited using the vacuum system UVN-2M equipped with the rotary and oil diffusion pumps. The residual pressure of the reactor was below $10^{-3}$ Pa. The plasma was ignited using radio frequency (RF) power supply Cito 1310-ACNA-N37A-FF (Comet, Flamatt, Switzerland) connected to the RFPG-128 disk generator (Beams & Plasmas) installed in the vacuum chamber. The duty cycle and the RF power were set to 5% and 500 W, respectively.

$CO_2$ (99.995%), Ar (99.998%), and $C_2H_4$ (99.95%) were fed into the vacuum chamber. The flows of the gases were controlled using a Multi-Gas Controller 647C (MKST, Newport, RI, USA). The flow rates of Ar, $CO_2$, and $C_2H_4$ were set to 50, 16.2, and 6.2 sccm, respectively. The pressure in the chamber was measured by a VMB-14 unit (Tokamak Company, Dubna, Russia) and D395-90-000 BOC Edwards controllers. The distance between RF-electrode and the substrate was set to 8 cm. The deposition time was 15 min and it led to the growth of ~100 nm thick plasma coatings. The plasma coated PCL-Cu nanofibers are referred to as PCL-Cu-COOH throughout the text.

### 2.4. Chemistry and Morphology Analysis

The microstructure of the nanofibers and the deposited plasma polymers was studied by scanning electron microscopy (SEM) using a Tescan Mira (Tescan, Brno, Czech Republic) device. The SEM micrographs were obtained in secondary emission mode with the accelerating voltage of 10 kV and working distance of 9 mm. Micrographs of 1024 × 1024 pixel were acquired. The elemental mappings were obtained using energy dispersive X-ray (EDX) detector (Oxford Instruments, High Wycombe, UK).

The chemical composition of the sample surfaces was determined by the X-ray photoelectron spectroscopy (XPS) using an Axis Supra spectrometer (Kratos Analytical, Manchester, UK) equipped with the monochromatic Al K$\alpha$ X-ray source. The maximum lateral resolution of the analyzed area was 0.7 mm. The spectra were fitted using the CasaXPS software after subtracting the Shirley-type background. The binding energies (BE) for all carbon and oxygen environments were taken from the literature [22]. The BE scale was calibrated by setting the $CH_x$ component at 285 eV.

### 2.5. The Ion Release Measuring

The High-Resolution Spectrometer iCAP 6500 (Thermo Fisher Scientific, Pittsburgh, PA, USA) was used. Samples size of 1 × 1 cm were placed in 5 mL of PBS and $H_2O$, respectively, and incubated at 37 °C. After 1, 2, 4, and 24 h, an aliquot of 1 of 500 µL was taken. The sample solution was injected into the plasma through a nebulizer of SeaSpray type using a peristaltic pump with a rate of 0.7 mL/min. Analysis conditions: cooling argon flow—12 L/min, secondary—0.5 L/min; registration time on the first slit—15 s; on the second slit—5 s. The power supplied to the ICP inductor was 1150 W (recommended by the manufacturer of the spectrometer). The registration of emission spectra was carried out at the axial observation of plasma. In the process of sample preparation, the following reagents were used: concentrated nitric acid extra pure, 69.0–71.0% (Sigma-Aldrich, Darmstadt, Germany), deionized water purified with the Direct-Q3 system (Millipore) >18 MΩ/cm; high purity argon; single component standard solution—copper (Cu) (Merck). Samples dissolution was performed using concentrated nitric acid with heating ~ 100–150 °C. Sample preparation was performed using disposable plastic tubes with a volume of 5–15 mL, polypropylene container with a volume of 10 mL, and automatic pipette with variable volume (1.00–5.00 mL, 100–1000 μL 10–100 μL,). To determine analytes of Cu, the most intense spectral lines were used (without the spectral influence of the matrix)—328.068, 338.289; 324.754, 327.396 nm, respectively. The validation of the technique by spike experiment was provided.

### 2.6. Modeling

The classical molecular dynamics method in the LAMMPS [23] software package was applied to the irradiation simulations of PCL by Cu atoms. All interatomic interactions in the system were described by ReaxFF potentials [24]. The dimer energies were calculated by the selected potential to estimate the parameters of the interaction of copper atoms with polymer atoms (see Table 1) and comparisons were made with similar calculations by the DFT method [25,26] in the Vienna Ab initio Simulation Package (VASP) [27,28]. Despite that the ReaxFF potentials underestimate the energies of individual dimers, they qualitatively describe changes in the energy of interactions of Cu atoms with Cu, H, C, O, since the energy decreases from Cu-Cu to Cu-O both in the case of DFT and in the case of ReaxFF calculations. In addition, the difference in link lengths between DFT and ReaxFF is negligible.

**Table 1.** Dimer binding energy calculated by DFT and ReaxFF potentials, the energy difference between DFT and ReaxFF, bond length in angstroms.

| Dimer | E, eV (PBE) | E, eV (ReaxFF) | ΔE, eV | R, Å (PBE) | R, Å (ReaxFF) |
|---|---|---|---|---|---|
| Cu-Cu | −2.75 | −1.38 | −1.36 | 2.22 | 2.33 |
| Cu-H | −4.27 | −2.58 | −1.69 | 1.46 | 1.53 |
| Cu-C | −4.28 | −2.65 | −1.63 | 1.76 | 1.61 |
| Cu-O | −5.25 | −4.05 | −1.21 | 1.69 | 1.75 |

### 2.7. Cell Tests

Cell viability was assayed by the MTT method and fluorescent microscopy. Human mesenchymal stromal cells were extracted from bone marrow using standard methods (the Ethics Committee approved the study of the RICEL-branch of ICG SB RAS (No 115 from 24.12.2015) and cultured in Dulbecco's modified Eagle's Medium (DMEM, Sigma Aldrich) that was supplemented with 10% fetal bovine serum (FBS, Gibco, Carlsbad, CA, USA). Cells were seeded in 96-well plates on scaffolds (round samples of diameter 0.5 cm) in concentration $7 \times 10^3$ cells/well. Additionally, cells were cultivated in 96-well plates in a medium in which the PCL-Cu and PCL-Cu-COOH were soaked (round samples of diameter 0.5 cm in 200 μL) for 1 h and then incubated for 72 h under 5% $CO_2$ atmosphere. A fresh culture medium was added to control cells. After that 5 μL of the MTT solution with the concentration of 5 mg/mL was added to each well, and the plates were incubated for 4 h and then solubilized with a dimethyl sulfoxide solution, as indicated in the manufacturer's instructions. The optical density was measured with a plate reader Multiskan FC (Thermo Fisher Scientific, Singapore) at the wavelength of 570 nm. The experiment was repeated three times on separate days. For fluorescent microscopic analysis, cells were incubated at 37 °C in the dark with stain solution (199 cell culture media with 5 μg/mL Hoechst 33,342 and 2 μM calcein AM (Thermo Fisher Scientific, St. Louis, MO, USA) for 30 min. Live cells are determined distinguished by the presence of ubiquitous intracellular intense uniform green fluorescence.

### 2.8. Microbiology

*E. coli* ATCC25922, *S. aureus* ATCC25923, *S. Typhimurium* ATCC14028, *P. aeruginosa* ATCC27853 strains were obtained from Remel™, Thermo Fisher Scientific, USA, and Becton Dickenson, France. The bacterial strains were grown in liquid Lysogeny broth (LB) medium at 37 °C for 24 h and then were diluted in saline to give concentrations of 0.675–2.5 × $10^5$ colony-forming units (CFU) $mL^{-1}$. Antibacterial activity against each strain was determined by the emersion of nanofibers (round samples in diameter 0.5 cm) in a medium volume of 300 µL with bacteria for 24 h. The number of viable microorganisms was estimated via counting of CFU after 24 h of cultivation. All experiments were performed in triplicate.

## 3. Results

### 3.1. Modeling of Cu Ions Penetration Depth

In order to correctly select the conditions where no destruction of such sensitive material as PCL occurs, the computational simulation was performed before the experiments of Cu deposition. The PCL unit cell (Figure 1) was taken from [29] and relaxed in VASP, after which it was used to create a slab supercell.

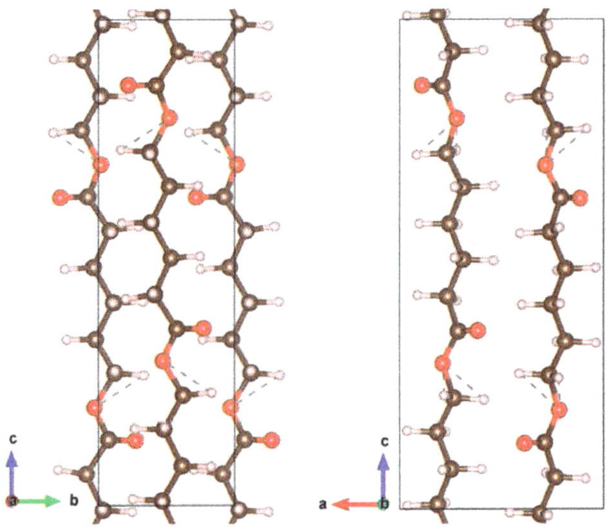


**Figure 1.** The PCL unit cell. The carbon, hydrogen and oxygen atoms are indicated by brown, gray and red colors.

To simulate the irradiation of a film of finite thickness, a PCL slab with a size of 22.7 × 3.94 × 5.31 nm, consisting of 64,800 atoms, was constructed. Periodic boundary conditions were applied in the direction of the y and z axes, and the irradiation was carried out along the x-axis. Before irradiation the slab was relaxed for 100 ps at constant pressure (NPT thermostat), then at a constant temperature (NVT thermostat) equal to T = 300 K.

The metal atom was placed randomly at a distance of 1.8 nm from the PCL slab surface, mainly in the center of the YZ plane of the supercell (Figure 2). The atom was given an initial velocity component normal to the slab plane following the energy under consideration. In the simulation, a variable time step was used, which was selected from the condition that the maximum displacement of atoms did not exceed 0.001 nm. This procedure made it possible to avoid unreasonably large approaches of atoms and kept the simulation stable. For example, at the largest energies of the metal atom considered, the minimum step was 0.012 fs.

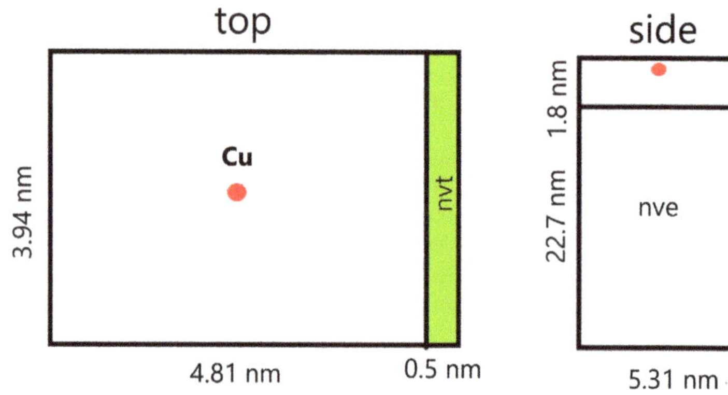


**Figure 2.** A diagram of a model with characteristic dimensions.

To simulate the dissipation of energy into an infinite volume of material, a 0.5 nm thick region was isolated from the side of the PCL slab to the atoms of which temperature control was applied (NVT thermostat, T = 300 K). An NVE thermostat was applied to the remaining atoms of the system, including the metal atom. The simulation continued until the energy of the metal atom exceeded the average thermal energy of the PCL slab.

The energies of a swooped Cu atom in the range from 500 eV to 2400 eV were considered. For each selected energy, a series of computer experiments were made, consisting of five simulations. The average values are shown in Figure 3.

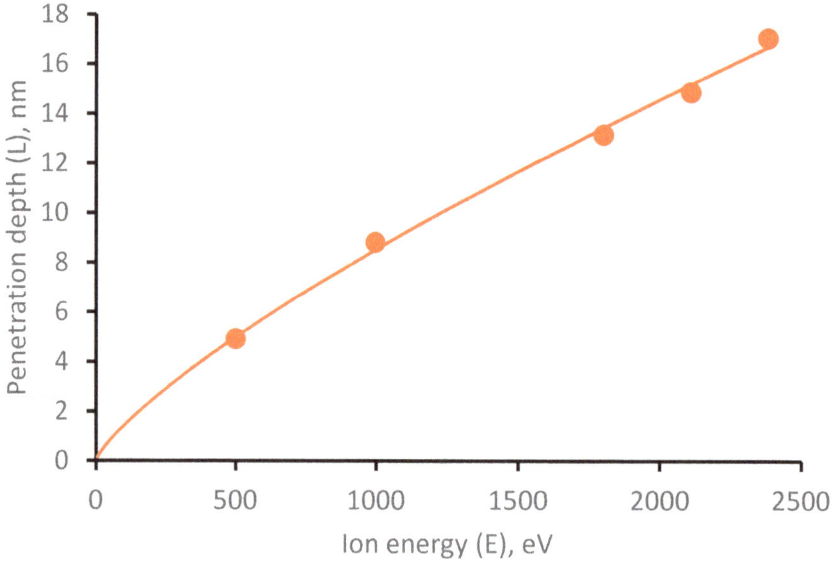


**Figure 3.** Dependence of the penetration depth of Cu atom into PCL depending on its initial energy.

In the energy range under consideration, a power-law dependence of the penetration depth on the atom's energy is obtained, described by the equation $L=4.07\cdot10^{-2}\cdot E^{0.7739}$. It is worth noting that in the process of penetration into the PCL, the copper atom has the ability to change the movement direction due to collisions with polymer atoms. Herewith, the penetration angle of the copper atom is random, and the average deviation angle for all simulations is 5.96 degrees.

### 3.2. PCL-Cu Nanofibers

The SEM micrograph of Cu-coated PCL nanofibrous mat is in Figure 4a. The fiber diameter was around 250 nm, and the structure of nanofibrous mesh was homogenous with no defects observed. The EDX mapping of the same sample (Figure 4b) revealed homogenous coverage of the nanofibers by copper. The XPS analysis also confirmed that the PCL-Cu sample is well covered by the copper layer. The percentages of all elements are shown in Table 2.

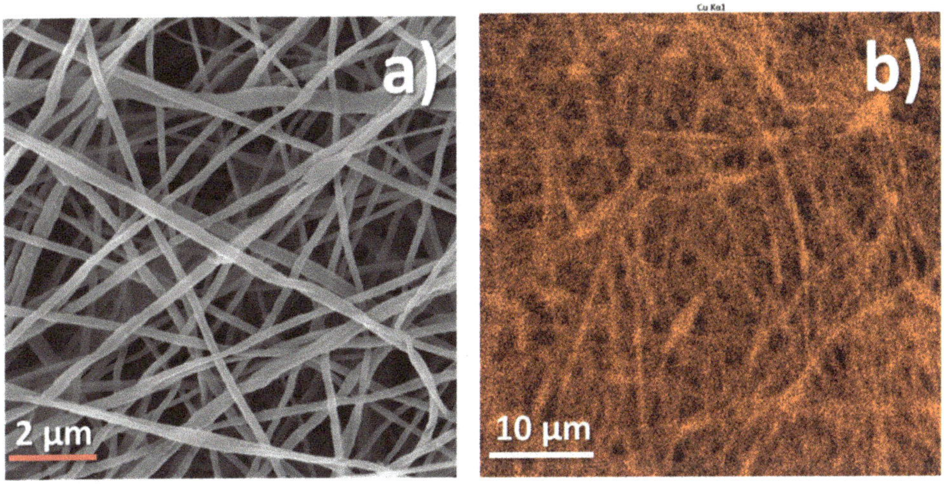


**Figure 4.** SEM micrograph (**a**) and EDS Cu signal cartography (**b**) of PCL-Cu sample.

**Table 2.** Composition of samples (in at. %) derived from XPS analysis.

| Sample Name | Cu | O | C |
|---|---|---|---|
| PCL-ref | 0.0 | 26.1 | 73.9 |
| PCL-Cu | 20.4 | 29.0 | 50.6 |
| PCL-Cu-PBS-2h | 17.2 | 35.7 | 47.1 |
| PCL-Cu-PBS-24h | 8.6 | 43.5 | 47.9 |
| PCL-COOH | 0.0 | 27.5 | 72.5 |
| PCL-Cu-COOH | 0.4 | 26.5 | 73.1 |

The Cu layer covering PCL nanofibers is relatively thick, as sputtering by Ar cluster gun revealed. As shown in Figure 5, the Cu concentration increased to 53.7 at.% after the sputtering for 3 min, and the sputtering for 100 min led to its decrease to 27 at.%. Hence, a deep penetration depth of Cu ions, i.e., below 20 nm is expected as predicted by the computational analysis.

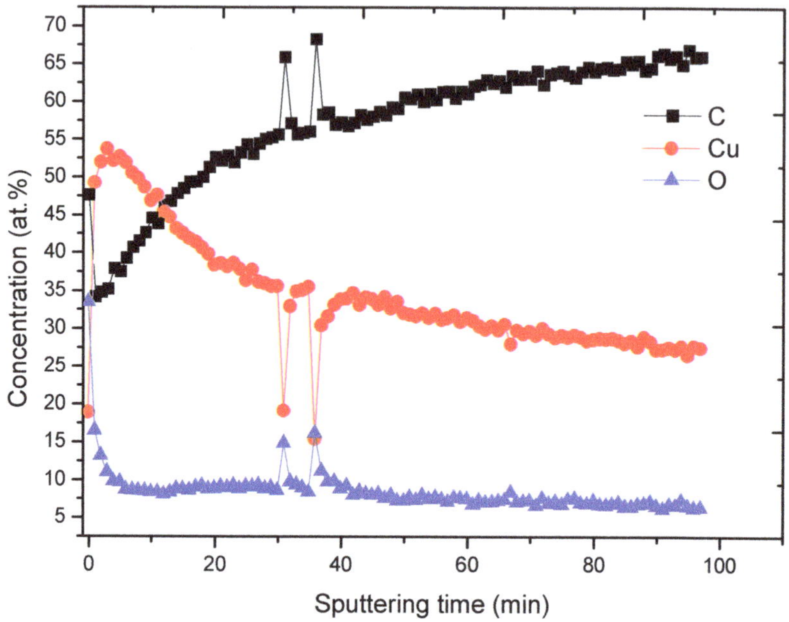


**Figure 5.** XPS depth profiling results for PCL-Cu sample.

To investigate the chemical bonds of copper and other elements presented in the samples, the fitting of high-resolution XPS Cu2p 3/2, C1s, and O1s signals was performed (Figures 6–8). We analyzed the copper environment solely by the fitting of Cu2p 3/2 signal (without the fitting of Cu2p 1/2) due to a higher signal of the Cu2p 3/2 line. The XPS Cu2p 3/2 signal was fitted by a sum of six peaks: metallic copper or copper oxide (I) $Cu^0/Cu^+$ (BE = 932.5 ± 0.1 eV, FWHM = 1.1 ± 0.1 eV), copper oxide (II) CuO (BE = 934.5 ± 0.1 eV, FWHM = 1.7 ± 0.2 eV), copper hydroxide $Cu(OH)_2$ (BE = 935.8 ± 0.2 eV, FWHM = 1.9 ± 0.2 eV) and three Cu(II) satellites centered at 940 ± 0.2 eV, 942.2 ± 0.2 eV and 944.3 ± 0.2 eV. The BE values for all Cu2p and O1s peaks were employed from the literature: $Cu^0/Cu^+$ from [30,31], $Cu(OH)_2$ from [31], and CuO 934 [32,33]. The curve fitting with percentages of each contribution is reported in Figure 6. The presence of copper oxide (II) was evident both by the high percentage of CuO peaks, high intensity of Cu(II) satellites as well by the peak of oxygen attributed to the metal oxide contribution Cu-O (BE = 530.5 eV, FWHM = 1.3 eV) in the O1s spectrum. The XPS O1s spectra also revealed peaks attributed to C-O (BE = 533.1 eV, FWHM = 1.6 eV) and OH (BE = 531.6 eV, FWHM = 1.6 eV) that most probably came from $Cu(OH)_2$. Hence, the XPS O1s spectrum of PCL-Cu is completely different as compared to the spectrum of PCL-ref that was fitted by a sum of two peaks: C-O (BE = 533.4 eV) and C = O (BE = 532.2 eV), as shown in Figure 8a,b.

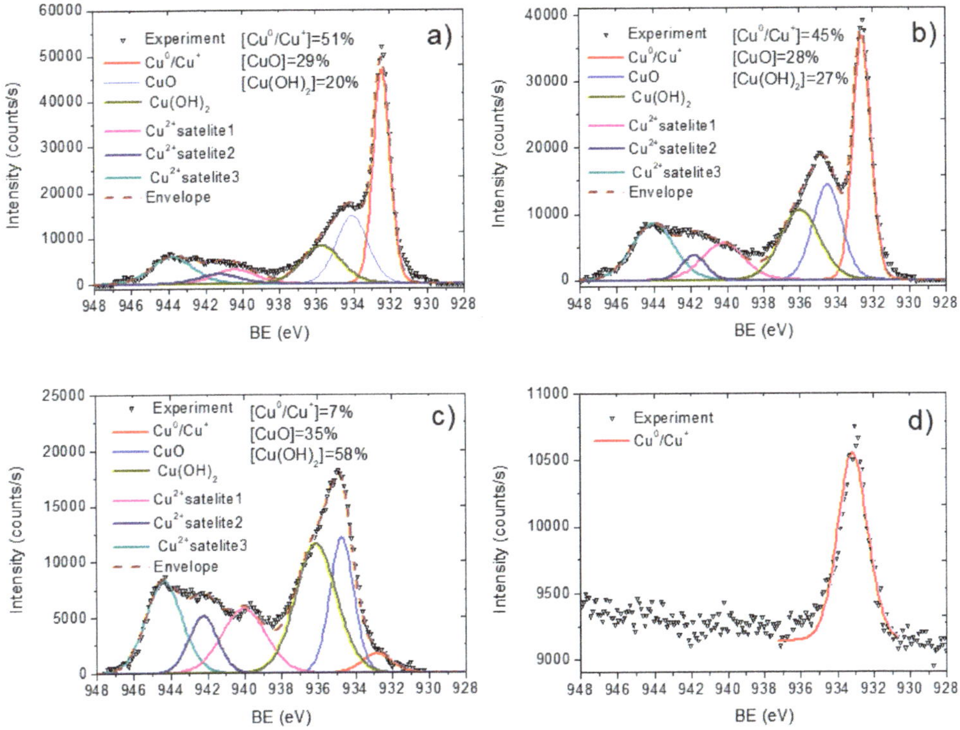


**Figure 6.** XPS Cu2p curve fitting of PCL-Cu as deposited (**a**), after 2h in PBS (**b**), after 24 h in PBS (**c**) and PCL-Cu-COOH (**d**).

The shape of C1s spectra of PCL-ref and PCL-Cu additionally confirmed significant changes in the surface chemistry after Cu deposition. The XPS C1s spectrum of PCL-ref (Figure 7a) was fitted by the sum of three components, namely hydrocarbons $CH_x$ (BE = 285 eV), carbon neighbored to ester group C-C(O)O (BE = 285.5 eV), ether group C-O (BE = 286.4 eV) and ester group C(O)O (BE = 289.0 eV). The FWHM for all peaks was $1.0 \pm 0.1$ eV. A sum of only three components fitted the XPS C1s spectrum of PCL-Cu: $CH_x$, C-O and C(O)O (Figure 7b), where $CH_x$ contribution with the concentration of 73% dominated over other environments. The nature of carbon presented at the surface of PCL-Cu related to the surface contaminations as well as to the signal coming from the area of nanofibers with thinner Cu coating.

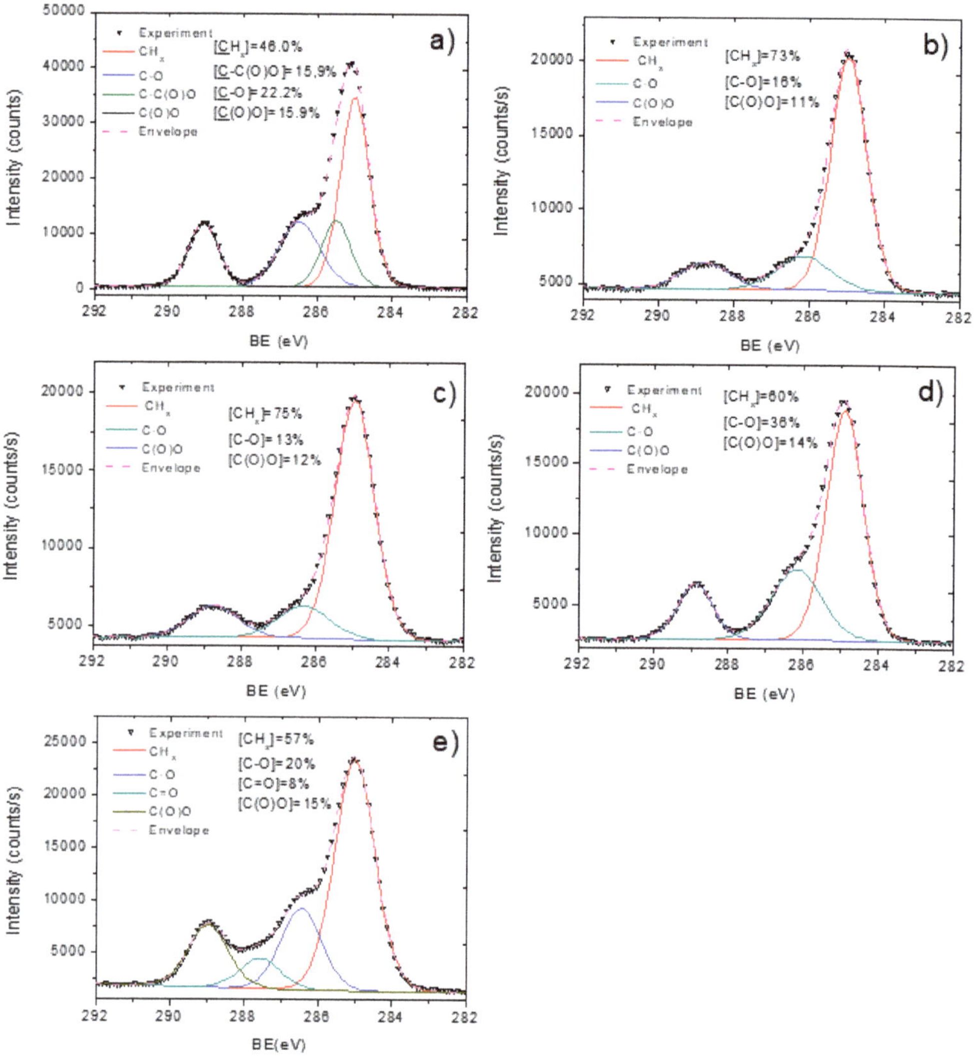


**Figure 7.** XPS C1s curve fitting of pristine PCL nanofibers PCL-ref (**a**), PCL-Cu as-deposited (**b**), PCL-Cu after two hours in PBS (**c**), PCL-Cu after 24 h in PBS (**d**) and PCL-Cu-COOH (**e**).

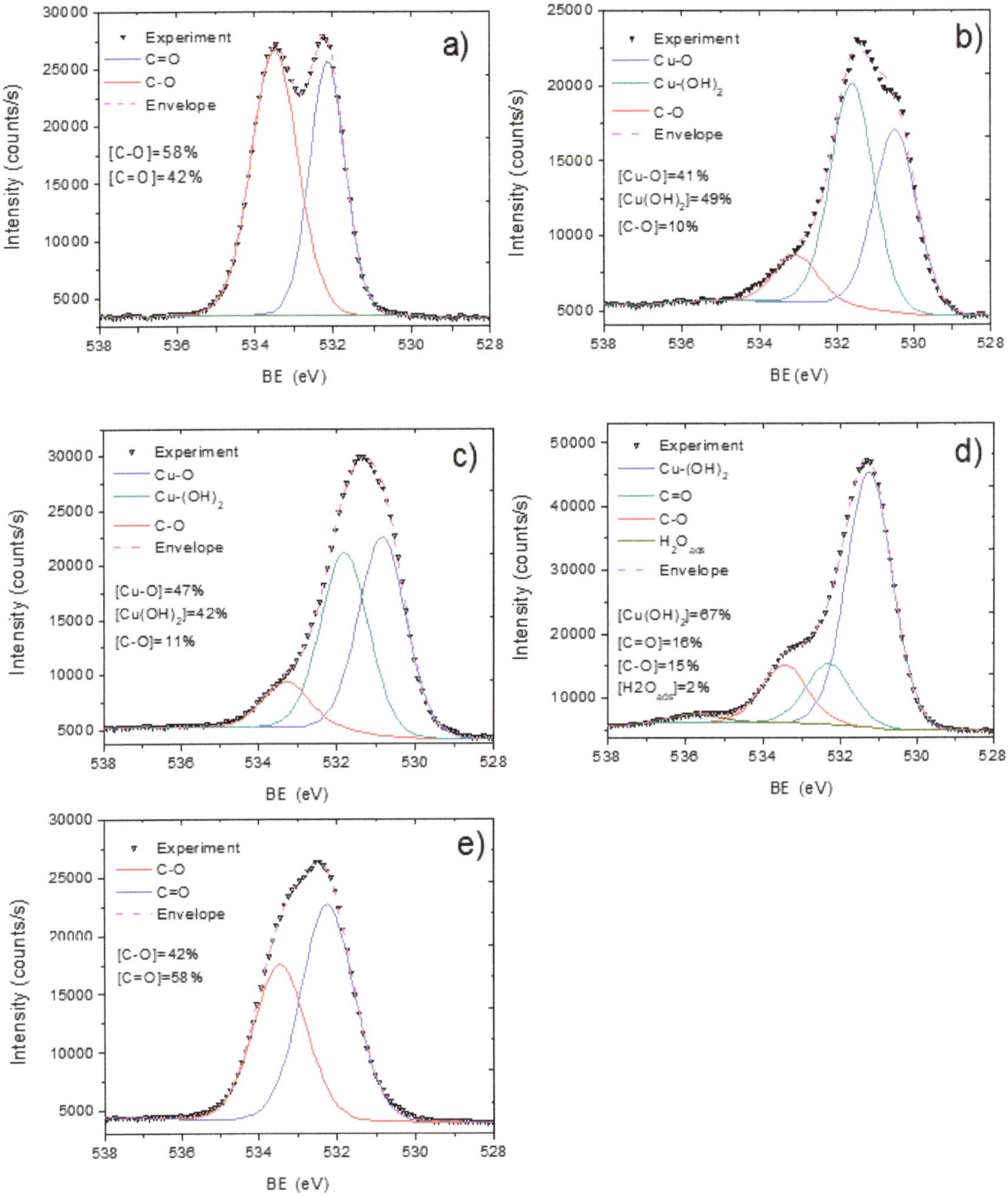


**Figure 8.** XPS O1s curve fitting of pristine PCL nanofibers PCL-ref (**a**), PCL-Cu as-deposited (**b**), PCL-Cu after two hours in PBS (**c**), PCL-Cu after 24 h in PBS (**d**), and PCL-Cu-COOH (**e**).

### 3.3. Cu-PCL Nanofibers Coated with Plasma Polymer

The deposition of the COOH layer of PCL-Cu allows improving the wettability of PCL-Cu nanofibers and the COOH groups play a role of active sites for grafting other compounds. The successful coating was evidenced by significantly different surface composition of COOH-plasma-coated samples (PCL-Cu-COOH) compared to PCL-Cu. As shown in Table 2, the composition of PCL-Cu-COOH exhibited a deficient Cu concentration (0.4 at.%), whereas C and O percentages were similar to PCL-COOH. Additionally, the C1s and O1s curve fitting also revealed significant changes in the carbon and oxygen environments (see Figures 7e and 8e). The high concentration of C(O)O environment of 15% confirmed efficient grafting of COOH moieties because C(O)O directly correlates with the concentration of carboxylic acid reactive groups [34]. The PCL-Cu-COOH was very hydrophilic, as its water contact angle decreased from 99° to 18°.

### *3.4. Stability in Water and Cu Ions Release*

The antibacterial properties of PCL-Cu and PCL-Cu-COOH samples are expected to be induced by the leaching of $Cu^{2+}$ ions. The stability of Cu layers was inspected visually and by XPS analysis of samples after soaking in PBS and by ICP-OES analyses of PBS and water after PCL-Cu and PCL-Cu-COOH soaking.

Very rapid dissolution of Cu layer was evident even by visual inspection as shown in Figure 9. The dissolution rate depends on the liquid's temperature and composition (culture media, PBS or deionized water). The highest dissolution rate was visually observed for PCL-Cu and PCL-COOH samples soaked in culture media at 37 °C, where no red/yellow was visible at the sample surface. It is worth noting that no flakes were floating in the liquid (Figure 9).

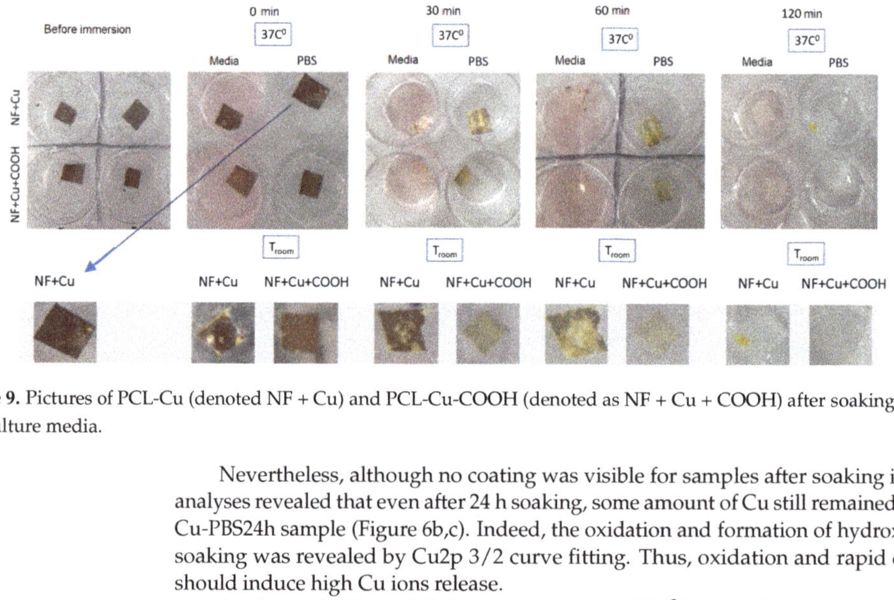


**Figure 9.** Pictures of PCL-Cu (denoted NF + Cu) and PCL-Cu-COOH (denoted as NF + Cu + COOH) after soaking in PBS and culture media.

Nevertheless, although no coating was visible for samples after soaking in PBS, XPS analyses revealed that even after 24 h soaking, some amount of Cu still remained at the PCL-Cu-PBS24h sample (Figure 6b,c). Indeed, the oxidation and formation of hydroxide during soaking was revealed by Cu2p 3/2 curve fitting. Thus, oxidation and rapid dissolution should induce high Cu ions release.

As shown in Figure 10, the concentration of $Cu^{2+}$ ions released in water or PBS at 37 °C is very high and recalculated concentration that is expected to be in the culture media while PCL-Cu cell tests (left Y-axis) exceeds 4.2 μg/mL.

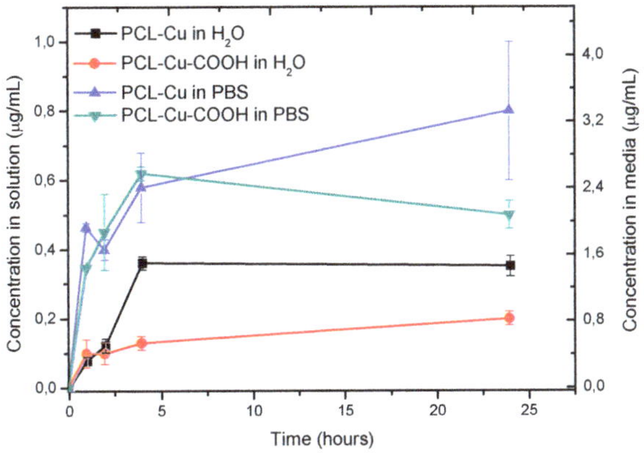


**Figure 10.** Cu ions release at 37 °C measured by ICP-OES.

### 3.5. Antibacterial Properties

The antimicrobial activity of nanofibers was studied on the most common pathogenic microorganisms, namely Gram-negative *E. coli, S. Typhimurium, P. aeruginosa*, and Gram-positive *S. aureus*. The PCL-ref was used as a negative control and did not have any antibacterial effect. The results are presented in Figure 11 and Table 3 as the percentage of colony-forming units (CFU) relative to the CFU of the control. The data was obtained by the emersion of nanofibers in a medium with bacteria for 24 h indicated that the PCL-Cu and PCL-Cu-COOH nanofiber reliably retard the rate of development of *E. coli* and *S. aureus*. In turn against *S. Typhimurium* and *P. aeruginosa* scaffolds had no significant antibacterial effect.

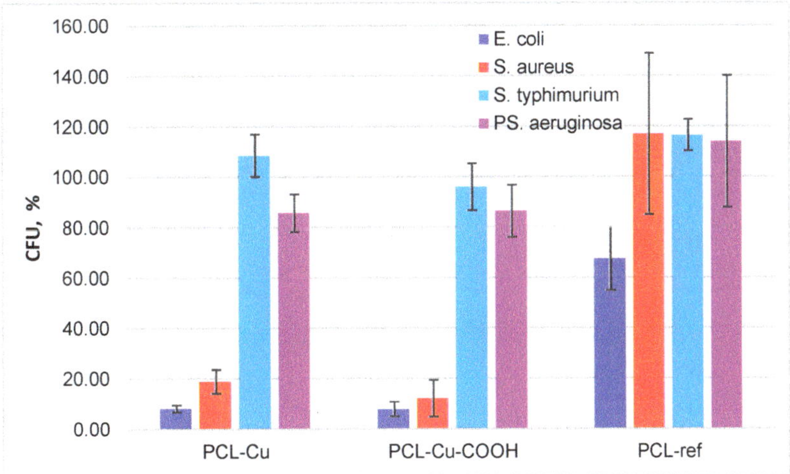


**Figure 11.** Antimicrobial effects of the PCL-ref, PCL-Cu and PCL-Cu-COOH as the percentage of colony forming units (CFU) relative to the CFU of the control.

**Table 3.** Antibacterial activity. The percentage is of CFU relative to the CFU of the control. The reliability of the differences was calculated according to the Student's criterion with the degree of freedom number 4 (3 + 3-2), ($p < 0.01$). Positive control was performed using PCL-COOH with covalent bonded of Penicillin-Streptomycin-Neomycin (PSN) Antibiotic Mixture under the same conditions.

| Microorganism \ Sample Name | PCL-Cu | PCL-Cu-COOH | PCL-Ref | Control |
|---|---|---|---|---|
| *E. coli* ATCC25922 | 7.9 ± 2.9 | 8.1 ± 1.4 | 67.4 ± 12.5 | 100 ± 7.9 |
| *S. aureus* ATCC25923 | 12.1 ± 7.2 | 18.8 ± 4.8 | 146.9 ± 92.0 | 100 ± 34.6 |
| *S. typhimurium* ATCC14028 | 95.9 ± 9.3 | 108.5 ± 8.5 | 116.4 ± 6.2 | 100 ± 8.1 |
| *PS. aeruginosa* ATCC27853 | 86.3 ± 10.3 | 85.6 ± 7.5 | 143.9 ± 76.2 | 100 ± 40.7 |

Gray—differences are significant compared to control.

### 3.6. Cytotoxicity

Cells seeded on PCL-Cu and PCL-Cu-COOH did not survive (data not shown). Figure 12 shows the results of the released copper ions formed during the soaking of nanomaterials in the medium effect on the MSCs viability. It has been determined that cell survival reaches 80% at copper concentrations of less than 0.87 μg/mL (Figure 12). Assessment of the viability of MSCs on nanofibers after their incubation in culture media for 1 h demonstrated an increase in their biocompatibility, the number of adhered cells on the surface of the nanomaterial was significantly higher (Figure 13). However, the functional activity of the cells, assessed with the calcein-AM fluorescent dye, was low.

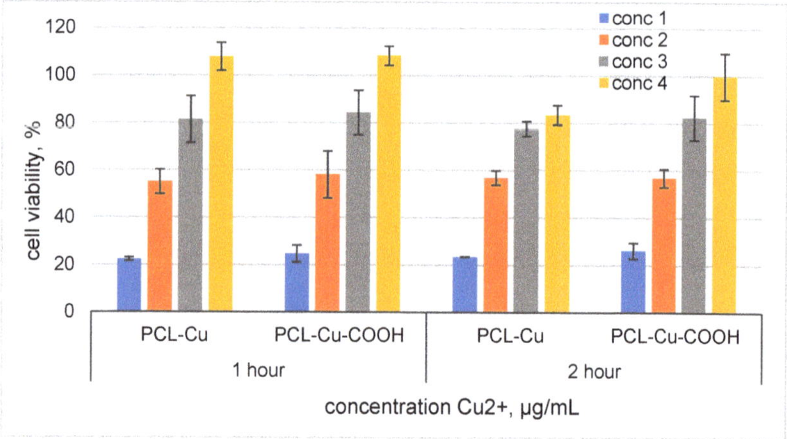


**Figure 12.** Cell viability effect of media after soaking PCL-Cu and PCL-Cu-COOH for 1 and 2 h. Cells were seeded in 96 well plates after 24 h media was replaced by medium soaked PCl-Cu and PCL-Cu-COOH. For **PCL-Cu 1** h: conc 1 = 2.3 μg/mL; conc 2 = 1.15 μg/mL, conc 3 = 0.57 μg/mL and conc 4 = 0.29 μg/mL. **PCL-Cu 2** h: conc 1 = 3.5 μg/mL; conc 2 = 1.75 μg/mL, conc 3 = 0.87 μg/mL and conc 4 = 0.44 μg/mL. For **PCL-Cu-COOH 1** h: conc 1 = 1.8 μg/mL; conc 2 = 0.9 μg/mL, conc 3 = 0.45 μg/mL and conc 4 = 0.23 μg/mL. **PCL-Cu-COOH 2** h: conc 1 = 2.9 μg/mL; conc 2 = 1.45 μg/mL, conc 3 = 0.73 μg/mL and conc 4 = 0.36 μg/mL.

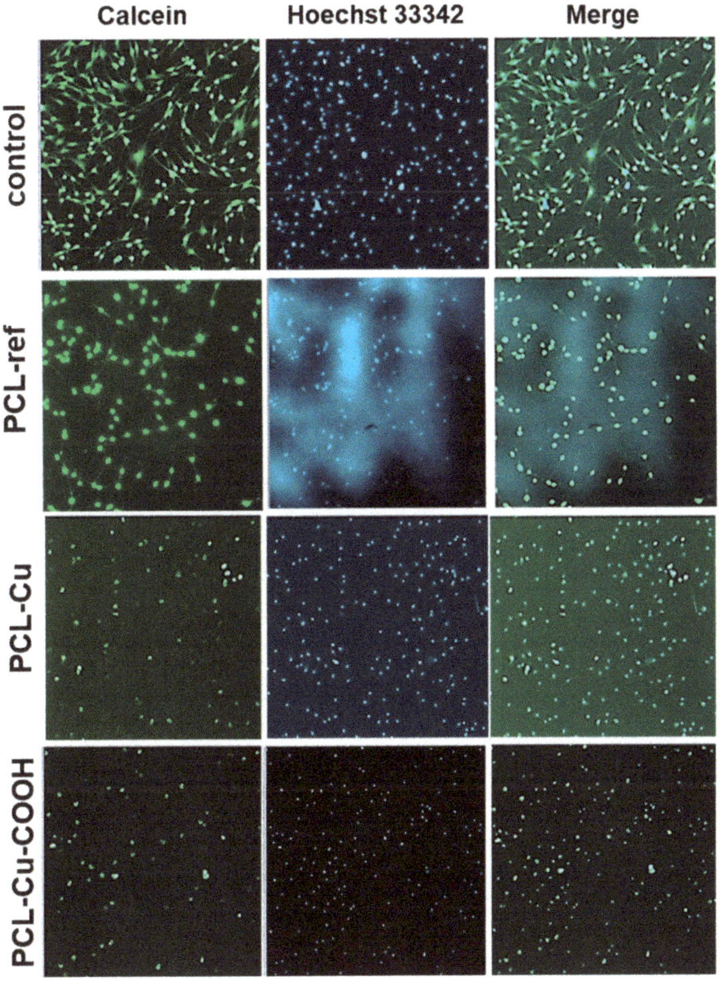


**Figure 13.** Mesenchymal stromal cells viability on the surface of PCL-ref, PLC-Cu, PCL-Cu-COOH, and on cultural plastic (control) after 3 days of cultivation. PCLs were preliminarily incubated in a culture medium for 1 h. Live cells were determined using calcein-AM, which changes from a non-fluorescent state to a fluorescent state (green) due to enzymatic conversion of intracellular esterases. The cell nuclei were stained by Hoechst 33342 (blue). Magnification 40x.

## 4. Discussion

The antibacterial properties of copper have been known since the ancient period. Now, in the era of antibiotic resistance, researchers again attract much attention, including the development of Cu-based antibacterial materials. Various forms of copper are used: nanoparticles [35], Cu containing wound dressing [6,35], soluble $Cu^{2+}$ salts, copper peroxide $CuO_2$, or copper hydroxide $Cu(OH)_2$. The antibacterial properties of the designed materials described in the literature differ significantly. Thus, when comparing the antibacterial action of $CuO_2$ and $Cu(OH)_2$, it was shown that $Cu(OH)_2$ has a significantly lower antibacterial activity compared to $CuO_2$ [36]. It is interesting to note that sometimes the antibacterial activity is not directly related to the concentration of copper ions. Thus, the authors demonstrated the antibacterial activity decreased with an increase of the Cu concentration [37]. In contrast, Lei et al. demonstrated that polyurethane nanofibers containing 5 wt% of CuO have no antibacterial effect and a minimum of 10 wt% of CuO is required to stimulate antibacterial effect against *E.coli* [38].

Our work demonstrated the very rapid dissolution of the Cu layer from Cu-coated PCL nanofibers and its antibacterial activity against Gram-negative *E. coli* and Gram-positive *S. aureus*. However, we did not find significant activity against the *S. typhimurium* and *Ps. aeruginosa* strains. This phenomenon is most probably related to different sensitivity towards copper for different strains. Bacteria have a number of defense mechanisms against the toxic effects of copper ions: the relative impermeability of the outer and inner membranes of the cell, which leads to a restriction of the intake (sequestration) of copper inside and the inner bacterial membranes to copper ions (intracellular sequestration), metallothionein-like proteins that absorb copper, in the cytoplasm and periplasm and energy-dependent efflux, precipitation as CuS, and extracellular complexation. Also in the cytoplasm and periplasm, some proteins actively remove copper from cells. For *S.typhimurium* and *Ps. aeruginosa* the following are known as the next copper-resistant proteins: copper detection proteins (SctR GolS, CueP for *S.typhimurium*), copper efflux proteins (GolT for *S.typhimurium* and CopA1, CopA2 for *PS. aeruginosa*) and copper sequestration proteins (CueP for *S.typhimurium*) [39,40]. The multicomponent copper efflux system CusCFBA and the multicopper oxidase CueO control the copper level and redox state in the periplasmic space, respectively [41,42]. It was shown that minimum inhibitory concentrations (MICs) of Cu against *S.typhimurium* are 8-16 times higher than for *E.coli* [42].

As a result, the MIC of *Ps. aeruginosa* to copper ions is reported at 44.8 μg/mL, significantly higher than the concentration released from nanomaterials (<10 μg/mL). Hence, possibly, the $Cu^{2+}$ concentration leached from our PCL-Cu layers is insignificant in inducing the antibacterial effects against *S.typhimurium* and *Ps. aeruginosa*.

Our work demonstrated that largely different amounts of copper are dissolved by distilled water, phosphate buffer, and culture media. We showed that the highest dissolution rate was visually observed for PCL-Cu and PCL-COOH samples soaked in culture media. The increased rate of copper release in the nutrient medium is most likely associated with the presence of chelating compounds in it, including low-density lipoproteins, proteins (for example, ceruloplasmin), which is the carrier of copper in the body. Using various media, it is possible to control the rate of release of copper ions and, accordingly, the antibacterial effect.

Molteni et al. compared the release of copper ions in various media [41]. The authors demonstrated that the highest copper release rate was in Tris-HCl and M17 media equal 2688 and 896 μg/mL, respectively. The $Cu^{2+}$ release in PBS and water was only 3.5 and 3.6 μg/mL, respectively. Hashmi et al. achieved $Cu^{2+}$ concentrations up to 33.98 ppm (33.98 μg/mL) for PAN nanofibers with CuO soaked in deionized water during 72 h [21]. Hence, the media that is used for $Cu^{2+}$ antibacterial analysis plays a significant role and affects the results.

It is worth noting that the dynamics of copper precipitation from the surface also differ. In some cases, there is a prolonged exposure for several days. In our work, similarly as in a majority of other reports [6], the rapid release of copper was observed in the first hours, followed by reaching a plateau.

Since the materials developed in our work have a great potential for biomedical use, including regenerative medicine, they must exhibit biocompatibility and activate the regenerative activity of cells. The PCL-Cu and PCL-Cu-COOH were tested for their biocompatibility towards the mesenchymal stromal cells. As mentioned earlier, the rate of copper release from the surface differs significantly depending on the medium. For MSCs cultivation we use standard culture media rich in ions, protein factors, and amino acids, which contribute to the copper extraction and binding. As a result, we have shown that seeded MSCs on PCL-Cu and PCL-Cu-COOH had a high death rate due to the high concentration of copper ions. Since the release of copper ions is fast enough, its concentration is high and cells seeded on fresh PCL-Cu/PCL-CU-COOH nanofibers have a high percentage of death. So in the reference [20], the concentration of copper released during the first 3 days is about 0.6 μg/mL, while in our work, in the first 2 h it varies from 2.9 to 3.5 μg/mL (for samples soaked in PBS). Accordingly, we experimented to assess the biocompatibility of these materials when they were previously soaked in a medium for 1 h. As a result, it was shown that cell survival increases significantly.

The influence of copper on tissue regeneration is enormous, thus the nanofibers developed in this work have excellent prospects for future research. Therefore, it will be necessary to check the effect of low pH environment, purulent wounds, with a high hydrogen peroxide content, which may increase the antibacterial activity. In this case, the high rate of copper release plays a positive role, since after cleaning the wound, $Cu^{2+}$ concentration will decrease to the optimal level for cell regeneration. It is known that copper ions are required for collagen synthesis, stimulate VEGF production, and enhance angiogenesis [43]. Moreover, angiogenin-bound copper is a potent inducer of blood vessel development, and it binds to endothelial cell receptors and extracellular matrix components [44]. Since in our work, we also developed PCL-Cu-COOH (containing active COOH groups), which promotes strong binding of protein factors and facilitates the proliferative activity of cells [45,46]. In the future, it is planned to test the effect of the developed nanofibers on angiogenesis. It is assumed that applying platelet reach plasma (PRP) with angiogenin will reduce the toxicity of copper while having regenerative activity.

## 5. Conclusions

The Cu-coated PCL nanofibrous mats were successfully prepared by using a robust and scalable approach based on Cu magnetron sputter-deposition onto electrospun polymer nanofibers. For the first time, the large-scale modeling of PCL films irradiation by molecular dynamics simulation was performed and allowed to predict the ions penetration depth and tune the deposition conditions. The copper-coated PCL nanofibers exhibited the antibacterial effect against *E. coli* and *S. aureus* due to a fast release of $Cu^{2+}$ ions (concentration up to 3.4 μg/mL), and sufficient biocompatibility. Thus, they may demonstrate an interesting synergistic effect when applied for wound healing. On the one hand, the rapid release of copper ions kills bacteria, while on the other hand, it stimulates the regeneration with the activation of immune cells, because copper ions are necessary for the bacteriostatic action of cells of the immune system [47]. The effect of Cu layer oxidation upon contact with liquid media was investigated by X-ray photoelectron spectroscopy revealing that, after two hours, 55% of Cu atoms are in form of CuO or $Cu(OH)_2$. The reactive COOH groups available at the surface of PCL-Cu coated with carboxyl plasma polymers may be used for grafting viable proteins, peptides, or drugs. It further explores the versatility of developed nanofibers for biomedical applications use.

**Author Contributions:** Conceptualization, A.M.M. and A.O.S.; methodology, A.O.S. and A.M.M.; biological testing N.A.S., A.R.T., A.Y.A. and L.S.A.; SEM and XPS experiments, L.B.; Cu coating, M.E.; plasma coatings, E.P.; draft preparation, A.M.M., A.O.S. and Z.I.P.; writing—review and editing, L.Z. and M.E.; modeling V.S.B. and Z.I.P. All authors have read and agreed to the published version of the manuscript.

**Funding:** The reported study was funded by Russian Foundation for Basic Research RFBR (project number 20-52-26020) and the Czech Science Foundation (project number GACR 21-12132J).

**Institutional Review Board Statement:** The use of human mesenchymal stromal cells was approved by the Ethics Committee of the RICEL-branch of ICG SB RAS (No 115 from 24 December 2015).

**Informed Consent Statement:** The data of this article were obtained using MSC cell line isolated back in 2015 ( the Ethics Committee approved the study of the RICEL-branch of ICG SB RAS (No 115 from 24 December 2015)) and frozen in liquid nitrogen. Humans did not involve in the work of this article.

**Data Availability Statement:** Not applicable.

**Acknowledgments:** Z.I.P. and V.S.B. are grateful to the Information Technology Center of Novosibirsk State University for providing access to the cluster computational resources. M.E., L.Z. and L.B. acknowledge CzechNanoLab Research Infrastructure supported by MEYS CR (LM2018110).

**Conflicts of Interest:** The authors declare no conflict of interest. The funders had no role in the design of the study; in the collection, analyses, or interpretation of data; in the writing of the manuscript, or in the decision to publish the results.

## References

1. Xie, L.; Li, Z.; Li, X.; Wenlong, W.; Hanjiang, Y. Electrospun copper oxide nanofibers and catalysis for combustion of ammonium perchlorate. *Ferroelectrics* **2019**, *549*, 23–28. [CrossRef]
2. Wang, H.; Fu, Q.; Zhang, G.; Sun, Y. The Synthesis of Ni–Cu Alloy Nanofibers via Vacuum Thermal Co-reduction Toward Hydrogen Generation from Hydrazine Decomposition. *Catal. Lett.* **2019**, *149*, 77–83. [CrossRef]
3. Zheng, X.; Shen, Z.; Cheng, C.; Shi, L.; Cheng, R.; Dong, J. Electrospinning Cu–$TiO_2$ nanofibers used for photocatalytic disinfection of bacteriophage f2: Preparation, optimization and characterization. *RSC Adv.* **2017**, *7*, 52172–52179. [CrossRef]
4. Cui, W.Y.; Yoo, H.J.; Li, Y.G.; Baek, C.; Min, J. Electrospun Nanofibers Embedded with Copper Oxide Nanoparticles to Improve Antiviral Function. *J. Nanosci. Nanotechnol.* **2021**, *21*, 4174–4178. [CrossRef]
5. Abedalwafa, M.A.; Zhang, H.; Mei, Q.; Li, Y.; Wang, F. A novel method to construct antimicrobial surface by decorating polyacrylonitrile nanofibrous membrane with nanoparticles. *J. Ind. Text.* **2020**. [CrossRef]
6. Ghasemian Lemraski, E.; Jahangirian, H.; Dashti, M.; Khajehali, E.; Sharafinia, M.S.; Rafiee-Moghaddam, R.; Webster, T.J. Antimicrobial Double-Layer Wound Dressing Based on Chitosan/Polyvinyl Alcohol/Copper: In vitro and in vivo Assessment. *Int. J. Nanomed.* **2021**, *16*, 223–235. [CrossRef]
7. Choi, S.-W.; Katoch, A.; Zhang, J.; Kim, S.S. Electrospun nanofibers of $CuOSnO_2$ nanocomposite as semiconductor gas sensors for $H_2S$ detection. *Sens. Actuators B Chem.* **2013**, *176*, 585–591. [CrossRef]
8. Ju, W.; Jiang, F.; Ma, H.; Pan, Z.; Zhao, Y.; Pagani, F.; Rentsch, D.; Wang, J.; Battaglia, C. Electrocatalytic Reduction of Gaseous $CO_2$ to CO on Sn/Cu-Nanofiber-Based Gas Diffusion Electrodes. *Adv. Energy Mater.* **2019**, *9*, 1901514. [CrossRef]
9. Nami-Ana, S.F.; Nasresfahani, S.; Tashkhourian, J.; Shamsipur, M.; Zargarpour, Z.; Sheikhi, M.H. Nanofibers of Polyaniline and Cu(II)–L-Aspartic Acid for a Room-Temperature Carbon Monoxide Gas Sensor. *ACS Appl. Mater. Interfaces* **2021**, *13*, 39791–39805. [CrossRef] [PubMed]
10. Khan, M.Q.; Kharaghani, D.; Nishat, N.; Sanaullah; Shahzad, A.; Hussain, T.; Kim, K.O.; Kim, I.S. The fabrications and characterizations of antibacterial PVA/Cu nanofibers composite membranes by synthesis of Cu nanoparticles from solution reduction, nanofibers reduction and immersion methods. *Mater. Res. Express* **2019**, *6*, 075051. [CrossRef]
11. Liu, X.; Xu, H.; Zhang, M.; Yu, D.-G. Electrospun Medicated Nanofibers for Wound Healing: Review. *Membranes* **2021**, *11*, 770. [CrossRef] [PubMed]
12. Mousavi, S.; Nejad, Z.M.; Hashemi, S.A.; Salari, M.; Gholami, A.; Ramakrishna, S.; Chiang, W.; Lai, C.W. Bioactive Agent-Loaded Electrospun Nanofiber Membranes for Accelerating Healing Process: A Review. *Membranes* **2021**, *11*, 702. [CrossRef] [PubMed]
13. Essa, W.; Yasin, S.; Saeed, I.; Ali, G. Nanofiber-Based Face Masks and Respirators as COVID-19 Protection: A Review. *Membrranes* **2021**, *11*, 250. [CrossRef]
14. Ullah, S.; Ullah, A.; Lee, J.; Jeong, Y.; Hashmi, M.; Zhu, C.; Joo, K., II; Cha, H.J.; Kim, I.S. Reusability Comparison of Melt-Blown vs Nanofiber Face Mask Filters for Use in the Coronavirus Pandemic. *ACS Appl. Nano Mater.* **2020**, *3*, 7231–7241. [CrossRef]
15. Mozaffari, A.; Gashti, M.P.; Mirjalili, M.; Parsania, M. Argon and Argon–Oxygen Plasma Surface Modification of Gelatin Nanofibers for Tissue Engineering Applications. *Membranes* **2021**, *11*, 31. [CrossRef] [PubMed]
16. Ramírez-Cedillo, E.; Ortega-Lara, W.; Rocha-Pizaña, M.; Gutierrez-Uribe, J.; Elías-Zúñiga, A.; Rodríguez, C. Electrospun Polycaprolactone Fibrous Membranes Containing Ag, $TiO_2$ and $Na_2Ti_6O_{13}$ Particles for Potential Use in Bone Regeneration. *Membranes* **2019**, *9*, 12. [CrossRef] [PubMed]
17. Haider, M.K.; Ullah, A.; Sarwar, M.N.; Yamaguchi, T.; Wang, Q.; Ullah, S.; Park, S.; Kim, I.S. Fabricating Antibacterial and Antioxidant Electrospun Hydrophilic Polyacrylonitrile Nanofibers Loaded with AgNPs by Lignin-Induced In-Situ Method. *Polymers* **2021**, *13*, 748. [CrossRef] [PubMed]
18. Haider, M.K.; Sun, L.; Ullah, A.; Ullah, S.; Suzuki, Y.; Park, S.; Kato, Y.; Tamada, Y.; Kim, I.S. Polyacrylonitrile/Carbon Black nanoparticle/Nano-Hydroxyapatite (PAN/nCB/HA) composite nanofibrous matrix as a potential biomaterial scaffold for bone regenerative applications. *Mater. Today Commun.* **2021**, *27*, 102259. [CrossRef]
19. Phan, D.-N.; Dorjjugder, N.; Saito, Y.; Khan, M.Q.; Ullah, A.; Bie, X.; Taguchi, G.; Kim, I.-S. Antibacterial mechanisms of various copper species incorporated in polymeric nanofibers against bacteria. *Mater. Today Commun.* **2020**, *25*, 101377. [CrossRef]
20. Haider, A.; Kwak, S.; Gupta, K.C.; Kang, I.-K. Antibacterial Activity and Cytocompatibility of PLGA/CuO Hybrid Nanofiber Scaffolds Prepared by Electrospinning. *J. Nanomater.* **2015**, *2015*, 1–10. [CrossRef]
21. Hashmi, M.; Ullah, S.; Kim, I.S. Copper oxide (CuO) loaded polyacrylonitrile (PAN) nanofiber membranes for antimicrobial breath mask applications. *Curr. Res. Biotechnol.* **2019**, *1*, 1–10. [CrossRef]
22. Permyakova, E.S.; Kiryukhantsev-Korneev, P.V.; Gudz, K.Y.; Konopatsky, A.S.; Polčak, J.; Zhitnyak, I.Y.; Gloushankova, N.A.; Shtansky, D.V.; Manakhov, A.M. Comparison of Different Approaches to Surface Functionalization of Biodegradable Polycaprolactone Scaffolds. *Nanomaterials* **2019**, *9*, 1769. [CrossRef]

23. Plimpton, S. Fast Parallel Algorithms for Short-Range Molecular Dynamics. *J. Comput. Phys.* **1995**, *117*, 1–19. [CrossRef]
24. Lloyd, A.; Cornil, D.; van Duin, A.C.T.; van Duin, D.; Smith, R.; Kenny, S.D.; Cornil, J.; Beljonne, D. Development of a ReaxFF potential for Ag/Zn/O and application to Ag deposition on ZnO. *Surf. Sci.* **2016**, *645*, 67–73. [CrossRef]
25. Kohn, W.; Sham, L.J. Self-Consistent Equations Including Exchange and Correlation Effects. *Phys. Rev.* **1965**, *140*, A1133–A1138. [CrossRef]
26. Hohenberg, P.; Kohn, W. Inhomogeneous Electron Gas. *Phys. Rev.* **1964**, *136*, B864–B871. [CrossRef]
27. Kresse, G.; Hafner, J. Ab initio molecular-dynamics simulation of the liquid-metal–amorphous-semiconductor transition in germanium. *Phys. Rev. B* **1994**, *49*, 14251–14269. [CrossRef] [PubMed]
28. Kresse, G.; Furthmüller, J. Efficiency of ab-initio total energy calculations for metals and semiconductors using a plane-wave basis set. *Comput. Mater. Sci.* **1996**, *6*, 15–50. [CrossRef]
29. Chatani, Y.; Okita, Y.; Tadokoro, H.; Yamashita, Y. Structural Studies of Polyesters. III. Crystal Structure of Poly-ε-caprolactone. *Polym. J.* **1970**, *1*, 555–562. [CrossRef]
30. Claros, M.; Setka, M.; Jimenez, Y.P.; Vallejos, S. AACVD Synthesis and Characterization of Iron and Copper Oxides Modified ZnO Structured Films. *Nanomaterials* **2020**, *10*, 471. [CrossRef]
31. Parmigiani, F.; Sangaletti, L. The Cu2p X-ray photoelectron core-lines in copper oxide based high temperature superconductors. *J. Electron Spectros. Relat. Phenom.* **1994**, *66*, 223–239. [CrossRef]
32. Biesinger, M.C. Advanced analysis of copper X-ray photoelectron spectra. *Surf. Interface Anal.* **2017**, 1325–1334. [CrossRef]
33. Nakamura, T.; Tomizuka, H.; Takahashi, M.; Hoshi, T. Methods of Powder Sample Mounting and Their Evaluations in XPS Analysis. *Hyomen. Kagaku* **1995**, *16*, 515–520. [CrossRef]
34. Manakhov, A.; Kiryukhantsev-Korneev, P.; Michlíček, M.; Permyakova, E.; Dvořáková, E.; Polčák, J.; Popov, Z.; Visotin, M.; Shtansky, D.V. Grafting of carboxyl groups using $CO_2/C_2H_4/Ar$ pulsed plasma: Theoretical modeling and XPS derivatization. *Appl. Surf. Sci.* **2018**, *435*, 1220–1227. [CrossRef]
35. Yang, Z.; Peng, H.; Wang, W.; Liu, T. Copper (II) Ions and Copper Nanoparticles-Loaded Chemically Modified Cotton Cellulose Fibers with Fair Antibacterial Properties. *J. Appl. Polym. Sci.* **2009**, *113*, 757–766.
36. Cui, H.; Liu, M.; Yu, W.; Cao, Y.; Zhou, H.; Yin, J.; Liu, H.; Que, S.; Wang, J.; Huang, C.; et al. Copper Peroxide-Loaded Gelatin Sponges for Wound Dressings with Antimicrobial and Accelerating Healing Properties. *ACS Appl. Mater. Interfaces* **2021**, *13*, 26800–26807. [CrossRef]
37. Fowler, L.; Engqvist, H.; Öhman-Mägi, C. Effect of Copper Ion Concentration on Bacteria and Cells. *Materials* **2019**, *12*, 3798. [CrossRef]
38. Lei, J.; Yao, G.; Sun, Z.; Wang, B.; Yu, C.; Zheng, S. Fabrication of a novel antibacterial TPU nanofiber membrane containing Cu-loaded zeolite and its antibacterial activity toward Escherichia coli. *J. Mater. Sci.* **2019**, *54*, 11682–11693. [CrossRef]
39. Chaturvedi, K.S.; Henderson, J.P. Pathogenic adaptations to host-derived antibacterial copper. *Front. Cell. Infect. Microbiol.* **2014**, *5*, 1–12. [CrossRef]
40. Achard, M.E.S.; Tree, J.J.; Holden, J.A.; Simpfendorfer, K.R.; Wijburg, O.L.C.; Strugnell, R.A.; Schembri, M.A.; Sweet, M.J.; Jennings, M.P.; McEwan, A.G. The multi-copper-ion oxidase CueO of Salmonella enterica serovar typhimurium is required for systemic virulence. *Infect. Immun.* **2010**, *78*, 2312–2319. [CrossRef]
41. Figueiredo, R.; Card, R.M.; Nunez-Garcia, J.; Mendonça, N.; Da Silva, G.J.; Anjum, M.F. Multidrug-Resistant Salmonella enterica Isolated from Food Animal and Foodstuff May Also Be Less Susceptible to Heavy Metals. *Foodborne Pathog. Dis.* **2019**, *16*, 166–172. [CrossRef]
42. Molteni, C.; Abicht, H.K.; Solioz, M. Killing of bacteria by copper surfaces involves dissolved copper. *Appl. Environ. Microbiol.* **2010**, *76*, 4099–4101. [CrossRef] [PubMed]
43. Harris, E.D.; Rayton, J.K.; Balthrop, J.E.; DiSilvestro, R.A.; Garcia-de-Quevedo, M. Copper and the synthesis of elastin and collagen. *Ciba Found. Symp.* **1980**, *79*, 163–182.
44. Cucci, L.M.; Satriano, C.; Marzo, T.; La Mendola, D. Angiogenin and copper crossing in wound healing. *Int. J. Mol. Sci.* **2021**, *22*, 704. [CrossRef]
45. Manakhov, A.; Permyakova, E.; Ershov, S.; Miroshnichenko, S.; Pykhtina, M.; Beklemishev, A.; Kovalskii, A.; Solovieva, A. XPS Modeling of Immobilized Recombinant Angiogenin and Apoliprotein A1 on Biodegradable Nanofibers. *Nanomaterials* **2020**, *10*, 879. [CrossRef]
46. Solovieva, A.; Miroshnichenko, S.; Kovalskii, A.; Permyakova, E.; Popov, Z.; Dvořáková, E.; Kiryukhantsev-Korneev, P.; Obrosov, A.; Polčak, J.; Zajíčková, L.; et al. Immobilization of Platelet-Rich Plasma onto COOH Plasma-Coated PCL Nanofibers Boost Viability and Proliferation of Human Mesenchymal Stem Cells. *Polymers* **2017**, *9*, 736. [CrossRef] [PubMed]
47. Chaturvedi, K.S.; Hung, C.S.; Crowley, J.R.; Stapleton, A.E.; Henderson, J.P. The siderophore yersiniabactin binds copper to protect pathogens during infection. *Nat. Chem. Biol.* **2012**, *8*, 731–736. [CrossRef]

membranes

Review

# Electrospun Antibacterial Nanomaterials for Wound Dressings Applications

Aysegul Gul [1], Izabela Gallus [2], Akshat Tegginamath [3], Jiri Maryska [2] and Fatma Yalcinkaya [2,*]

1 Institute for Nanomaterials, Advanced Technology and Innovation, Technical University of Liberec, Studentska 1402/2, 46117 Liberec, Czech Republic; aysegul.gul@tul.cz
2 Faculty of Mechatronics, Informatics and Interdisciplinary Studies, Technical University of Liberec, Studentska 1402/2, 46117 Liberec, Czech Republic; izabela.gallus@tul.cz (I.G.); jiri.maryska@tul.cz (J.M.)
3 Faculty of Mechanical Engineering, Technical University of Liberec, Studentska 1402/2, 46117 Liberec, Czech Republic; akshattm93@gmail.com
* Correspondence: fatma.yalcinkaya@tul.cz; Tel.: +420-485353389

**Abstract:** Chronic wounds are caused by bacterial infections and create major healthcare discomforts; to overcome this issue, wound dressings with antibacterial properties are to be utilized. The requirements of antibacterial wound dressings cannot be fulfilled by traditional wound dressing materials. Hence, to improve and accelerate the process of wound healing, an antibacterial wound dressing is to be designed. Electrospun nanofibers offer a promising solution to the management of wound healing, and numerous options are available to load antibacterial compounds onto the nanofiber webs. This review gives us an overview of some recent advances of electrospun antibacterial nanomaterials used in wound dressings. First, we provide a brief overview of the electrospinning process of nanofibers in wound healing and later discuss electrospun fibers that have incorporated various antimicrobial agents to be used in wound dressings. In addition, we highlight the latest research and patents related to electrospun nanofibers in wound dressing. This review also aims to concentrate on the importance of nanofibers for wound dressing applications and discuss functionalized antibacterial nanofibers in wound dressing.

**Keywords:** nanofiber; nanomaterial; wound dressing; antibacterial; tissue engineering; biomedical; electrospinning

**Citation:** Gul, A.; Gallus, I.; Tegginamath, A.; Maryska, J.; Yalcinkaya, F. Electrospun Antibacterial Nanomaterials for Wound Dressings Applications. *Membranes* **2021**, *11*, 908. https://doi.org/10.3390/membranes11120908

Academic Editor: Andrea Ehrmann

Received: 26 October 2021
Accepted: 19 November 2021
Published: 23 November 2021



## 1. Introduction

The skin is the body's largest organ, covering the entire external surface, which shields the internal organs from germs and thus aids in the prevention of infections. However, cuts, burns, surgical incisions, and illnesses such as diabetes can affect the structure and function of this organ.

Skin is divided into two layers, the epidermis and dermis. The epidermis is responsible for the healing process of the skin. A major part of the epidermal barrier is the stratum corneum, which plays an important role in this process. Several factors influence the health of the epidermal barrier, including the individual and the environment. The pH of the skin, the epidermal hydration, trans-epidermal water loss, and sebum excretion are the most important biophysical parameters that characterize the status of this barrier. In addition, the thickness of the epidermis's outer layer, the size of corneocytes, and the composition of superficial lipids all impact the regenerative properties of the skin, which contributes to the various courses of dermatological diseases during the healing process [1–3]. The understanding of biophysical skin processes could be useful in the development of wound dressing materials to restore barrier functionality.

Wound dressings serve three functions: (a) Absorption of wound secretions, (b) protection of the wound from injury, and (c) protection of the wound from bacterial contamination [4]. There are high rates of morbidity and mortality associated with skin and soft tissue infections (SSTIs). Although some SSTIs can be successfully treated with medication, those that affect the subcutaneous tissue, fascia, or muscle can delay the healing process and lead to life-threatening conditions resulting from the delayed healing process. This necessitates the use of more effective treatments [5].

Over the last few decades, a wide range of wound barrier materials have been studied, such as films, hydrogel, emulsions, composites, nano/microfibers, and so on [6–12]. Among them, nano/microfibers, in particular, have shown a promising future in wound dressing applications in recent years, making them very appealing to researchers. Figure 1 shows the growing number of publications in electrospinning for wound healing applications from the Web of Science database.

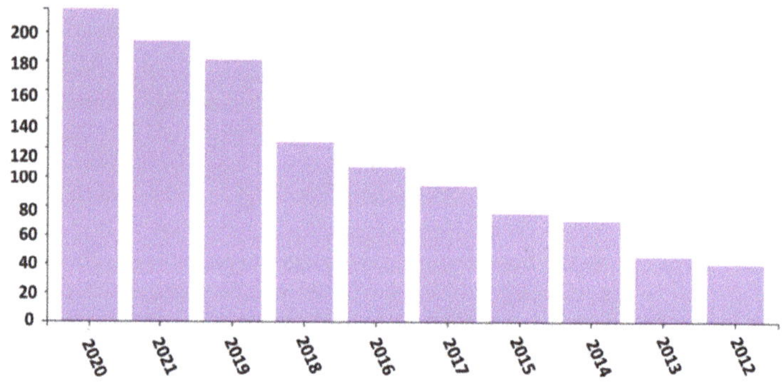


**Figure 1.** Recent publications related to electrospinning for wound dressing (September 2021).

Nano-sized materials have a high surface area/volume ratio, facilitating efficient drug encapsulation and controlled release kinetics. Furthermore, the physicochemical properties of nanomaterials, such as hydrophobicity, surface charge, or particle size, can easily be modified and can be specifically designed to mimic the extracellular matrix (ECM) or other cellular components while avoiding natural clearance mechanisms such as the immune system [13–15]. The ECM is crucial in controlling cell behavior and regulates the cells and sends environmental signals to them for site-specific cellular regulation and distinguishes one tissue area from another [16]. In the early 1960s, researchers speculated that nanometer-sized features influence cell behavior [17]. According to recent studies, cells attach better to fibers that are smaller in diameter than the diameters of the cells [18,19]. Hence, it is critical to replicate the natural ECM size to create an ideal dressing that functions as a synthetic ECM to guide the wound healing process. The application of nanometer-sized fibers in wound dressings has been demonstrated over and over again of for its value in medical healing treatments.

It is critical to figure out how to create an in-vivo-like architecture that supports cell growth and re-creation as closely as possible. Due to the various parameters that can be controlled, the process of electrospinning is of paramount importance in the production of nanofibers. Using the process of electrospinning, nanofibrous wound dressing materials can be produced that have diameters ranging from a few nanometers to hundreds of nanometers, along with specified pore size, porosity, and patterns and alignments to meet various requirements.

Basic wound dressing properties include absorbency, bacterial barrier, oxygen permeability (gas transfer), non-adhesion to healing tissue, and bioactivity, all provided by electrospun nanofiber structures [20,21]. Abrigo et al. [22] gave an evolution of electrospun wound dressings. This classification is based on the previous commercial dressing classifications: Passive, interactive, advanced, and bioactive. Passive meshes in wound dressings provide physical (i.e., water and gas permeability) and morphological (i.e., adequate porosity and nanometer-scale) properties. Interactive electrospun meshes combine the necessary morphological and physical requirements for wound healing with the value-added capability to address optimal cell responses and limit bacterial proliferation in the wound bed. The primary strategy used to develop interactive systems is a combination of synthetic polymers and biopolymers with antibacterial properties and an affinity towards ECM components. Multicomponent systems are more similar to the ECM. Many researchers are currently developing drug-loaded nanofibrous meshes to manufacture interactive dressings capable of treating bacterial infection. The goal of bioactive electrospun meshes is to be a multifunctional system that combines various properties that are capable of treating all aspects of the wound. Adequate mechanical and physicochemical properties protect the wound, stimulate the healing process, and control the bacterial load in the wound bed [23].

Researchers are currently experimenting with various strategies to create electrospun meshes that can support wound healing while preventing infection. Table S1 in the supplementary information lists the most recent patents for wound dressing materials using electrospinning methods.

In this review, the main process associated with electrospinning are described, wound dressings which are currently available are presented; the advances in the fabrication of electrospun meshes as wound dressings are highlighted, focusing on the current strategies for developing effective antibacterial nanofibrous wound dressing. Compared to previous papers, this review highlighted the most recent, up to date literature about functional nanofibers and their application in the wound healing process. Furthermore, the recent achievements, developments and current challenges in antibacterial nanofiber webs for the purpose of wound dressings are discussed.

## 2. Electrospinning Process (Parameters and Biomedical Applications)

Electrospinning is a voltage-driven technique in which a liquid droplet is electrified to create a jet, which is then stretched and elongated to create fibers. The main setup for electrospinning, shown in Figure 2, includes a spinneret (syringe needle) connected to a high-voltage (5 to 50 kV) supplier, a syringe pump, and a grounded or oppositely charged collector.

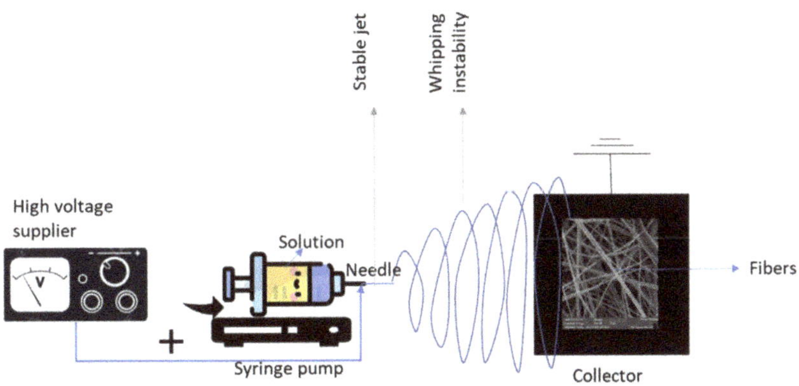


**Figure 2.** Electrospinning setup.

The liquid is extruded from the spinneret during electrospinning, producing a pendant droplet due to surface tension. When a droplet is electrified, electrostatic repulsion between surface charges with the same sign deforms it into a conical shape known as the Taylor cone, from which a charged jet is released. As soon as the electric field reaches a critical value (where the repulsive electric forces overcome the surface tension forces), a charged solution jet is ejected from the tip of the Taylor cone. Because of bending instabilities, the jet initially extends in a straight line and subsequently undergoes severe whipping motions.

An electric field can control the route of the jet as the jet is charged. As the jet flies in the air, the solvent evaporates, leaving behind a charged polymer fiber [23–26].

Certain factors have an impact on the electrospinning process. These factors are divided into three groups, as shown in Table 1. Researchers studied the effect of the controlling parameters, voltage, solution flow rate, concentration, molecular weight, distance, and solvent grade on the polymer jet's electric current and charge density during electrospinning. The viscosity of the solution has been found to influence the fiber diameter linked to the polymer concentration and molecular weight. Increasing the solution viscosity has been linked to the formation of larger-diameter fibers [24,25]. Solution conductivity is also linked to the voltage and effect on fiber diameter; the high solution conductivity results from thin fibers [26,27]. The molecular weight is linked to viscosity, surface tension, and conductivity, which affects fiber diameter; if it is low, bead structures form [28,29]. The applied voltage is linked to the tip-to-collector distance, conductivity, and feed rate. Higher voltage results in thinner fibers, but jet instabilities occur if the voltage is too high, resulting in thicker fibers [30]. Temperature is linked to viscosity, and an increase in temperature results in a decrease in fiber diameter thanks to a decrease in viscosity [31].

**Table 1.** Effecting parameters of electrospinning.

| Parameters | Effect on Fibers | References |
| --- | --- | --- |
| | Solution Parameters | |
| Viscosity | A higher viscosity results in a large fiber diameter. If the viscosity is very low, there will be no continuous fiber formation; if the viscosity is too high, the jet will be difficult to eject from the needle tip. | [24,32,33] |
| Solution Concentration | A minimum solution concentration is required for fiber formation in the electrospinning process. Increased concentration leads to larger diameters. | [34] |
| Molecular weight | Low molecular weight solutions tend to form beads rather than fibers, whereas high molecular weight nanofiber solutions produce fibers with a larger average diameter. | [25] |
| Solution electrical conductivity | When the electrical conductivity of the solution increases, the diameter of the electrospun nanofibers decreases significantly. Beads may also be observed due to the solution's low conductivity, which results in insufficient elongation of a jet by electrical force to produce uniform fiber. | [27,35] |
| Surface tension | The surface tension of the solution can drive droplets, beads, and fibers and the solution's low surface tension ensures that spinning occurs with a lower electric field requirement. | [36–38] |
| | Process Parameters | |
| Applied voltage | It has been discovered that increasing the electrostatic potential leads to thinner fibers. However, if too much voltage is applied, the jet may become unstable, and the fiber diameters may increase. | [39] |
| Distance from needle to the collector | The traveling time of the polymeric jet is affected. Traveling time should be long enough for complete evaporation of the solvent. | [40,41] |
| Volume feed rate | Increasing the feed rate resulted in an increase in fiber diameter and the formation of a bead structure. | [36,37] |
| | Environmental Parameters | |
| Humidity | High humidity can cause pores on the surface of the fiber. | [33,42] |
| Temperature | Temperature increases cause a decrease in fiber diameter due to a decrease in viscosity. | [42] |

Electrospun nanofibers are widely used in biomedical applications such as tissue-engineered scaffolds (vascular implants) [43], drug-delivery systems [44–46], and medical treatments in healthcare to improve wound healing [8,12,47]. Figure 3 shows the medical applications of electrospinning. Wound dressings are one of the most well-known of these applications. Thanks to the electrospinning technique, the fibers can be patterned or aligned to increase the contact efficiency of the cells. Furthermore, nanofibrous scaffolds have been shown to improve cell adhesion, protein adsorption, and cell growth and differentiation.

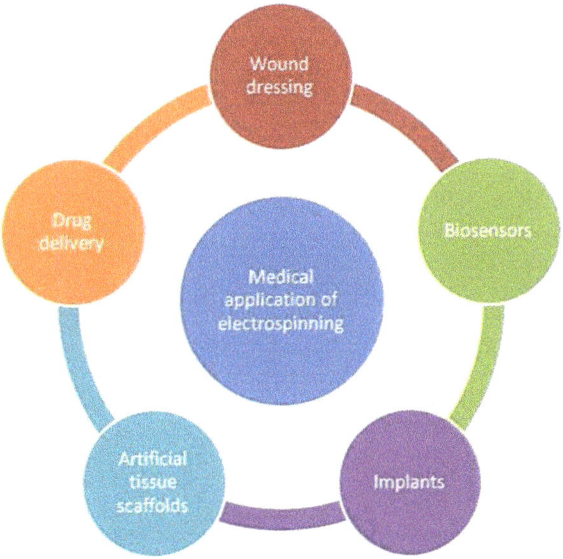


**Figure 3.** Medical applications of electrospinning.

Antibacterial nanofibers have received special attention. With the incorporation of antimicrobial agents, the design goal of wound dressing materials has been to avoid or reduce infection, which is the cause of bacteria. Antimicrobial nanofibrous wound dressings have recently emerged as a viable technique to decrease infection and wound bacterial colonization to improve the healing process (Table S2 shows recent studies for antibacterial electrospun wound dressings).

## 3. Antibacterial Nanofibers for Wound Dressing

One of the major causes of chronic infections can be linked to bacterial infections [48–51], which fester at a very high rate in existing wounds; thus, the need to use antibacterial materials is of paramount importance. With a large surface area, antibacterial nanofibers allow for the efficient integration of antibacterial agents [52]. In recent years, nanotechnology has advanced at a blistering pace. The areas of research under nanotechnology are also expanding at an exponential rate. One of the research areas under this revolution is nanomedicine, and over recent decades, this field has shown great potential of becoming a major field of research. Research in this field has led to drastic improvement of human health [53]. Several techniques have been utilized to produce nanofibers, such as melt spinning, chemical vapor deposition, sinter technology, solution spinning, and electrospinning. Among these techniques, the electrospinning technique has been determined as the most cost-effective method in producing continuous nanofibers from numerous polymers or compounds [54–58]. The nanofibers produced via electrospinning have a large specific area, a high porosity, and huge interest in applications in tissue engineering, regenerative medicine, and wound dressing.

The foremost function of the skin is to protect the internal organs, muscles, and bones, which can be affected by burns, cuts, or illnesses. The process of healing a wound starts instantly when the skin is affected. The presence of bacteria will reduce the efficacy of healing the wound and increase the chances for an infection to occur and fester. The absence of Gram-positive organisms such as staphylococcus aureus and streptococcus pyogenes would exponentially increase the wound's healing rate. Thus, these microorganisms must be eliminated quickly. After surgery or an injury, the exposed tissues may be in danger of contracting an infection, which may lead to diseases, and in severe cases, it may even lead to death [48]. Thus, the dressing of the wounds would help prevent infections and maintain an environment conducive to healing wounds [59]. For wound healing, electrospun nanofibers have the following features that are imperative for their usage:

- Mimicry of the composition.
- Mimicry of the structure.
- Incorporation of bioactive materials.
- Mechanical mimicry.
- Regulation of the skin cell response [48].

### 3.1. Mimicry of the Composition

Various materials have been used in the field of wound healing, such as hydrogels, gas-foaming formed scaffolds, or decellularized porcine dermal matrices [60–62]. However, these materials cannot reproduce the skin's extracellular matrix (ECM) [63]. Electrospinning has found traction in recent years for wound healing, as it can be used to produce biomimetic nanofibers with the required features from numerous synthetic and natural polymers [64]. Collagens, laminins, elastins, proteoglycans, and polysaccharides are some of the proteins present in the ECM of skin [65]. Due to electrospinning's multifaceted nature, nanofibers of type I and III, which make up a major portion of the dermal matrix, can be produced [66]. By direct electrospinning, surface immobilization, or blending, electrospun nanofibers can be produced that have a high degree of similarities with the ECM of the skin. Table 2 shows the various electrospun nanofibers that can be utilized to recreate the ECM of the skin.

**Table 2.** Electrospun nanofibers mimicking the ECM of the skin.

| Composition | Approximate Diameter | Reference |
|---|---|---|
| Collagen | 460 nm | [66,67] |
| Collagen/chitosan | (134 ± 42) nm | [68] |
| Collagen/PCL | (170 ± 0.075) nm | [69] |
| Collagen/Zein | (423–910) nm | [70] |
| Collagen/elastin/PEO | (220–600) nm | [71] |
| Laminin I | (90–300) nm | [72] |
| PCL/gelatin | (470 ± 120) nm;<br>(409 ± 88) nm | [73,74] |
| Gelatin | (570 ± 10) nm | [75] |
| Polyurethane/gelatin | (0.4–2.1) µm | [76,77] |
| HA/PEO | (70–110) nm | [78] |
| Silk fibroin/chitosan | (185.5–249.7) nm | [48] |
| Silk fibroin/PEO | (414 ± 73) nm; 1 µm | [79,80] |
| Chitin | 163 nm | [81] |
| Carboxyethyl chitosan/PVA | (131–456) nm | [82] |
| Chitosan/gelatin | (120–220) nm | [83] |

**Table 2.** *Cont.*

| Composition | Approximate Diameter | Reference |
|---|---|---|
| PLGA | (150–225) nm | [84] |
| PLGA/collagen | (170–650) nm | [85] |
| Chitosan/PEO | (130–150) nm | [86] |
| Hyperbranched polyglycerol | (58–80) nm | [87] |

### *3.2. Mimicry of the Structure*

Upon observation under an electron microscope [88], human skin was found to have three zones (papillary, mid, and deep zones), which are composed of a fine layer of fibers near the epidermis with a thick layer of fiber bundles and a loosely arranged fiber bundle layer. The fiber bundles consist of parallelly aligned fibrils. It was later found that the collagen present in the skin has a basket-weave structure [89–91]. To achieve this structure, numerous attempts were made to produce electrospun nanofibers similar to it [92–94]; using the weaving techniques present in the industry, forays have been made to produce nanofiber yarns with a basket-weave [95,96]. Using a process called 'noobing', 3D nanofiber scaffolds with a basket-weave structure were produced [97].

### *3.3. Incorporation of Bioactive Materials*

With the introduction of therapeutic agents, the process of wound healing can be accelerated at the site of a wound. The local delivery of therapeutic agents such as antioxidants, anesthetics, enzymes, growth factors, and antimicrobial agents can be comprehensively achieved with the help of electrospun nanofibers [98]. The advantage of using electrospun nanofibers to deliver these agents over the commonly used drug delivery system is that the nanofibers have a fast response rate with greater control over the release rate [99,100]. The therapeutic agents can be introduced into the electrospun nanofibers via co-axial electrospinning or emulsion electrospinning [101,102]. The process of CO2 impregnation or infusion or surface immobilization can be utilized to introduce the therapeutic agents into the electrospun nanofibers [103,104]. Table S3 in Supplementary Materials lists therapeutic agents that can be incorporated with electrospun nanofibers.

### *3.4. Mechanical Mimicry*

The parameters of the scaffold can influence the process of tissue regeneration, and cellular behavior used [105]. Thus, the mechanical properties of late have come into the limelight [106]. Due to a low degree of orientation and extension of polymer chains, electrospun nanofibers have low tensile strength and Young's modulus [107]. Thus, it is of preponderant importance to select the appropriate raw material that can encompass the desired properties [108]. Surface coating, mechanical treatments, and thermal treatments can be utilized to introduce the required properties into the electrospun nanofibers [109–111]. In Table 3, a collection of nanofibers that come close to the mechanical properties of the human skin and a comparison with the mechanical properties is presented.

**Table 3.** Nanofibers mimic the mechanical properties of the human skin with a comparison.

| | Human Skin | PCL/Collagen | HA/PLGA | PLGA/Collagen |
|---|---|---|---|---|
| **Tensile modulus (MPa)** | 15–150 | 21.42 ± 0.04 | 28.0 | 40.43 ± 3.53 |
| **Ultimate tensile stress (MPa)** | 1–32 | 8.63 ± 1.44 | 1.52 | 1.22 ± 0.12 |
| **Ultimate tensile strain (%)** | 35–115 | 24.0 ± 7.16 | 60.07 | 96 ± 13 |
| **Reference** | [112] | [113] | [114] | [112] |

### 3.5. Regulation of the Skin Cell Response

For a wound to heal ECM deposition, skin cell proliferation and migration must take place. It was found that a spreading morphology was shown by cells when electrospun nanofibers with type I collagen, laminin, and integrin ligands were used for wound dressings [67]. In a study conducted by Yoo et al., it was found that the mRNA levels for loricin and keratin 1 were higher when the PCL nanofibers were cultured with keratinocytes and chemically conjugated human epidermal growth factors were utilized [115]. To increase re-epithelialization when scaffolds are used, aligned PVA nanofibers can be used, as they would assist the keratinocytes in the wound healing process [116].

## 4. Biopolymeric Nanofibrous Antibacterial Wound Dressings

The ecofriendly nature and biocompatibility of biopolymers are some of the characteristics due to which biopolymers are extensively studied to create wound dressings with the desired characteristics. Since the biopolymers show a high degree of similarity to the ECM structure, high bioactivity, and are biodegradable, polysaccharide biopolymers, among the many biopolymers utilized, are comprehensively studied. The following are the various biopolymers used in the study of wound healing:

- Collagen.
- Alginate.
- Chitosan.
- Gelatin [117].
- Fibronectin and fibrin [118].

Collagen is used in wound dressing due to the following reasons:

- Low antigenicity and inherent biocompatibility.
- Increase in fibroblast production and permeation.
- Helps to preserve leukocytes, macrophages, fibroblasts, and epithelial cells.
- Attracts fibroblasts and encourages the deposition of new collagen to the wound bed.

Collagen nanofiber webs are similar to native tissue architecture and are easily remodeled due to their simple structure, easy preparation, availability, and relative uniformity. Collagen nanofiber helps the healing process but does not show anti-bacterial properties. An antibacterial additive or treatment is needed. On the other hand, chitosan not only shows biocompatibility and biofunctionality but also antibacterial, analgesic, antioxidant, and neuroprotective properties. Electrospun chitosan nanofiber webs are promising candidates for wound healing.

Gelatin nanofibers are interesting for use in the wound healing process due to their biodegradable, easy to spin, controllable thickness, and physical stability properties. Gelatin nanofiber does not show antibacterial properties. However, mixing with antibacterial materials such as chitosan, curcumin, or nanoparticles can improve the antibacterial property of gelatin nanofibers [119–121].

To treat burn injuries, a cellulose nanofibril wood-based wound dressing has been developed. Cellulose is a very commonly available polysaccharide that helps speed up wound healing by providing assistance in the processes of epithelialization, granulation, and tissue regeneration [122]. Cellulose can be obtained from bacteria (Acetobacter xylinum) and plants. The cellulose obtained from the bacteria is called bacterial cellulose; this cellulose has great mechanical characteristics, biocompatibility, biodegradability, and physicochemical properties required to produce a wound dressing material [123]. Bacterial cellulose can be used to regenerate blood vessels, reconstruct the damaged tissues, and wound healing since it can mimic the structure of the ECM with great ease and similarities [124]. The cellulose-based wound dressing properties can be elevated by introducing antimicrobial drugs, hormones, antioxidants, and enzymes [125]. Gallic acid can be used to functionalize cellulose acetate nanofibers, as it is a polyphenol compound with antioxidant, anti-inflammatory, and antibacterial characteristics [126]. The ECM of vertebrates contains Hyaluronic acid (HA), a naturally occurring nonimmunogenic linear polysaccharide [127].

For wound healing, numerous hydrogels based on HA were examined. The HA was functionalized with thiol [128], glycidyl methacrylate [129], and DNA [130], to help with networking. The HA used in wound dressing materials mostly supports cellular migration, proliferation, and absorbing exudates, hence, leading to the regeneration of tissues and healing of the wound [131]. Shell fibers of HA core-poly (lactic-co-glycolic acid, PLGA) with epigallocatechin-3-0-gallate (EGCG) were produced and developed by Shin et al. [100] and used as a wound dressing on diabetic rats, and it was found that the HA/PLGA-E fibers used helped to increase the rate of the wound healing process. A blend of HA/poly (vinyl alcohol) (PVA) nanofibers was also developed for wound dressing [132]; here, the HA is carried by the PVA polymer along with the addition of hydroxypropyl-βcyclodextrin (HP βCD), which is used a stabilizing agent in electrospinning, to allow a water-based fabrication process. Due to a high degree of biocompatibility and biodegradability, chitosan (CS) and chitin are good options in developing wound dressing materials, with chitin being one of the most available natural amino polysaccharides whose production is equal to that of cellulose and can be found in fungi cell walls as well as the exoskeletons of crustaceans, insects, and invertebrates. For the purpose of wound dressing, PVA/CS/tetracycline hydrochloride (TCH) [133], honey/PVA/CS [134], and bacterial cellulose/CS/polyethylene oxide (PEO) CS-based antibacterial nanofibers have been suggested. With very low toxicity levels, good biocompatibility, and inexpensive cost, alginate, an anionic polymer derived naturally [132], can be utilized to produce wound dressings made from collagen alginate, gelatin alginate calcium alginate, and calcium sodium alginate [125]. When used for wound healing, alginate maintains appropriate levels of moisture and greatly reduces bacterial activity at the wound site and accelerates the wound's healing process [135]. It is blended with various synthetic polymers to produce an electrospun nanofibrous wound dressing based on alginate [136]. Collagen nanofibers were used to produce wound dressing materials by Zhou et al. [137], which were used to vitalize epidermal differentiation and human keratinocytes and increase the rate of healing of the wound. The collagen fibrils were paired with synthetic and natural polymers, which would help maintain the moisture and help absorb the exudate from the wound and accelerate the process of wound healing [138]. Yao et al. [139] developed a gelatin/keratin blended nanofiber wound dressing material, which enhanced the migration, adhesion, and cell proliferation leading to vascularization and healing of the wound, which was observed in the animal test model.

Natural biopolymers such as silk fibroin (SF) obtained from the mulberry silkworm, Bombyx mori, are utilized in biomedical applications due to their inexpensiveness, biocompatibility, biodegradability, green processing, and very low inflammatory response [140]. Along with these properties, SF has great exudate absorption capacity, pliability, and adherence. This can be used as a stand alone or combined with alginate, multiwalled carbon nanotubes, chitosan, etc. [125]. The skin's environment can be mimicked to a high degree by the SF, leading to an accelerated wound healing process and minimized scarring [141]. Thus, wound dressing materials based on SF are being researched and developed [142]. Antioxidant Fenugreek/SF nanofiber wound dressing material was fabricated by Selvaraj and Fathima [143], which, along with wound healing characteristics, also helps with collagen deposition and complete re-epithelialization. The potential for wound healing using biopolymeric nanofibers is excellent, but the properties offered are seldom enough to fulfill both disinfection and wound healing. Functional agents that help in accelerating the wound healing rates must be used, hence hybridizing the biopolymeric nanofibers [118].

## 5. Nanoparticle Containing Nanocomposite Antibacterial Nanofibers

The wound healing process is continually put on the line and tested with the presence of bacteria. When bacteria are present, they may lead to inflammation of the wound and delay the process of healing. Bioactive wound dressings are a new field of wound dressing and show great potential in displacing the conventional wound dressing methods [118]. Wound dressing materials can be modified with surface-functionalized agents, bio blends, and antibacterial nanocomposites or nanoparticles to have antibacterial action. In recent years, silver nanoparticles were used in polymeric nanofibers due to their ability to resist bacterial activity [144]. The wound dressing material physically shields the wound from bacterial activities and helps with the differentiation of fibroblasts and their migration at the wound site. According to the mode of loading and the type of antibacterial agent utilized, various types of wound dressings are present such as hydrogels, films, foams, or sponges [118].

An open wound is open for bacterial attacks, increasing inflammation and leading to long periods of wound healing. As a result, it would lead to impeding the production of new granulation tissues and damage the ECM's constituents. When an antimicrobial dressing is applied at the site of the wound, pathogens cannot enter the wound as their pathway is blocked, and those that entered prior to applying the wound dressing will be eliminated efficiently. Moreover, the immune system is induced to promote the migration of keratinocytes/fibroblasts, leading to faster wound healing [145].

Nanoparticles such as zinc oxide, silver, iron oxide, and gold are used for biodetection, medical devices, drug delivery, and wound healing [146]. Because of their ability to fight human pathogens, they can be used to design wound dressings. For this reason, metallic nanoparticles have recently attracted much interest from researchers. Silver nanoparticles are of particular interest. They have strong toxicity and a large surface area, increasing contact with pathogens [147]. Silver nanoparticles and silver complexes have already found wide use in producing antimicrobial materials and wound healing [6]. The incorporation of metal nanoparticles and metal oxide into the polymeric membrane structure is considered one of the better solutions for developing dressings with antimicrobial properties. Materials such as hydrogels, nanocomposites, and nanofibers have high porosity, excellent gas permeability, and a high surface-area-to-volume ratio. These are required in wound healing as they ensure proper cellular respiration, hemostasis, exudate removal, improved skin regeneration, and hydration [148]. In the design of wound healing materials, it is believed that the best strategy is to combine various non-conventional antimicrobial formulations in order to harness their synergistic effects to overcome microbial resistance [146]. Combining hydrogels, nanocomposites, or nanofibers with nanoparticles seems to be the optimal solution for creating wound healing materials. The introduction of Ag nanoparticles into the polymer structure can be carried out by different methods such as electrospinning, chemical modification, or hydrogel formation [149,150]. Hongli et al. managed to obtain porous silver nanoparticle/chitosan composites with wound healing activity by in situ reductions of silver nanoparticles with gelatin [6]. Kumar et al. created a chitin hydrogel/nano ZnO composite bandage [148], while Jatoi et al. obtained poly(vinyl alcohol) composite nanofibers embedded with silver-anchored silica nanoparticles [150]. Table 4 summarizes various research studies using nanocomposites, nanofibers, hydrogels, and nanoparticles to produce materials suitable for wound healing.

**Table 4.** An overview of recent wound dressing materials constructed from nanoparticles and nanomaterials.

| Material | Nanoparticles | Bacterial Species | Ref. |
|---|---|---|---|
| Carboxymethyl Chitosan/Polyethylene Oxide Nanofibers (CMCTS–PEO) | Ag (12–18 nm) | *S. aureus, P. aeruginosa, E. coli, fungus Candida albicans* | [149] |
| Alginate/Nicotinamide Nanocomposites | Ag (20–80 nm) | *S. aureus* and *E. coli* | [150] |
| Nanofibrous Poly vinyl alcohol, chitosan | Ag | *S. aureus* and *E. coli.* | [151] |
| Nanofibrous mats from cellulose acetate | Ag | *S. aureus* and *E. coli.* | [152] |
| Nanofibrous membrane from Gum Arabic, polycaprolactone, polyvinyl alcohol | Ag | *S. aureus, E. coli, P. aeruginosa* and *C. albicans* | [153] |
| PVA-co-PE nanofibrous membrane | Ag | *S. aureus* and *E. coli.* | [154] |
| Electrospun peppermint oil on polyethylene oxide/Graphene oxide | $CeO_2$ | *S. aureus* and *E. coli.* | [155] |

Table 4. *Cont.*

| Material | Nanoparticles | Bacterial Species | Ref. |
|---|---|---|---|
| Hyaluronic acid | ZnO | *S. aureus, B. subtilis, E. coli, P. aeruginosa,* and *V. cholerae* | [156] |
| Chitosan/cellulose acetate | $CeO_2$ | *S. aureus* and *E. coli.* | [157] |
| Chitosan/poly(N-vinylpyrrolidone) | $TiO_2$ | *E. coli, S. aureus, B. subtilis* and *P. aeruginosa* | [158] |
| Chitosan/pectin | $TiO_2$ | *E. coli, S. aereus, A. niger, B. subtilis, P. aeruginosa* | [159] |
| Electrospun Chitosan/Gelatin | $Fe_3O_4$ | *S. aureus* and *E. coli.* | [160] |
| B-Chitin Hydrogel | Ag (4–8 nm) | *S. aureus* and *E. coli.* | [161] |
| Chitosan/Polyvinyl Alcohol Hydrogel, Collagen | Ag (4–19 nm) | *P. aeruginosa* and *S. aureus* | [162] |
| Linseed hydrogel | Ag (10–35 nm) | *E. coli, S. mutans, A. niger, S. epidermidis, P. aeruginosa, S. aureus, acillus subtilis, Actinomyces odontolyticus* | [163] |

## 6. Biofunctionalized Antibacterial Nanofibers

Biofunctionalized antibacterial nanofibers are a type of wound dressing material where the biopolymeric nanofibers are surface functionalized with amino acids and antimicrobial peptides [118]. The two most important biopolymers are chitosan and silk fibroin when dealing with biofunctionalized nanomaterials, since they allow various antimicrobial agents to be attached via the numerous functional groups present. The antimicrobial peptides (AMPs) bound to the surface of the nanofibers are studied widely [164–166]. Due to the biocompatibility offered by the AMPs, they have now become one of the most utilized antimicrobial additives for wound dressings. This wound dressing system produced is a hybrid system, and the type of AMP tailors the antimicrobial activity of these hybrid systems utilized [118]. For the AMP to be immobilized on the surface of the nanofiber, numerous approaches are utilized. Co-spinning and covalent binding are approaches implemented in producing the nanofibers with AMP immobilized onto the surface [166]. The process of covalent immobilization provides the best process, as this leads to negligible leaching of the AMP and long-term stability and nontoxicity [167]. Various antibacterial biohybrid nanofibrous wound dressings are produced based on the surface functionality of silk fibroin (SF). On the SF nanofibers, various functional groups such as carboxyl, hydroxyl, phenol, and amines are loaded [168]. It has been observed that SF biohybrid nanofibers do not allow the growth of bacteria [118]. If the immobilized factor amount is higher, the antibacterial activity is higher. Over a period of three weeks, the effect of biofunctionalized nanofibers remains constant, disregarding the temperature of storage. It has been found that the bacteria S. Aureus can reduce the efficiency of AMPs by lowering the negative surface charge, changing the fluidity of their membrane, or using their pumps to keep the AMPs away [169].

Another biopolymer that is biocompatible and biodegradable is chitosan, and this has excellent antimicrobial properties against various microorganisms such as fungi, algae, viruses, and bacteria [170–174]. The electrostatic interactions of the amine groups present in chitosan undergo electrostatic interactions on the cell wall [172]; due to this, the permeability of the cell wall is altered. Hence, the osmotic balance is disrupted, which leads to the restriction of the growth of the microorganism. In addition, the leakage of intracellular electrolytes occurs due to the hydrolysis of peptidoglycans [175]. Due to this, several blends of functionalized chitosan nanofibers have been suggested [118]. The positive charge of the amino acids is the key factor in the protection against microorganisms; to this end, L-asparagine [176], L-arginine [177], and L-lysine [118] have been grafted onto the nanofibers of chitosan to increase the density of positive charge present. A wound dressing made from deacetylated/arginine functionalized chitosan has been developed [178]. The bio functional component helps with the higher deposition ability of collagen; this, in turn, helps with the healing of the wound at a greater rate [118]. Table 5 below shows examples of biofunctionalized antibacterial materials.

**Table 5.** Biofuntionalized antibacterial materials with proteins.

| Protein | Co-Polymer | Antimicrobial Agent | Bacterial Species | Ref. |
|---|---|---|---|---|
| Zein | PU | Ag NPs | *E. coli, S. aureus* | [151] |
| Zein | PU/CA | Streptomycin | *V. vulnificus, S. aureus, B. subtilis* | [177] |
| Keratin | PVA, PEO | Ag NPs | *E. coli, S. aureus* | [178] |
| Collagen | CS | ZnO | *S. aureus, E. coli* | [179] |
| α-lactoglobulin | PEO | Ampicillin | *E. coli, P. aeruginosa, B. thailandensis* | [180] |
| Silk fibroin | PEO | $TiO_2$ NPs | *E. coli* | [181] |
| Silk fibroin | - | Ag NP coating | *S. aureus, P. aeruginosa* | [182] |
| Silk fibroin | PEO | $Cu_2O$ NPs | *S. aureus, E. coli* | [47] |
| Lactoferrin | Gelatin | - | *E. coli, S. aureus* | [183] |
| Gelatin | Alginatedialdehyde | Ciprofloxacin, gentamicin | *P. aeruginosa, S. epidermidis* | [184] |

Proteins can be combined with polymer structures using electrospinning. However, this is a very challenging process due to their molecular weight, the ionic, hydrogen, and disulfide bonds present, and the complexity of their structure. During the electrospinning of proteins, the most important factor is the proper choice of solvent. One must consider their solubility in a given solvent and the degree of unfolding and entanglement of the protein chain. In addition, the solvent affects the fiber size, crystallinity, morphology, and mechanical properties of the protein. Therefore, adding a synthetic polymer during electrospinning is necessary for this to occur continuously and without interference. The production of wound healing dressings during electrospinning uses animal or plant-based proteins [185–188]. The activity, degradation, and stability of the material are determined by the proteins' size, chemical structure, purification process, and protein isolation. The purity and composition of the obtained raw material affect the reproducibility of the electrospinning process and the properties of the final product [47,188].

Silver-based compounds have been used since the early 1970s for wound care applications [189] and hence the combination of silver with sulphadiazine was established, which led to the usage of silver in wound dressings [190]. The active antimicrobial entity in wound dressings that makes use of silver is the silver ion, and these ions react with the thiol (-SH) groups, leading to the generation of reactive oxygen species (ROS), and this is the major contributor to the antibacterial efficacy of the wound dressing used. Silver ions, when released, have the potential to cross various biological partitions [189].

Due to the history of the usage of silver in therapeutic agents, the potential toxicity of silver is a well-documented fact. The ingestion or dermal exposure or inhalation of salts of silver in sufficient amounts lead to Argyria and Argyrosis, which is blue–grey discoloration of the skin and eyes. This occurs mainly due to the deposition of the silver precipitates. Although argyria is not toxic in nature, it leads to disfigurement and, hence, this is considered an undesirable effect [191,192]. Historical studies have shown that high dosages of silver nitrate lead to gastrointestinal damage and rarely lead to fatalities [191]. There has been little to no evidence to support the fact that silver in any form might be toxic in nature to the cardiovascular or immune, reproductive, or nervous systems in humans [193,194]. A threshold limit value of 0.01 mg/m$^3$ for metallic silver in soluble form and 0.1 mg/m$^3$ for metallic silver has been set by the American Conference of Governmental Industrial Hygienists (ACGIH). These values have been set based on the limit values for protection against Argyria [195].

Liu et al. [196] conducted a cytotoxicity study of the nanofibrous membranes produced from PEU and CA for 3 days in in vitro conditions with rat skin fibroblast cells according to DS/EN ISO10993-5 [197]. The results of these tests showed that the pure PEU and co-spun PEU/CA nanofibers containing PHMB had no toxicity towards the fibroblast cells of the rats, as the cells showed adhesion to the nanofibrous membranes and showed growth. Thus, this led to the conclusion that the polymers used were biocompatible and safe to use as wound dressing materials.

To determine the biocompatibility of nanofiber-based wound dressing materials, clinical trials have to be undertaken to gain extensive knowledge, but the number of studies being undertaken at the clinical phase is very limited [198]. The number of clinical trials to determine the effects of electrospun nanofibers can be found on the clinical trial website [199].

## 7. Conclusions

The interest in electrospun nanofiber mats has risen drastically due to their unique properties such as high specific surface area, highly porous structure, tight pore size and pore size distribution, interconnected pores, and good chemical and biological activity.

Herein, we have briefly reviewed the role of the nanofiber web in wound dressing applications. For an ideal wound dressing, a future perspective, the requirements are:

- Nontoxic to mammal cells.
- Nonantigenic.
- Good mechanical resistance.
- Elastic and flexible.
- Antibacterial.
- Permeable for gas exchange.
- Inexpensive.
- Long shelf-life.

Incorporating functional nanoparticles or bioactive agents into nanofibers improves the antibacterial property of wound dressing materials. There is no doubt that the nanofiber web has provided a promising wound dressing material in biomedical applications for its unique properties. In recent years, the limitation of low production behind the electrospinning process has been due to industrial production devices. On the other hand, bringing nanofiber webs into the clinical field still needs to be improved. With more clinical research and improved functional nanofiber web, the electrospun nanomaterials can offer an unprecedented breakthrough in biomedical applications.

**Supplementary Materials:** The following are available online at https://www.mdpi.com/article/10.3390/membranes11120908/s1, Table S1: Patents from 2020 to 2021 for antibacterial electrospun wound dressings materials, Table S2: Studies for antibacterial electrospun wound dressing materials in 2020–2021, and Table S3. Therapeutic agents used in nanofibers.

**Author Contributions:** Conceptualization, A.G., I.G., A.T., J.M. and F.Y., software, A.G. and F.Y.; resources, J.M.; writing—original draft preparation, A.G., I.G., A.T. and F.Y.; writing—review and editing, F.Y.; supervision, F.Y.; project administration, J.M. All authors have read and agreed to the published version of the manuscript.

**Funding:** The APC was funded by the Ministry of Industry and Trade of the Czech Republic (Research and development of laboratory workplace with antibacterial nanosurface based on a thin surface sol-gel) grant number FV30148.

**Institutional Review Board Statement:** Not applicable.

**Informed Consent Statement:** Not applicable.

**Data Availability Statement:** Not applicable.

**Acknowledgments:** The research presented in this paper was supported by the Internal Grant of the Technical University of Liberec, Czech Republic (Project No. SGS-2021-3027). The authors acknowledge the assistance provided by the Ministry of Industry and Trade of the Czech Republic (Research and development of laboratory workplace with antibacterial nanosurface based on a thin surface sol-gel, Project No. FV30148).

**Conflicts of Interest:** The authors declare no conflict of interest. The funders had no role in the design of the study; in the collection, analyses, or interpretation of data; in the writing of the manuscript, or in the decision to publish the results.

## References

1. Boer, M.; Duchnik, E.; Maleszka, R.; Marchlewicz, M. Structural and Biophysical Characteristics of Human Skin in Maintaining Proper Epidermal Barrier Function. *Postepy Dermatol. Alergol.* **2016**, *33*, 1–5. [CrossRef] [PubMed]
2. Wysocki, A.B. Skin Anatomy, Physiology, and Pathophysiology. *Nurs. Clin. N. Am.* **1999**, *34*, 777–797.
3. Kolarsick, P.A.J.; Kolarsick, M.A.; Goodwin, C. Anatomy and Physiology of the Skin. *J. Dermatol. Nurses' Assoc.* **2011**, *3*, 203–213. [CrossRef]
4. Chrintz, H.; Vibits, H.; Cordtz, T.O.; Harreby, J.S.; Waaddegaard, P.; Larsen, S.O. Need for Surgical Wound Dressing. *Br. J. Surg.* **1989**, *76*, 204–205. [CrossRef]
5. Simões, D.; Miguel, S.P.; Ribeiro, M.P.; Coutinho, P.; Mendonça, A.G.; Correia, I.J. Recent Advances on Antimicrobial Wound Dressing: A Review. *Eur. J. Pharm. Biopharm.* **2018**, *127*, 130–141. [CrossRef]
6. Chen, K.; Wang, F.; Liu, S.; Wu, X.; Xu, L.; Zhang, D. In Situ Reduction of Silver Nanoparticles by Sodium Alginate to Obtain Silver-Loaded Composite Wound Dressing with Enhanced Mechanical and Antimicrobial Property. *Int. J. Biol. Macromol.* **2020**, *148*, 501–509. [CrossRef]
7. Wang, S.; Wang, Z.; Xu, C.; Cui, L.; Meng, G.; Yang, S.; Wu, J.; Liu, Z.; Guo, X. PEG-α-CD/AM/Liposome @amoxicillin Double Network Hydrogel Wound Dressing—Multiple Barriers for Long-Term Drug Release. *J. Biomater. Appl.* **2021**, *35*, 1085–1095. [CrossRef]
8. Li, H.; Wei, X.; Yi, X.; Tang, S.; He, J.; Huang, Y.; Cheng, F. Antibacterial, Hemostasis, Adhesive, Self-Healing Polysaccharides-Based Composite Hydrogel Wound Dressing for the Prevention and Treatment of Postoperative Adhesion. *Mater. Sci. Eng. C* **2021**, *123*, 111978. [CrossRef]
9. Zhang, J.; Fang, W.; Zhang, F.; Gao, S.; Guo, Y.; Li, J.; Zhu, Y.; Zhang, Y.; Jin, J. Ultrathin Microporous Membrane with High Oil Intrusion Pressure for Effective Oil/Water Separation. *J. Membr. Sci.* **2020**, *608*, 118201. [CrossRef]
10. Li, D.; Fei, X.; Wang, K.; Xu, L.; Wang, Y.; Tian, J.; Li, Y. A Novel Self-Healing Triple Physical Cross-Linked Hydrogel for Antibacterial Dressing. *J. Mater. Chem. B* **2021**, *9*, 6844–6855. [CrossRef]
11. Shalumon, K.T.; Sheu, C.; Chen, C.-H.; Chen, S.-H.; Jose, G.; Kuo, C.-Y.; Chen, J.-P. Multi-Functional Electrospun Antibacterial Core-Shell Nanofibrous Membranes for Prolonged Prevention of Post-Surgical Tendon Adhesion and Inflammation. *Acta Biomater.* **2018**, *72*, 121–136. [CrossRef]
12. Pan, H.; Fan, D.; Duan, Z.; Zhu, C.; Fu, R.; Li, X. Non-Stick Hemostasis Hydrogels as Dressings with Bacterial Barrier Activity for Cutaneous Wound Healing. *Mater. Sci. Eng. C* **2019**, *105*, 110118. [CrossRef]
13. Alberti, T.; Coelho, D.S.; Voytena, A.; Pitz, H.; de Pra, M.; Mazzarino, L.; Kuhnen, S.; Ribeiro-do-Valle, R.M.; Maraschin, M.; Veleirinho, B. Nanotechnology: A Promising Tool towards Wound Healing. *Curr. Pharm. Des.* **2017**, *23*, 3515–3528. [CrossRef]
14. Newman, M.D.; Stotland, M.; Ellis, J.I. The Safety of Nanosized Particles in Titanium Dioxide–and Zinc Oxide–Based Sunscreens. *J. Am. Acad. Dermatol.* **2009**, *61*, 685–692. [CrossRef]
15. Kalashnikova, I.; Das, S.; Seal, S. Nanomaterials for Wound Healing: Scope and Advancement. *Nanomedicine* **2015**, *10*, 2593–2612. [CrossRef]
16. Hubbell, J.A. Chapter 21—Matrix Effects. In *Principles of Tissue Engineering*, 4th ed.; Lanza, R., Langer, R., Vacanti, J., Eds.; Academic Press: Boston, MA, USA, 2014; pp. 407–421; ISBN 978-0-12-398358-9.
17. Rosenberg, M.D. Cell Guidance by Alterations in Monomolecular Films. *Science* **1963**, *139*, 411–412. [CrossRef]
18. Laurencin, C.T.; Ambrosio, A.M.A.; Borden, M.D.; Cooper, J.A., Jr. Tissue Engineering: Orthopedic Applications. *Annu. Rev. Biomed. Eng.* **1999**, *1*, 19–46. [CrossRef]
19. Tanzli, E.; Ehrmann, A. Electrospun Nanofibrous Membranes for Tissue Engineering and Cell Growth. *Appl. Sci.* **2021**, *11*, 6929. [CrossRef]
20. Younes, P.-S.; Mehdi, D.; Abbas, M.; Amir, F.; Roghayeh, S.; Nosratollah, Z. An Overview on Application of Natural Substances Incorporated with Electrospun Nanofibrous Scaffolds to Development of Innovative Wound Dressings. *Mini-Rev. Med. Chem.* **2018**, *18*, 414–427.
21. Tiyek, I.; Gunduz, A.; Yalcinkaya, F.; Chaloupek, J. Influence of Electrospinning Parameters on the Hydrophilicity of Electrospun Polycaprolactone Nanofibres. *J. Nanosci. Nanotechnol.* **2019**, *19*, 7251–7260. [CrossRef]
22. Abrigo, M.; McArthur, S.L.; Kingshott, P. Electrospun Nanofibers as Dressings for Chronic Wound Care: Advances, Challenges, and Future Prospects. *Macromol. Biosci.* **2014**, *14*, 772–792. [CrossRef] [PubMed]

23. Liu, Y.; Li, T.; Han, Y.; Li, F.; Liu, Y. Recent Development of Electrospun Wound Dressing. *Curr. Opin. Biomed. Eng.* **2021**, *17*, 100247. [CrossRef]
24. Dodero, A.; Brunengo, E.; Alloisio, M.; Sionkowska, A.; Vicini, S.; Castellano, M. Chitosan-Based Electrospun Membranes: Effects of Solution Viscosity, Coagulant and Crosslinker. *Carbohydr. Polym.* **2020**, *235*, 115976. [CrossRef] [PubMed]
25. Olkhov, A.A.; Staroverova, O.V.; Kuherenko, E.L.; Iordanskii, A.L. Effect of Electrospinning Solution Parameters on the Properties of Nonvolven Fibrous Material Based on Polyhydroxibutyrate. *J. Phys. Conf. Ser.* **2020**, *1431*, 012029. [CrossRef]
26. Fallahi, D.; Rafizadeh, M.; Mohammadi, N.; Vahidi, B. Effects of Feed Rate and Solution Conductivity on Jet Current and Fiber Diameter in Electrospinning of Polyacrylonitrile Solutions. *E-Polymers* **2009**, *9*. [CrossRef]
27. Angammana, C.; Jayaram, S. Analysis of the Effects of Solution Conductivity on Electrospinning Process and Fiber Morphology. *IEEE Trans. Ind. Appl.* **2011**, *47*, 1109–1117. [CrossRef]
28. Afifi, A.M.; Yamane, H.; Kimura, Y. Effect of Polymer Molecular Weight on the Electrospinning of Polylactides in Entangled and Aligned Fiber Forms. *Sen'i Gakkaishi* **2010**, *66*, 35–42. [CrossRef]
29. Park, B.K.; Um, I.C. Effect of Molecular Weight on Electro-Spinning Performance of Regenerated Silk. *Int. J. Biol. Macromol.* **2018**, *106*, 1166–1172. [CrossRef]
30. Liu, Y.; Dong, L.; Fan, J.; Wang, R.; Yu, J.-Y. Effect of Applied Voltage on Diameter and Morphology of Ultrafine Fibers in Bubble Electrospinning. *J. Appl. Polym. Sci.* **2011**, *120*, 592–598. [CrossRef]
31. Yang, G.Z.; Li, H.P.; Yang, J.H.; Wan, J.; Yu, D.G. Influence of Working Temperature on the Formation of Electrospun Polymer Nanofibers. *Nanoscale Res. Lett.* **2017**, *12*, 1–10. [CrossRef]
32. Levitt, A.; Vallett, R.; Dion, G.; Schauer, C. Effect of Electrospinning Processing Variables on Polyacrylonitrile Nanoyarns. *J. Appl. Polym. Sci.* **2018**, *135*, 46404. [CrossRef]
33. Nezarati, R.M.; Eifert, M.B.; Cosgriff-Hernandez, E. Effects of Humidity and Solution Viscosity on Electrospun Fiber Morphology. *Tissue Eng. Part C Methods* **2013**, *19*, 810–819. [CrossRef]
34. Baykara, T.; Taylan, G. Coaxial Electrospinning of PVA/Nigella Seed Oil Nanofibers: Processing and Morphological Characterization. *Mater. Sci. Eng. B* **2021**, *265*, 115012. [CrossRef]
35. Daenicke, J.; Lämmlein, M.; Steinhübl, F.; Schubert, D.W. Revealing Key Parameters to Minimize the Diameter of Polypropylene Fibers Produced in the Melt Electrospinning Process. *E-Polymers* **2019**, *19*, 330–340. [CrossRef]
36. Deitzel, J.M.; Kleinmeyer, J.; Harris, D.; Beck Tan, N.C. The Effect of Processing Variables on the Morphology of Electrospun Nanofibers and Textiles. *Polymer* **2001**, *42*, 261–272. [CrossRef]
37. Geng, X.; Kwon, O.-H.; Jang, J. Electrospinning of Chitosan Dissolved in Concentrated Acetic Acid Solution. *Biomaterials* **2005**, *26*, 5427–5432. [CrossRef]
38. Thompson, C.J.; Chase, G.G.; Yarin, A.L.; Reneker, D.H. Effects of Parameters on Nanofiber Diameter Determined from Electrospinning Model. *Polymer* **2007**, *48*, 6913–6922. [CrossRef]
39. Beachley, V.; Wen, X. Effect of Electrospinning Parameters on the Nanofiber Diameter and Length. *Mater. Sci. Eng. C Mater. Biol. Appl.* **2009**, *29*, 663–668. [CrossRef]
40. Jabur, A.R.; Aldain, S.M.M. Effects of Ambient Temperature and Needle to Collector Distance on PVA Nanofibers Diameter Obtained from Electrospinning Technique. Available online: https://www.semanticscholar.org/paper/Effects-of-Ambient-Temperature-and-Needle-to-on-PVA-Jabur-Aldain/0104988ab6c79c6e95c15d1b3ea589858caa1ed7 (accessed on 18 October 2021).
41. Zhang, S.; Liu, H.; Tang, N.; Yu, J.; Ding, B. Chapter 8—Electronetting. In *Electrospinning: Nanofabrication and Applications*; Ding, B., Wang, X., Yu, J., Eds.; Micro and Nano Technologies; William Andrew Publishing: Norwich, NY, USA, 2019; pp. 249–282; ISBN 978-0-323-51270-1.
42. De Vrieze, S.; Van Camp, T.; Nelvig, A.; Hagström, B.; Westbroek, P.; De Clerck, K. The Effect of Temperature and Humidity on Electrospinning. *J. Mater. Sci.* **2009**, *44*, 1357–1362. [CrossRef]
43. Karaman, O.; Şen, M.; Demirci, E.A. 11—Electrospun scaffolds for vascular tissue engineering. In *Electrospun Materials for Tissue Engineering and Biomedical Applications*; Uyar, T., Kny, E., Eds.; Woodhead Publishing: Sawston, UK, 2017; pp. 261–287; ISBN 978-0-08-101022-8.
44. Torres-Martínez, E.J.; Bravo, J.M.C.; Medina, A.S.; González, G.L.P.; Gómez, L.J.V. A Summary of Electrospun Nanofibers as Drug Delivery System: Drugs Loaded and Biopolymers Used as Matrices. *Curr. Drug Deliv.* **2018**, *15*, 1360–1374. [CrossRef]
45. Eren Boncu, T.; Ozdemir, N.; Uskudar Guclu, A. Electrospinning of Linezolid Loaded PLGA Nanofibers: Effect of Solvents on Its Spinnability, Drug Delivery, Mechanical Properties, and Antibacterial Activities. *Drug Dev. Ind. Pharm.* **2020**, *46*, 109–121. [CrossRef] [PubMed]
46. Khalf, A.; Madihally, S.V. Recent Advances in Multiaxial Electrospinning for Drug Delivery. *Eur. J. Pharm. Biopharm.* **2017**, *112*, 1–17. [CrossRef] [PubMed]
47. Akhmetova, A.; Heinz, A. Electrospinning Proteins for Wound Healing Purposes: Opportunities and Challenges. *Pharmaceutics* **2021**, *13*, 4. [CrossRef] [PubMed]
48. Bjarnsholt, T. The role of bacterial biofilms in chronic infections. *Apmis* **2013**, *121*, 1–58. [CrossRef]
49. Wang, L.; Hu, C.; Shao, L. The Antimicrobial Activity of Nanoparticles: Present Situation and Prospects for the Future. *Int. J. Nanomed.* **2017**, *12*, 1227–1249. [CrossRef]
50. Inzana, J.A.; Schwarz, E.M.; Kates, S.L.; Awad, H.A. Biomaterials Approaches to Treating Implant-Associated Osteomyelitis. *Biomaterials* **2016**, *81*, 58–71. [CrossRef]

51. Tan, X.W.; Goh, T.W.; Saraswathi, P.; Nyein, C.L.; Setiawan, M.; Riau, A.; Lakshminarayanan, R.; Liu, S.; Tan, D.; Beuerman, R.W.; et al. Effectiveness of Antimicrobial Peptide Immobilization for Preventing Perioperative Cornea Implant-Associated Bacterial Infection. *Antimicrob. Agents Chemother.* **2014**, *58*, 5229–5238. [CrossRef]
52. Wei, Z.; Wang, L.; Zhang, S.; Chen, T.; Yang, J.; Long, S.; Wang, X. Electrospun Antibacterial Nanofibers for Wound Dressings and Tissue Medicinal Fields: A Review. *J. Innov. Opt. Health Sci.* **2020**, *13*, 2030012. [CrossRef]
53. Ragelle, H.; Danhier, F.; Préat, V.; Langer, R.; Anderson, D.G. Nanoparticle-Based Drug Delivery Systems: A Commercial and Regulatory Outlook as the Field Matures. *Expert Opin. Drug Deliv.* **2017**, *14*, 851–864. [CrossRef]
54. Huang, Z.-M.; Zhang, Y.-Z.; Kotaki, M.; Ramakrishna, S. A Review on Polymer Nanofibers by Electrospinning and Their Applications in Nanocomposites. *Compos. Sci. Technol.* **2003**, *63*, 2223–2253. [CrossRef]
55. Jones, J.R. Review of Bioactive Glass: From Hench to Hybrids. *Acta Biomater.* **2013**, *9*, 4457–4486. [CrossRef]
56. Reneker, D.H.; Yarin, A.L.; Zussman, E.; Xu, H. Electrospinning of Nanofibers from Polymer Solutions and Melts. In *Advances in Applied Mechanics*; Elsevier: Amsterdam, The Netherlands, 2007; Volume 41, pp. 43–346; ISBN 978-0-12-002057-7.
57. Liu, Z.-Y.; Wei, Z.-M.; Wang, X.-J.; Zhang, G.; Long, S.-R.; Yang, J. Preparation and Characterization of Multi-Layer Poly (Arylene Sulfide Sulfone) Nanofibers Membranes for Liquid Filtration. *Chin. J. Polym. Sci.* **2019**, *37*, 1248–1256. [CrossRef]
58. Wehlage, D.; Blattner, H.; Sabantina, L.; Böttjer, R.; Grothe, T.; Rattenholl, A.; Gudermann, F.; Lütkemeyer, D.; Ehrmann, A. Sterilization of PAN/Gelatin Nanofibrous Mats for Cell Growth. *Tekstilec* **2019**, *62*, 78–88. [CrossRef]
59. Percival, N.J. Classification of Wounds and Their Management. *Surgery* **2002**, *20*, 114–117. [CrossRef]
60. Chen, S.; Zhang, M.; Shao, X.; Wang, X.; Zhang, L.; Xu, P.; Zhong, W.; Zhang, L.; Xing, M.; Zhang, L. A Laminin Mimetic Peptide SIKVAV-Conjugated Chitosan Hydrogel Promoting Wound Healing by Enhancing Angiogenesis, Re-Epithelialization and Collagen Deposition. *J. Mater. Chem. B* **2015**, *3*, 6798–6804. [CrossRef]
61. Kuna, V.K.; Padma, A.M.; Håkansson, J.; Nygren, J.; Sjöback, R.; Petronis, S.; Sumitran-Holgersson, S. Significantly Accelerated Wound Healing of Full-Thickness Skin Using a Novel Composite Gel of Porcine Acellular Dermal Matrix and Human Peripheral Blood Cells. *Cell Transplant.* **2017**, *26*, 293–307. [CrossRef]
62. Murali, R.; Thanikaivelan, P. Bionic, Porous, Functionalized Hybrid Scaffolds with Vascular Endothelial Growth Factor Promote Rapid Wound Healing in Wistar Albino Rats. *RSC Adv.* **2016**, *6*, 19252–19264. [CrossRef]
63. Meyer, U.; Handschel, J.; Wiesmann, H.P.; Meyer, T. (Eds.) *Fundamentals of Tissue Engineering and Regenerative Medicine*; Springer: Berlin/Heidelberg, Germany, 2009; ISBN 978-3-540-77754-0.
64. Greiner, A.; Wendorff, J.H. Electrospinning: A Fascinating Method for the Preparation of Ultrathin Fibers. *Angew. Chem. Int. Ed.* **2007**, *46*, 5670–5703. [CrossRef]
65. Schultz, G.; Ladwig, G.; Wysocki, A. Extracellular Matrix: Review of Its Roles in Acute and Chronic Wounds. *World Wide Wounds* **2005**, *2005*, 1–18.
66. Matthews, J.A.; Wnek, G.E.; Simpson, D.G.; Bowlin, G.L. Electrospinning of Collagen Nanofibers. *Biomacromolecules* **2002**, *3*, 232–238. [CrossRef]
67. Rho, K.S.; Jeong, L.; Lee, G.; Seo, B.-M.; Park, Y.J.; Hong, S.-D.; Roh, S.; Cho, J.J.; Park, W.H.; Min, B.-M. Electrospinning of Collagen Nanofibers: Effects on the Behavior of Normal Human Keratinocytes and Early-Stage Wound Healing. *Biomaterials* **2006**, *27*, 1452–1461. [CrossRef] [PubMed]
68. Chen, J.-P.; Chang, G.-Y.; Chen, J.-K. Electrospun Collagen/Chitosan Nanofibrous Membrane as Wound Dressing. *Colloids Surf. A Physicochem. Eng. Asp.* **2008**, *313–314*, 183–188. [CrossRef]
69. Venugopal, J.R.; Zhang, Y.; Ramakrishna, S. In Vitro Culture of Human Dermal Fibroblasts on Electrospun Polycaprolactone Collagen Nanofibrous Membrane. *Artif. Organs* **2006**, *30*, 440–446. [CrossRef] [PubMed]
70. Lin, J.; Li, C.; Zhao, Y.; Hu, J.; Zhang, L.-M. Co-Electrospun Nanofibrous Membranes of Collagen and Zein for Wound Healing. *ACS Appl. Mater. Interfaces* **2012**, *4*, 1050–1057. [CrossRef]
71. Buttafoco, L.; Kolkman, N.G.; Engbers-Buijtenhuijs, P.; Poot, A.A.; Dijkstra, P.J.; Vermes, I.; Feijen, J. Electrospinning of Collagen and Elastin for Tissue Engineering Applications. *Biomaterials* **2006**, *27*, 724–734. [CrossRef]
72. Neal, R.A.; McClugage, S.G.; Link, M.C.; Sefcik, L.S.; Ogle, R.C.; Botchwey, E.A. Laminin Nanofiber Meshes that Mimic Morphological Properties and Bioactivity of Basement Membranes. *Tissue Eng. Part C Methods* **2009**, *15*, 11–21. [CrossRef]
73. Chong, E.J.; Phan, T.T.; Lim, I.J.; Zhang, Y.Z.; Bay, B.H.; Ramakrishna, S.; Lim, C.T. Evaluation of Electrospun PCL/Gelatin Nanofibrous Scaffold for Wound Healing and Layered Dermal Reconstitution. *Acta Biomater.* **2007**, *3*, 321–330. [CrossRef]
74. Duan, H.; Feng, B.; Guo, X.; Wang, J.; Zhao, L.; Zhou, G.; Liu, W.; Cao, Y.; Zhang, W.J. Engineering of Epidermis Skin Grafts Using Electrospun Nanofibrous Gelatin/Polycaprolactone Membranes. *Int. J. Nanomed.* **2013**, *8*, 2077. [CrossRef]
75. Powell, H.M.; Boyce, S.T. Fiber Density of Electrospun Gelatin Scaffolds Regulates Morphogenesis of Dermal–Epidermal Skin Substitutes. *J. Biomed. Mater. Res.* **2008**, *84*, 1078–1086. [CrossRef]
76. Kim, S.E.; Heo, D.N.; Lee, J.B.; Kim, J.R.; Park, S.H.; Jeon, S.H.; Kwon, I.K. Electrospun Gelatin/Polyurethane Blended Nanofibers for Wound Healing. *Biomed. Mater.* **2009**, *4*, 044106. [CrossRef]
77. Heo, D.N.; Yang, D.H.; Lee, J.B.; Bae, M.S.; Kim, J.H.; Moon, S.H.; Chun, H.J.; Kim, C.H.; Lim, H.-N.; Kwon, I.K. Burn-Wound Healing Effect of Gelatin/Polyurethane Nanofiber Scaffold Containing Silver-Sulfadiazine. *J. Biomed. Nanotechnol.* **2013**, *9*, 511–515. [CrossRef]
78. Ji, Y.; Ghosh, K.; Shu, X.Z.; Li, B.; Sokolov, J.C.; Prestwich, G.D.; Clark, R.A.F.; Rafailovich, M.H. Electrospun Three-Dimensional Hyaluronic Acid Nanofibrous Scaffolds. *Biomaterials* **2006**, *27*, 3782–3792. [CrossRef]

79. Chutipakdeevong, J.; Ruktanonchai, U.R.; Supaphol, P. Process Optimization of Electrospun Silk Fibroin Fiber Mat for Accelerated Wound Healing. *J. Appl. Polym. Sci.* **2013**, *130*, 3634–3644. [CrossRef]
80. Schneider, A.; Wang, X.Y.; Kaplan, D.L.; Garlick, J.A.; Egles, C. Biofunctionalized Electrospun Silk Mats as a Topical Bioactive Dressing for Accelerated Wound Healing. *Acta Biomater.* **2009**, *5*, 2570–2578. [CrossRef]
81. Noh, H.K.; Lee, S.W.; Kim, J.-M.; Oh, J.-E.; Kim, K.-H.; Chung, C.-P.; Choi, S.-C.; Park, W.H.; Min, B.-M. Electrospinning of Chitin Nanofibers: Degradation Behavior and Cellular Response to Normal Human Keratinocytes and Fibroblasts. *Biomaterials* **2006**, *27*, 3934–3944. [CrossRef]
82. Zhou, Y.; Yang, D.; Chen, X.; Xu, Q.; Lu, F.; Nie, J. Electrospun Water-Soluble Carboxyethyl Chitosan/Poly(Vinyl Alcohol) Nanofibrous Membrane as Potential Wound Dressing for Skin Regeneration. *Biomacromolecules* **2008**, *9*, 349–354. [CrossRef]
83. Dhandayuthapani, B.; Krishnan, U.M.; Sethuraman, S. Fabrication and Characterization of Chitosan-Gelatin Blend Nanofibers for Skin Tissue Engineering. *J. Biomed. Mater. Res.* **2010**, *94*, 264–272. [CrossRef]
84. Kumbar, S.G.; Nukavarapu, S.P.; James, R.; Nair, L.S.; Laurencin, C.T. Electrospun Poly (Lactic Acid-Co-Glycolic Acid) Scaffolds for Skin Tissue Engineering. *Biomaterials* **2008**, *29*, 4100–4107. [CrossRef]
85. Liu, S.-J.; Kau, Y.-C.; Chou, C.-Y.; Chen, J.-K.; Wu, R.-C.; Yeh, W.-L. Electrospun PLGA/Collagen Nanofibrous Membrane as Early-Stage Wound Dressing. *J. Membr. Sci.* **2010**, *355*, 53–59. [CrossRef]
86. Xie, Z.; Paras, C.B.; Weng, H.; Punnakitikashem, P.; Su, L.-C.; Vu, K.; Tang, L.; Yang, J.; Nguyen, K.T. Dual Growth Factor Releasing Multi-Functional Nanofibers for Wound Healing. *Acta Biomater.* **2013**, *9*, 9351–9359. [CrossRef]
87. Vargas, E.A.T.; do Vale Baracho, N.C.; de Brito, J.; de Queiroz, A.A.A. Hyperbranched Polyglycerol Electrospun Nanofibers for Wound Dressing Applications. *Acta Biomater.* **2010**, *6*, 1069–1078. [CrossRef] [PubMed]
88. Carr, K.E. Scanning Electron Microscope Studies of Human Skin. *Br. J. Plast. Surg.* **1970**, *23*, 66–72. [CrossRef]
89. Van Zuijlen, P.P.M.; Ruurda, J.J.B.; van Veen, H.A.; van Marle, J.; van Trier, A.J.M.; Groenevelt, F.; Kreis, R.W.; Middelkoop, E. Collagen Morphology in Human Skin and Scar Tissue: No Adaptations in Response to Mechanical Loading at Joints. *Burns* **2003**, *29*, 423–431. [CrossRef]
90. Rawlins, J.M.; Lam, W.L.; Karoo, R.O.; Naylor, I.L.; Sharpe, D.T. Quantifying Collagen Type in Mature Burn Scars: A Novel Approach Using Histology and Digital Image Analysis. *J. Burn. Care Res.* **2006**, *27*, 60–65. [CrossRef] [PubMed]
91. Osman, O.S.; Selway, J.L.; Harikumar, P.E.; Stocker, C.J.; Wargent, E.T.; Cawthorne, M.A.; Jassim, S.; Langlands, K. A Novel Method to Assess Collagen Architecture in Skin. *BMC Bioinform.* **2013**, *14*, 260. [CrossRef] [PubMed]
92. Abbasipour, M.; Khajavi, R. Nanofiber Bundles and Yarns Production by Electrospinning: A Review. *Adv. Polym. Technol.* **2013**, *32*. [CrossRef]
93. Yousefzadeh, M.; Latifi, M.; Teo, W.-E.; Amani-Tehran, M.; Ramakrishna, S. Producing Continuous Twisted Yarn from Well-Aligned Nanofibers by Water Vortex. *Polym. Eng. Sci.* **2011**, *51*, 323–329. [CrossRef]
94. Ali, U.; Niu, H.; Abbas, A.; Shao, H.; Lin, T. Online Stretching of Directly Electrospun Nanofiber Yarns. *RSC Adv.* **2016**, *6*, 30564–30569. [CrossRef]
95. Khil, M.-S.; Bhattarai, S.R.; Kim, H.-Y.; Kim, S.-Z.; Lee, K.-H. Novel Fabricated Matrix via Electrospinning for Tissue Engineering. *J. Biomed. Mater. Res. Part B Appl. Biomater.* **2005**, *72*, 117–124. [CrossRef]
96. Ravandi, S.A.H.; Tork, R.B.; Dabirian, F.; Gharehaghaji, A.A.; Sajjadi, A. Characteristics of Yarn and Fabric Made out of Nanofibers. *Mater. Sci. Appl.* **2015**, *6*, 103. [CrossRef]
97. Xie, J.; Ma, B.; Michael, P.L. Fabrication of Novel 3D Nanofiber Scaffolds with Anisotropic Property and Regular Pores and Their Potential Applications. *Adv. Healthc. Mater.* **2012**, *1*, 674–678. [CrossRef]
98. Chakraborty, S.; Liao, I.-C.; Adler, A.; Leong, K.W. Electrohydrodynamics: A Facile Technique to Fabricate Drug Delivery Systems. *Adv. Drug Deliv. Rev.* **2009**, *61*, 1043–1054. [CrossRef]
99. Weng, L.; Xie, J. Smart Electrospun Nanofibers for Controlled Drug Release: Recent Advances and New Perspectives. *Curr. Pharm. Des.* **2015**, *21*, 1944–1959. [CrossRef]
100. Said, S.S.; El-Halfawy, O.M.; El-Gowelli, H.M.; Aloufy, A.K.; Boraei, N.A.; El-Khordagui, L.K. Bioburden-Responsive Antimicrobial PLGA Ultrafine Fibers for Wound Healing. *Eur. J. Pharm. Biopharm.* **2012**, *80*, 85–94. [CrossRef]
101. Chou, S.-F.; Carson, D.; Woodrow, K.A. Current Strategies for Sustaining Drug Release from Electrospun Nanofibers. *J. Control. Release* **2015**, *220*, 584–591. [CrossRef]
102. Kim, K.; Luu, Y.K.; Chang, C.; Fang, D.; Hsiao, B.S.; Chu, B.; Hadjiargyrou, M. Incorporation and Controlled Release of a Hydrophilic Antibiotic Using Poly(Lactide-Co-Glycolide)-Based Electrospun Nanofibrous Scaffolds. *J. Control. Release* **2004**, *98*, 47–56. [CrossRef]
103. Ayodeji, O.; Graham, E.; Kniss, D.; Lannutti, J.; Tomasko, D. Carbon Dioxide Impregnation of Electrospun Polycaprolactone Fibers. *J. Supercrit. Fluids* **2007**, *41*, 173–178. [CrossRef]
104. Geiger, B.C.; Nelson, M.T.; Munj, H.R.; Tomasko, D.L.; Lannutti, J.J. Dual Drug Release from $CO_2$-Infused Nanofibers via Hydrophobic and Hydrophilic Interactions. *J. Appl. Polym. Sci.* **2015**, *132*. [CrossRef]
105. Brandl, F.; Sommer, F.; Goepferich, A. Rational Design of Hydrogels for Tissue Engineering: Impact of Physical Factors on Cell Behavior. *Biomaterials* **2007**, *28*, 134–146. [CrossRef]
106. Discher, D.E.; Janmey, P.; Wang, Y. Tissue Cells Feel and Respond to the Stiffness of Their Substrate. *Science* **2005**, *310*, 1139–1143. [CrossRef]
107. Yao, J.; Bastiaansen, C.W.M.; Peijs, T. High Strength and High Modulus Electrospun Nanofibers. *Fibers* **2014**, *2*, 158–186. [CrossRef]

108. Mohammadzadehmoghadam, S.; Dong, Y.; Davies, I.J. Recent Progress in Electrospun Nanofibers: Reinforcement Effect and Mechanical Performance. *J. Polym. Sci. Part B Polym. Phys.* **2015**, *53*, 1171–1212. [CrossRef]
109. Ravandi, S.A.H.; Sadrjahani, M. Mechanical and Structural Characterizations of Simultaneously Aligned and Heat Treated PAN Nanofibers. *J. Appl. Polym. Sci.* **2012**, *124*, 3529–3537. [CrossRef]
110. Xie, J.; Michael, P.L.; Zhong, S.; Ma, B.; MacEwan, M.R.; Lim, C.T. Mussel Inspired Protein-Mediated Surface Modification to Electrospun Fibers and Their Potential Biomedical Applications. *J. Biomed. Mater. Res. Part A* **2012**, *100*, 929–938. [CrossRef]
111. Xie, J.; Zhong, S.; Ma, B.; Shuler, F.D.; Lim, C.T. Controlled Biomineralization of Electrospun Poly (ε-Caprolactone) Fibers to Enhance Their Mechanical Properties. *Acta Biomater.* **2013**, *9*, 5698–5707. [CrossRef]
112. Ma, K.; Chan, C.K.; Liao, S.; Hwang, W.Y.K.; Feng, Q.; Ramakrishna, S. Electrospun Nanofiber Scaffolds for Rapid and Rich Capture of Bone Marrow-Derived Hematopoietic Stem Cells. *Biomaterials* **2008**, *29*, 2096–2103. [CrossRef]
113. Gümüşderelioğlu, M.; Dalkıranoğlu, S.; Aydın, R.S.T.; Çakmak, S. A Novel Dermal Substitute Based on Biofunctionalized Electrospun PCL Nanofibrous Matrix. *J. Biomed. Mater. Res. Part A* **2011**, *98*, 461–472. [CrossRef]
114. Lee, E.J.; Lee, J.H.; Jin, L.; Jin, O.S.; Shin, Y.C.; Oh, S.J.; Lee, J.; Hyon, S.-H.; Han, D.-W. Hyaluronic Acid/Poly (Lactic-Co-Glycolic Acid) Core/Shell Fiber Meshes Loaded with Epigallocatechin-3-O-Gallate as Skin Tissue Engineering Scaffolds. *J. Nanosci. Nanotechnol.* **2014**, *14*, 8458–8463. [CrossRef]
115. Choi, J.S.; Leong, K.W.; Yoo, H.S. In Vivo Wound Healing of Diabetic Ulcers Using Electrospun Nanofibers Immobilized with Human Epidermal Growth Factor (EGF). *Biomaterials* **2008**, *29*, 587–596. [CrossRef]
116. Pelipenko, J.; Kocbek, P.; Govedarica, B.; Rošic, R.; Baumgartner, S.; Kristl, J. The Topography of Electrospun Nanofibers and Its Impact on the Growth and Mobility of Keratinocytes. *Eur. J. Pharm. Biopharm.* **2013**, *84*, 401–411. [CrossRef]
117. Ehrmann, A. Non-Toxic Crosslinking of Electrospun Gelatin Nanofibers for Tissue Engineering and Biomedicine—A Review. *Polymers* **2021**, *13*, 1973. [CrossRef]
118. Homaeigohar, S.; Boccaccini, A.R. Antibacterial Biohybrid Nanofibers for Wound Dressings. *Acta Biomater.* **2020**, *107*, 25–49. [CrossRef]
119. Hakkarainen, T.; Koivuniemi, R.; Kosonen, M.; Escobedo-Lucea, C.; Sanz-Garcia, A.; Vuola, J.; Valtonen, J.; Tammela, P.; Mäkitie, A.; Luukko, K.; et al. Nanofibrillar Cellulose Wound Dressing in Skin Graft Donor Site Treatment. *J. Control. Release* **2016**, *244*, 292–301. [CrossRef]
120. Czaja, W.; Krystynowicz, A.; Bielecki, S.; Brown, R.M. Microbial Cellulose—the Natural Power to Heal Wounds. *Biomaterials* **2006**, *27*, 145–151. [CrossRef]
121. Czaja, W.K.; Young, D.J.; Kawecki, M.; Brown, R.M. The Future Prospects of Microbial Cellulose in Biomedical Applications. *Biomacromolecules* **2007**, *8*, 1–12. [CrossRef]
122. Mogoşanu, G.D.; Grumezescu, A.M. Natural and Synthetic Polymers for Wounds and Burns Dressing. *Int. J. Pharm.* **2014**, *463*, 127–136. [CrossRef]
123. Wutticharoenmongkol, P.; Hannirojram, P.; Nuthong, P. Gallic Acid-Loaded Electrospun Cellulose Acetate Nanofibers as Potential Wound Dressing Materials. *Polym. Adv. Technol.* **2019**, *30*, 1135–1147. [CrossRef]
124. Liu, Y.; Ma, G.; Fang, D.; Xu, J.; Zhang, H.; Nie, J. Effects of Solution Properties and Electric Field on the Electrospinning of Hyaluronic Acid. *Carbohydr. Polym.* **2011**, *83*, 1011–1015. [CrossRef]
125. Eng, D.; Caplan, M.; Preul, M.; Panitch, A. Hyaluronan Scaffolds: A Balance between Backbone Functionalization and Bioactivity. *Acta Biomater.* **2010**, *6*, 2407–2414. [CrossRef]
126. Leach, J.B.; Bivens, K.A.; Patrick, C.W., Jr.; Schmidt, C.E. Photocrosslinked Hyaluronic Acid Hydrogels: Natural, Biodegradable Tissue Engineering Scaffolds. *Biotechnol. Bioeng.* **2003**, *82*, 578–589. [CrossRef]
127. Chen, W. Preparation of Hyaluronan-DNA Matrices and Films. *Cold Spring Harb. Protoc.* **2012**, *2012*, pdb.prot071522. [CrossRef] [PubMed]
128. Uppal, R.; Ramaswamy, G.N.; Arnold, C.; Goodband, R.; Wang, Y. Hyaluronic Acid Nanofiber Wound Dressing—Production, Characterization, and in Vivo Behavior. *J. Biomed. Mater. Res. Part B Appl. Biomater.* **2011**, *97*, 20–29. [CrossRef] [PubMed]
129. Séon-Lutz, M.; Couffin, A.-C.; Vignoud, S.; Schlatter, G.; Hébraud, A. Electrospinning in Water and in Situ Crosslinking of Hyaluronic Acid/Cyclodextrin Nanofibers: Towards Wound Dressing with Controlled Drug Release. *Carbohydr. Polym.* **2019**, *207*, 276–287. [CrossRef] [PubMed]
130. Alavarse, A.C.; de Oliveira Silva, F.W.; Colque, J.T.; da Silva, V.M.; Prieto, T.; Venancio, E.C.; Bonvent, J.-J. Tetracycline Hydrochloride-Loaded Electrospun Nanofibers Mats Based on PVA and Chitosan for Wound Dressing. *Mater. Sci. Eng. C* **2017**, *77*, 271–281. [CrossRef]
131. Sarhan, W.A.; Azzazy, H.M.E.; El-Sherbiny, I.M. Honey/Chitosan Nanofiber Wound Dressing Enriched with Allium Sativum and Cleome Droserifolia: Enhanced Antimicrobial and Wound Healing Activity. *ACS Appl. Mater. Interfaces* **2016**, *8*, 6379–6390. [CrossRef]
132. Homaeigohar, S.; Tsai, T.-Y.; Young, T.-H.; Yang, H.J.; Ji, Y.-R. An Electroactive Alginate Hydrogel Nanocomposite Reinforced by Functionalized Graphite Nanofilaments for Neural Tissue Engineering. *Carbohydr. Polym.* **2019**, *224*, 115112. [CrossRef]
133. Lee, K.Y.; Mooney, D.J. Alginate: Properties and Biomedical Applications. *Prog. Polym. Sci.* **2012**, *37*, 106–126. [CrossRef]
134. Tang, Y.; Lan, X.; Liang, C.; Zhong, Z.; Xie, R.; Zhou, Y.; Miao, X.; Wang, H.; Wang, W. Honey Loaded Alginate/PVA Nanofibrous Membrane as Potential Bioactive Wound Dressing. *Carbohydr. Polym.* **2019**, *219*, 113–120. [CrossRef]

135. Zhou, T.; Wang, N.; Xue, Y.; Ding, T.; Liu, X.; Mo, X.; Sun, J. Electrospun Tilapia Collagen Nanofibers Accelerating Wound Healing via Inducing Keratinocytes Proliferation and Differentiation. *Colloids Surf. B Biointerfaces* **2016**, *143*, 415–422. [CrossRef]
136. Wen, X.; Zheng, Y.; Wu, J.; Wang, L.-N.; Yuan, Z.; Peng, J.; Meng, H. Immobilization of Collagen Peptide on Dialdehyde Bacterial Cellulose Nanofibers via Covalent Bonds for Tissue Engineering and Regeneration. *Int. J. Nanomed.* **2015**, *10*, 4623–4637. [CrossRef]
137. Yao, C.-H.; Lee, C.-Y.; Huang, C.-H.; Chen, Y.-S.; Chen, K.-Y. Novel Bilayer Wound Dressing Based on Electrospun Gelatin/Keratin Nanofibrous Mats for Skin Wound Repair. *Mater. Sci. Eng. C* **2017**, *79*, 533–540. [CrossRef]
138. Melke, J.; Midha, S.; Ghosh, S.; Ito, K.; Hofmann, S. Silk Fibroin as Biomaterial for Bone Tissue Engineering. *Acta Biomater.* **2016**, *31*, 1–16. [CrossRef]
139. Fan, L.; Cai, Z.; Zhang, K.; Han, F.; Li, J.; He, C.; Mo, X.; Wang, X.; Wang, H. Green Electrospun Pantothenic Acid/Silk Fibroin Composite Nanofibers: Fabrication, Characterization and Biological Activity. *Colloids Surf. B Biointerfaces* **2014**, *117*, 14–20. [CrossRef]
140. Song, D.W.; Kim, S.H.; Kim, H.H.; Lee, K.H.; Ki, C.S.; Park, Y.H. Multi-Biofunction of Antimicrobial Peptide-Immobilized Silk Fibroin Nanofiber Membrane: Implications for Wound Healing. *Acta Biomater.* **2016**, *39*, 146–155. [CrossRef]
141. Selvaraj, S.; Fathima, N.N. Fenugreek Incorporated Silk Fibroin Nanofibers-A Potential Antioxidant Scaffold for Enhanced Wound Healing. *ACS Appl. Mater. Interfaces* **2017**, *9*, 5916–5926. [CrossRef]
142. GhavamiNejad, A.; Rajan Unnithan, A.; Kurup Sasikala, A.R.; Samarikhalaj, M.; Thomas, R.G.; Jeong, Y.Y.; Nasseri, S.; Murugesan, P.; Wu, D.; Hee Park, C.; et al. Mussel-Inspired Electrospun Nanofibers Functionalized with Size-Controlled Silver Nanoparticles for Wound Dressing Application. *ACS Appl. Mater. Interfaces* **2015**, *7*, 12176–12183. [CrossRef]
143. Wang, J.; Windbergs, M. Functional Electrospun Fibers for the Treatment of Human Skin Wounds. *Eur. J. Pharm. Biopharm.* **2017**, *119*, 283–299. [CrossRef]
144. Kalantari, K.; Mostafavi, E.; Afifi, A.M.; Izadiyan, Z.; Jahangirian, H.; Rafiee-Moghaddam, R.; Webster, T.J. Wound Dressings Functionalized with Silver Nanoparticles: Promises and Pitfalls. *Nanoscale* **2020**, *12*, 2268–2291. [CrossRef]
145. Shankar, P.D.; Shobana, S.; Karuppusamy, I.; Pugazhendhi, A.; Ramkumar, V.S.; Arvindnarayan, S.; Kumar, G. A Review on the Biosynthesis of Metallic Nanoparticles (Gold and Silver) Using Bio-Components of Microalgae: Formation Mechanism and Applications. *Enzym. Microb. Technol.* **2016**, *95*, 28–44. [CrossRef]
146. Kumar, P.T.S.; Lakshmanan, V.-K.; Biswas, R.; Nair, S.V.; Jayakumar, R. Synthesis and Biological Evaluation of Chitin Hydrogel/Nano ZnO Composite Bandage as Antibacterial Wound Dressing. *J. Biomed. Nanotechnol.* **2012**, *8*, 891–900. [CrossRef]
147. Stoica, E.; Chircov, C.; Grumezescu, A. Nanomaterials for Wound Dressings: An Up-to-Date Overview. *Molecules* **2020**, *25*, 2699. [CrossRef] [PubMed]
148. Jatoi, A.W.; Jo, Y.K.; Lee, H.; Oh, S.-G.; Hwang, D.S.; Khatri, Z.; Cha, H.J.; Kim, I.S. Antibacterial Efficacy of Poly (Vinyl Alcohol) Composite Nanofibers Embedded with Silver-Anchored Silica Nanoparticles. *J. Biomed. Mater. Res. Part B Appl. Biomater.* **2018**, *106*, 1121–1128. [CrossRef] [PubMed]
149. Fouda, M.M.G.; El-Aassar, M.R.; Al-Deyab, S.S. Antimicrobial Activity of Carboxymethyl Chitosan/Polyethylene Oxide Nanofibers Embedded Silver Nanoparticles. *Carbohydr. Polym.* **2013**, *92*, 1012–1017. [CrossRef] [PubMed]
150. Hebeish, A.A.; Ramadan, M.A.; Montaser, A.S.; Farag, A.M. Preparation, Characterization and Antibacterial Activity of Chitosan-g-Poly Acrylonitrile/Silver Nanocomposite. *Int. J. Biol. Macromol.* **2014**, *68*, 178–184. [CrossRef]
151. Maharjan, B.; Joshi, M.K.; Tiwari, A.P.; Park, C.H.; Kim, C.S. In-Situ Synthesis of AgNPs in the Natural/Synthetic Hybrid Nanofibrous Scaffolds: Fabrication, Characterization and Antimicrobial Activities. *J. Mech. Behav. Biomed. Mater.* **2017**, *65*, 66–76. [CrossRef]
152. Hashmi, M.; Ullah, S.; Ullah, A.; Akmal, M.; Saito, Y.; Hussain, N.; Ren, X.; Kim, I.S. Optimized Loading of Carboxymethyl Cellulose (CMC) in Tri-Component Electrospun Nanofibers Having Uniform Morphology. *Polymers* **2020**, *12*, 2524. [CrossRef]
153. Eghbalifam, N.; Shojaosadati, S.A.; Hashemi-Najafabadi, S.; Khorasani, A.C. Synthesis and Characterization of Antimicrobial Wound Dressing Material Based on Silver Nanoparticles Loaded Gum Arabic Nanofibers. *Int. J. Biol. Macromol.* **2020**, *155*, 119–130. [CrossRef]
154. Alven, S.; Buyana, B.; Feketshane, Z.; Aderibigbe, B.A. Electrospun Nanofibers/Nanofibrous Scaffolds Loaded with Silver Nanoparticles as Effective Antibacterial Wound Dressing Materials. *Pharmaceutics* **2021**, *13*, 964. [CrossRef]
155. Bharathi, B.S.; Stalin, T. Cerium Oxide and Peppermint Oil Loaded Polyethylene Oxide/Graphene Oxide Electrospun Nanofibrous Mats as Antibacterial Wound Dressings. *Mater. Today Commun.* **2019**, *21*, 100664. [CrossRef]
156. One-Pot Synthesis of ZnO Nanobelt-like Structures in Hyaluronan Hydrogels for Wound Dressing Applications—ScienceDirect. Available online: https://www.sciencedirect.com/science/article/pii/S014486171930791X?via%3Dihub (accessed on 17 October 2021).
157. Kalaycıoğlu, Z.; Kahya, N.; Adımcılar, V.; Kaygusuz, H.; Torlak, E.; Akın-Evingür, G.; Erim, F.B. Antibacterial Nano Cerium Oxide/Chitosan/Cellulose Acetate Composite Films as Potential Wound Dressing. *Eur. Polym. J.* **2020**, *133*, 109777. [CrossRef]
158. In Vivo Evaluation of Chitosan–PVP–Titanium Dioxide Nanocomposite as Wound Dressing Material—ScienceDirect. Available online: https://www.sciencedirect.com/science/article/pii/S0144861713002725?via%3Dihub (accessed on 17 October 2021).
159. Archana, D.; Dutta, J.; Dutta, P.K. Evaluation of Chitosan Nano Dressing for Wound Healing: Characterization, in Vitro and in Vivo Studies. *Int. J. Biol. Macromol.* **2013**, *57*, 193–203. [CrossRef]

160. Cai, N.; Li, C.; Han, C.; Luo, X.; Shen, L.; Xue, Y.; Yu, F. Tailoring Mechanical and Antibacterial Properties of Chitosan/Gelatin Nanofiber Membranes with $Fe_3O_4$ Nanoparticles for Potential Wound Dressing Application. *Appl. Surf. Sci.* **2016**, *369*, 492–500. [CrossRef]
161. Madhumathi, K.; Sudheesh Kumar, P.T.; Abhilash, S.; Sreeja, V.; Tamura, H.; Manzoor, K.; Nair, S.V.; Jayakumar, R. Development of Novel Chitin/Nanosilver Composite Scaffolds for Wound Dressing Applications. *J. Mater. Sci. Mater. Med.* **2010**, *21*, 807–813. [CrossRef]
162. Nguyen, N.; Nguyen, L.; Thanh, N.; Vo, T.; Quyen, T.; Tran, P.; Nguyen, T.H. Stabilization of Silver Nanoparticles in Chitosan and Gelatin Hydrogel and Its Applications. *Mater. Lett.* **2019**, *248*, 241–245. [CrossRef]
163. Haseeb, M.T.; Hussain, M.A.; Abbas, K.; Youssif, B.G.; Bashir, S.; Yuk, S.H.; Bukhari, S.N.A. Linseed Hydrogel-Mediated Green Synthesis of Silver Nanoparticles for Antimicrobial and Wound-Dressing Applications. *Int. J. Nanomed.* **2017**, *12*, 2845–2855. [CrossRef]
164. Felgueiras, H.P.; Amorim, M.T.P. Functionalization of Electrospun Polymeric Wound Dressings with Antimicrobial Peptides. *Colloids Surf. B Biointerfaces* **2017**, *156*, 133–148. [CrossRef]
165. Costa, F.; Carvalho, I.F.; Montelaro, R.C.; Gomes, P.; Martins, M.C.L. Covalent Immobilization of Antimicrobial Peptides (AMPs) onto Biomaterial Surfaces. *Acta Biomater.* **2011**, *7*, 1431–1440. [CrossRef]
166. Deshmukh, K.; Sankaran, S.; Basheer Ahamed, M.; Khadheer Pasha, S.K. Biomedical Applications of Electrospun Polymer Composite Nanofibres. In *Polymer Nanocomposites in Biomedical Engineering*; Sadasivuni, K.K., Ponnamma, D., Rajan, M., Ahmed, B., Al-Maadeed, M.A.S.A., Eds.; Lecture Notes in Bioengineering; Springer International Publishing: Cham, Switzerland, 2019; pp. 111–165; ISBN 978-3-030-04741-2.
167. Peschel, A. How Do Bacteria Resist Human Antimicrobial Peptides? *Trends Microbiol.* **2002**, *10*, 179–186. [CrossRef]
168. Goy, R.C.; Morais, S.T.B.; Assis, O.B.G. Evaluation of the Antimicrobial Activity of Chitosan and Its Quaternized Derivative on *E. Coli* and *S. Aureus* Growth. *Rev. Bras. Farmacogn.* **2016**, *26*, 122–127. [CrossRef]
169. Saharan, V.; Mehrotra, A.; Khatik, R.; Rawal, P.; Sharma, S.S.; Pal, A. Synthesis of Chitosan Based Nanoparticles and Their in Vitro Evaluation against Phytopathogenic Fungi. *Int. J. Biol. Macromol.* **2013**, *62*, 677–683. [CrossRef]
170. Arkoun, M.; Daigle, F.; Heuzey, M.-C.; Ajji, A. Mechanism of Action of Electrospun Chitosan-Based Nanofibers against Meat Spoilage and Pathogenic Bacteria. *Molecules* **2017**, *22*, 585. [CrossRef] [PubMed]
171. Park, S.-C.; Choi, C.; Jeong, G.-W.; Lee, H.-S.; Choi, S.-J.; Kim, W.-S.; Nah, J.-W. Algicidal Effects of Free-Amine Water-Soluble Chitosan to Marine Harmful Algal Species. *J. Ind. Eng. Chem.* **2016**, *34*, 139–145. [CrossRef]
172. Chirkov, S.N. The Antiviral Activity of Chitosan. *Appl. Biochem. Microbiol.* **2002**, *38*, 1–8. [CrossRef]
173. Chen, S.; Liu, B.; Carlson, M.A.; Gombart, A.F.; Reilly, D.A.; Xie, J. Recent Advances in Electrospun Nanofibers for Wound Healing. *Nanomedicine* **2017**, *12*, 1335–1352. [CrossRef] [PubMed]
174. Jeon, Y.; Kim, S. Effect of Antimicrobial Activity by Chitosan Oligosaccharide N-Conjugated with Asparagine. *J. Microbiol. Biotechnol.* **2001**, *11*, 281–286.
175. Xiao, B.; Wan, Y.; Zhao, M.; Liu, Y.; Zhang, S. Preparation and Characterization of Antimicrobial Chitosan-N-Arginine with Different Degrees of Substitution. *Carbohydr. Polym.* **2011**, *83*, 144–150. [CrossRef]
176. Antunes, B.P.; Moreira, A.F.; Gaspar, V.M.; Correia, I.J. Chitosan/Arginine–Chitosan Polymer Blends for Assembly of Nanofibrous Membranes for Wound Regeneration. *Carbohydr. Polym.* **2015**, *130*, 104–112. [CrossRef]
177. Unnithan, A.R.; Gnanasekaran, G.; Sathishkumar, Y.; Lee, Y.S.; Kim, C.S. Electrospun Antibacterial Polyurethane–Cellulose Acetate–Zein Composite Mats for Wound Dressing. *Carbohydr. Polym.* **2014**, *102*, 884–892. [CrossRef]
178. He, M.; Chen, M.; Dou, Y.; Ding, J.; Yue, H.; Yin, G.; Chen, X.; Cui, Y. Electrospun Silver Nanoparticles-Embedded Feather Keratin/Poly(Vinyl Alcohol)/Poly(Ethylene Oxide) Antibacterial Composite Nanofibers. *Polymers* **2020**, *12*, 305. [CrossRef]
179. Sun, L.; Han, J.; Liu, Z.; Wei, S.; Su, X.; Zhang, G. The Facile Fabrication of Wound Compatible Anti-Microbial Nanoparticles Encapsulated Collagenous Chitosan Matrices for Effective Inhibition of Poly-Microbial Infections and Wound Repairing in Burn Injury Care: Exhaustive in Vivo Evaluations. *J. Photochem. Photobiol. B Biol.* **2019**, *197*, 111539. [CrossRef]
180. Stie, M.B.; Corezzi, M.; Juncos Bombin, A.D.; Ajalloueian, F.; Attrill, E.; Pagliara, S.; Jacobsen, J.; Chronakis, I.S.; Nielsen, H.M.; Fodera, V. Waterborne Electrospinning of $\alpha$-Lactalbumin Generates Tunable and Biocompatible Nanofibers for Drug Delivery. *ACS Appl. Nano Mater.* **2020**, *3*, 1910–1921. [CrossRef]
181. Jao, W.-C.; Yang, M.-C.; Lin, C.-H.; Hsu, C.-C. Fabrication and Characterization of Electrospun Silk Fibroin/$TiO_2$ Nanofibrous Mats for Wound Dressings. *Polym. Adv. Technol.* **2012**, *23*, 1066–1076. [CrossRef]
182. Antimicrobial Electrospun Silk Fibroin Mats with Silver Nanoparticles for Wound Dressing Application. Available online: https://www.semanticscholar.org/paper/Antimicrobial-electrospun-silk-fibroin-mats-with-Uttayarat-Jetawattana/bc67376 5a402c9ed556fad65274d13493864488a (accessed on 17 October 2021).
183. Padrão, J.; Machado, R.; Casal, M.; Lanceros-Méndez, S.; Rodrigues, L.R.; Dourado, F.; Sencadas, V. Antibacterial Performance of Bovine Lactoferrin-Fish Gelatine Electrospun Membranes. *Int. J. Biol. Macromol.* **2015**, *81*, 608–614. Available online: https://ro.uow.edu.au/eispapers/4614/ (accessed on 17 October 2021). [CrossRef]
184. Chen, J.; Liu, Z.; Chen, M.; Zhang, H.; Li, X. Electrospun Gelatin Fibers with a Multiple Release of Antibiotics Accelerate Dermal Regeneration in Infected Deep Burns. *Macromol. Biosci.* **2016**, *16*, 1368–1380. [CrossRef]
185. Yıldız, A.; Kara, A.A.; Acartürk, F. Peptide-Protein Based Nanofibers in Pharmaceutical and Biomedical Applications. *Int. J. Biol. Macromol.* **2020**, *148*, 1084–1097. [CrossRef]

186. Azimi, B.; Maleki, H.; Zavagna, L.; De la Ossa, J.G.; Linari, S.; Lazzeri, A.; Danti, S. Bio-Based Electrospun Fibers for Wound Healing. *J. Funct. Biomater.* **2020**, *11*, 67. [CrossRef]
187. Khabbaz, B.; Solouk, A.; Mirzadeh, H. Polyvinyl alcohol/soy protein isolate nanofibrous patch for wound-healing applications. *Prog. Biomater.* **2019**, *8*, 185–186. [CrossRef]
188. Akshaykumar, K.P.; Zare, E.N.; Torres-Mendieta, R.; Wacławek, S.; Makvandi, P.; Černík, M.; Padil, V.V.; Varma, R.S. Electrospun fibers based on botanical, seaweed, microbial, and animal sourced biomacromolecules and their multidimensional applications. *Int. J. Biol. Macromol.* **2021**, *171*, 130–149. [CrossRef]
189. Wilkinson, L.J.; White, R.J.; Chipman, J.K. Silver and Nanoparticles of Silver in Wound Dressings: A Review of Efficacy and Safety. *J. Wound Care* **2011**, *20*, 543–549. [CrossRef]
190. White, R.; Cooper, R. Silver Sulphadiazine: A Review of the Evidence. *Wounds UK* **2005**, *1*, 51.
191. Lansdown, A.B.G. A Pharmacological and Toxicological Profile of Silver as an Antimicrobial Agent in Medical Devices. *Adv. Pharmacol. Sci.* **2010**, *2010*, 910686. [CrossRef] [PubMed]
192. Samberg, M.E.; Oldenburg, S.J.; Monteiro-Riviere, N.A. Evaluation of Silver Nanoparticle Toxicity in Skin in Vivo and Keratinocytes in Vitro. *Environ. Health Perspect.* **2010**, *118*, 407–413. [CrossRef]
193. Wijnhoven, S.W.P.; Peijnenburg, W.J.G.M.; Herberts, C.A.; Hagens, W.I.; Oomen, A.G.; Heugens, E.H.W.; Roszek, B.; Bisschops, J.; Gosens, I.; Van De Meent, D.; et al. Nano-Silver—A Review of Available Data and Knowledge Gaps in Human and Environmental Risk Assessment. *Nanotoxicology* **2009**, *3*, 109–138. [CrossRef]
194. Chen, X.; Schluesener, H.J. Nanosilver: A Nanoproduct in Medical Application. *Toxicol. Lett.* **2008**, *176*, 1–12. [CrossRef] [PubMed]
195. Drake, P.L.; Hazelwood, K.J. Exposure-Related Health Effects of Silver and Silver Compounds: A Review. *Ann. Occup. Hyg.* **2005**, *49*, 575–585. [CrossRef] [PubMed]
196. Liu, X.; Lin, T.; Gao, Y.; Xu, Z.; Huang, C.; Yao, G.; Jiang, L.; Tang, Y.; Wang, X. Antimicrobial Electrospun Nanofibers of Cellulose Acetate and Polyester Urethane Composite for Wound Dressing. *J. Biomed. Mater. Res. Part B Appl. Biomater.* **2012**, *100*, 1556–1565. [CrossRef]
197. Tang, Y.; Wong, C.; Wang, H.; Sutti, A.; Kirkland, M.; Wang, X.; Lin, T. Three-Dimensional Tissue Scaffolds from Interbonded Poly (ε-Caprolactone) Fibrous Matrices with Controlled Porosity. *Tissue Eng. Part C Methods* **2011**, *17*, 209–218. [CrossRef]
198. Balusamy, B.; Senthamizhan, A.; Uyar, T. In Vivo Safety Evaluations of Electrospun Nanofibers for Biomedical Applications. In *Electrospun Materials for Tissue Engineering and Biomedical Applications*; Uyar, T., Kny, E., Eds.; Woodhead Publishing: Sawston, UK, 2017; pp. 101–113; ISBN 978-0-08-101022-8.
199. Home—ClinicalTrials.Gov. Available online: https://clinicaltrials.gov/ (accessed on 15 November 2021).

*membranes*

Article

# *Zingiber cassumunar* Roxb. Essential Oil-Loaded Electrospun Poly(lactic acid)/Poly(ethylene oxide) Fiber Blend Membrane for Antibacterial Wound Dressing Application

**Pattawika Sinsup [1], Veerawat Teeranachaideekul [2], Arthit Makarasen [3], Laemthong Chuenchom [4], Pongthep Prajongtat [1], Supanna Techasakul [3], Peerada Yingyuad [3,5,*] and Decha Dechtrirat [1,3,6,*]**

1 Department of Materials Science, Faculty of Science, Kasetsart University, Bangkok 10900, Thailand; pattawika.s@ku.th (P.S.); fscipop@ku.ac.th (P.P.)
2 Department of Pharmacy, Faculty of Pharmacy, Mahidol University, Bangkok 10400, Thailand; veerawat.tee@mahidol.edu
3 Laboratory of Organic Synthesis, Chulabhorn Research Institute, Bangkok 10210, Thailand; arthit@cri.or.th (A.M.); supanna@cri.or.th (S.T.)
4 Division of Physical Science, Center of Excellence for Innovation in Chemistry, Faculty of Science, Prince of Songkla University, Songkhla 90110, Thailand; laemthong.c@psu.ac.th
5 Department of Chemistry, Faculty of Science, Kasetsart University, Bangkok 10900, Thailand
6 Specialized Center of Rubber and Polymer Materials for Agriculture and Industry (RPM), Faculty of Science, Kasetsart University, Bangkok 10900, Thailand
* Correspondence: fscipdy@ku.ac.th (P.Y.); fscidcd@ku.ac.th (D.D.)

**Citation:** Sinsup, P.; Teeranachaideekul, V.; Makarasen, A.; Chuenchom, L.; Prajongtat, P.; Techasakul, S.; Yingyuad, P.; Dechtrirat, D. *Zingiber cassumunar* Roxb. Essential Oil-Loaded Electrospun Poly(lactic acid)/Poly(ethylene oxide) Fiber Blend Membrane for Antibacterial Wound Dressing Application. *Membranes* **2021**, *11*, 648. https://doi.org/10.3390/membranes11090648

Academic Editor: Andrea Ehrmann

Received: 29 July 2021
Accepted: 20 August 2021
Published: 24 August 2021



**Abstract:** The essential oil from *Zingiber cassumunar* Roxb. (*Plai*) has long been used in Thai herbal remedies to treat inflammation, pains, sprains, and wounds. It was therefore loaded into an electrospun fibrous membrane for use as an analgesic and antibacterial dressing for wound care. The polymer blend between poly(lactic acid) and poly(ethylene oxide) was selected as the material of choice because its wettability can be easily tuned by changing the blend ratio. Increasing the hydrophilicity and water uptake ability of the material while retaining its structural integrity and porosity provides moisture balance and removes excess exudates, thereby promoting wound healing. The effect of the blend ratio on the fiber morphology and wettability was investigated using scanning electron microscopy (SEM) and contact angle measurement, respectively. The structural determination of the prepared membranes was conducted using Fourier-transform infrared spectroscopy (FTIR). The release behavior of (E)-1-(3,4-dimethoxyphenyl) butadiene (DMPBD), a marker molecule with potent anti-inflammatory activity from the fiber blend, showed a controlled release characteristic. The essential oil-loaded electrospun membrane also showed antibacterial activity against *S. aureus* and *E. coli*. It also exhibited no toxicity to both human fibroblast and keratinocyte cells, suggesting that the prepared material is suitable for wound dressing application.

**Keywords:** controlled release; electrospinning; essential oil; fibrous membrane; wound dressing; *Zingiber cassumunar* Roxb.

## 1. Introduction

*Zingiber cassumunar* Roxb., named *Plai* in Thailand, is a medicinal plant that has long been used in Thai traditional herbal remedies for pain, sprains, inflammation, wounds, skin diseases, asthma, and rheumatism [1–3]. The essential oil and extract from the fresh rhizome of *Zingiber cassumunar* Roxb. have been reported to exhibit local anesthetic and analgesic effects [4,5], reduce pain and inflammation [6–9], and inhibit the growth of fungi and bacteria [10–12]. Due in part to its reported medical uses and pharmacological activities, *Zingiber cassumunar* Roxb. is a good candidate for development as a medical or health care product.

Wound dressing plays an essential role in accelerating the wound healing process. The ideal dressing should prevent bacterial infection, remove excess exudates, allow the ease of gas exchange, and maintain moisture balance on the wound bed [13,14]. Compared with typical dressings such as gauzes, lint, bandages, and cotton wool, nanofibers are more attractive, owing to their large specific surface area, high porosity, and excellent pore interconnectivity. These extraordinary characteristics exert several advantages for wound healing, such as assisting cell attachment and proliferation as well as facilitating the permeability of moisture and gas, which are beneficial for cell growth and allowing the absorption of additional exudates containing nutrients for bacterial growth. In addition, nanofiber scaffolds closely mimic the structure of the extracellular matrix (ECM) [15–17]. Therefore, nanofibers are promising candidates for wound dressing materials as they provide a suitable environment for wound healing.

In recent years, both natural and synthetic polymers have been used to prepare nanofiber-based wound dressings via the electrospinning process. Among these different polymers, poly(lactic acid) (PLA), an FDA-approved synthetic biodegradable polymer derived from renewable resources, is of particular interest. Due to its biodegradability, biocompatibility, good mechanical properties, and low cost, PLA has become a candidate of choice in bio-related applications [18–20]. The hydrophobic nature of PLA could render better interaction with lipophilic drugs or plant essential oils. However, this property could also reduce cellular interactions and limit exudate uptake capability. Therefore, PLA needs to be modified in order to adjust its wettability property while retaining its structural integrity and porous morphology. In bulk modification, hydrophilic polymers such as polyethylene oxide (PEO) are usually incorporated into the electrospinning solution [21–24]. PEO is biodegradable, biocompatible, non-toxic, and FDA-approved. Thus, it is considered a suitable choice to produce PLA-based composite nanofibers. It has been reported that PLA mixed with PEO could successfully be electrospun, resulting in nanofibers with a smooth surface, high porosity, and enhanced hydrophilicity while maintaining structural integrity when compared with pure PLA nanofibers [21]. In a previous study, the incorporation of rapamycin, a water-insoluble antibiotic and antiproliferative agent, into the PLA and PEO blend solution in the appropriate ratios produced smooth and uniform nanofibers with high encapsulation efficiency [25]. The blended PLA/PEO nanofiber was also shown to manipulate and control the release of loaded natural and synthetic compounds [25,26]. In addition, the PLA/PEO fiber blend loaded with grape seed extract (GSE) enhanced fibroblast cell adhesion and proliferation in comparison with the PLA/GSE nanofiber as a result of increased hydrophilicity [27].

The electrospun nanofiber loaded with *Plai* oil has been studied previously by Tonglairoum et al. using polyvinylpyrrolidone (PVP), a hydrophilic polymer, blended with 2-hydroxypropyl-β-cyclodextrin (HPβCD) [28]. It was found that a maximum of 20% of *Plai* oil could be loaded into the nanofiber with high entrapment efficiency. However, entrapment efficiency decreased markedly when the incorporated amount of *Plai* oil was increased to 30%. These findings were primarily due to less hydrophilic-hydrophobic interaction between the PVP polymer and *Plai* oil. In addition, the PVP nanofiber showed a relatively short shelf life of up to one week before they fused together, which could be a result of the hygroscopic nature of PVP. PLA has been proven to be a promising matrix to improve the entrapment efficiency of *Plai* oil as well as prolong nanofiber shelf life [29]. Therefore, this work incorporated *Plai* essential oil into a PLA/PEO fiber blend. PEO was blended with PLA to produce a matrix with enhanced hydrophilicity, resulting in the increased water uptake capacity of the fiber. The *Plai* essential oil-loaded fiber blend was prepared through electrospinning (Scheme 1). Characterization of the prepared materials was carried out using various techniques. In vitro release, antibacterial, and cytotoxicity tests were also performed to demonstrate the potential of using the prepared materials as wound dressings.

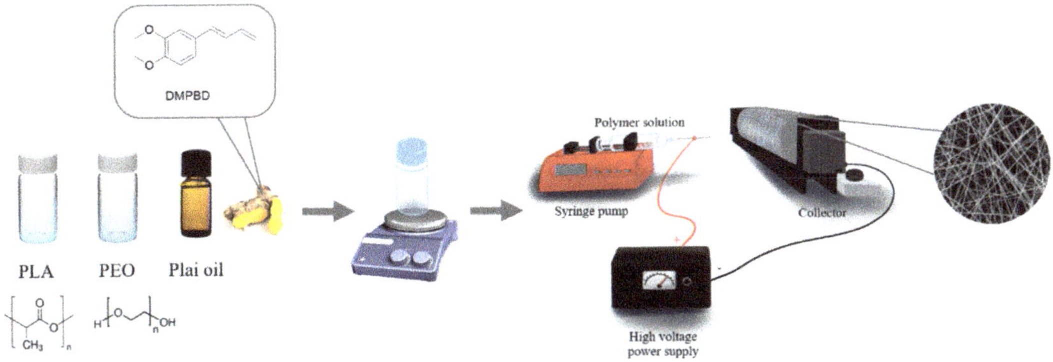


**Scheme 1.** Schematic representation for the fabrication of the *Plai* essential oil-loaded electrospun PLA/PEO fibrous membrane.

## 2. Materials and Methods

### 2.1. Materials

*Plai* essential oil was obtained from Thai China Flavours & Fragrances Industry Co., Ltd. (Bangkok, Thailand). PEO (Mw ~ 100,000) and PLA (Mw ~ 60,000) were acquired from Sigma Aldrich. Dulbecco's modified Eagle's medium (DMEM, Gibco®, MD, USA) with 10% fetal bovine serum (FBS, Gibco®), 1 mM sodium pyruvate (Gibco®), and penicillin/streptomycin (Gibco®) were supplied by Life Technologies. In addition, (3-(4,5-dimethylthiazol-2-yl)-5-(3-carboxymethoxyphenyl)-2-(4-sulfophenyl)-2H-tetrazolium inner salt was from obtained Promega Corporation. All buffer salts and organic solvents were of analytical grade from Merck.

### 2.2. Analysis of Plai Essential Oil

The chemical constituents and DMPBD content in *Plai* essential oil were determined using GC/MS (QP2020, Shimadzu, Japan). Briefly, the analysis was performed using a capillary column (SH-Rxi-5Sil MS; 30 m length; 0.25 mm ID; 0.25 μm film thickness, Shimadzu, Japan) with an injection volume of 1 μL. The temperature of the injection port was fixed at 200 °C. The temperature of the oven was increased from 60 °C to 150 °C within 18 min, from 150 °C to 180 °C within 3 min, and then kept at 180 °C for 5 min. The mass scan was operated using the electron impact ionization mode over the mass range of 45–400 amu. DMPBD was isolated and characterized as previously described by Wongkanya et al. [29]. Briefly, *Plai* oil was loaded onto the silica gel column and then eluted with hexane/ethyl acetate (95:5 *v*/*v*). DMPBD, as a colorless oil, was obtained after solvent evaporation under reduced pressure.

### 2.3. Fabrication of Electrospun Fiber Blend Membranes

PLA/PEO solutions were prepared in DCM/DMSO (8:2 *v*/*v*) at 8%, 10%, and 12% *w*/*v*. The weight ratios of PLA and PEO were 9:1, 8:2, and 7:3 *w*/*w*. The effects of polymer concentration and PLA/PEO weight ratio on fiber morphology were studied. *Plai* oil at 30% (*w*/*w*, to the polymer content) was added into the polymer solutions and then thoroughly mixed. The prepared solutions were filled in a 10 mL glass syringe connected with a blunt needle (20-gauge). The syringe was assembled to the infusion pump, and the polymer solution was delivered at 0.5 mL/h. The electrospinning was operated using a high-voltage power supply (ES30P-5W, GAMMA, Ormond Beach, FL, USA) at 20 kV with the collector/needle tip distance of 15 cm. The experiments were carried out at 25 °C and 40% RH. The electrospun fibers were collected on a spinning drum and thereafter stored in a desiccator in the dark until further study.

### *2.4. Materials Characterization*

#### 2.4.1. Scanning Electron Microscopy

The electrospun membranes were mounted to the stubs. The samples were subsequently coated with gold using a sputter coater. Then, the fiber morphology was visualized using a scanning electron microscope (Quanta 450, FEI, Eindhoven, The Netherlands) at an accelerating voltage of 15 kV. Fiber diameters from the SEM micrographs were analyzed using the ImageJ software (NIH), and the distribution and average of fiber diameters were determined from 100 random fibers.

#### 2.4.2. Contact Angle Measurement

Water contact angle measurement was conducted by an optical contact angle measuring system (Dataphysics OCA 20). A 5 μL water droplet was dispensed onto the fiber mat, and the corresponding image of the water droplet was taken. In total, 10 different positions on the fiber mat were tested, and the average contact angle was calculated.

#### 2.4.3. ATR-FTIR

The pristine fiber, fiber blend, and *Plai* oil-loaded fiber blend membranes were analyzed by a Fourier-transform infrared spectrometer (Perkin-Elmer Spectrum One FTIR, USA). The spectra were collected from 4000 $cm^{-1}$ to 650 $cm^{-1}$ with 64 scans and a resolution of 4 $cm^{-1}$.

#### 2.4.4. Entrapment Efficiency of DMPBD in the Fiber Blend

Solvent extraction of DMPBD from the fiber blend was performed by placing the fiber membrane in a sealed glass vial containing 20 mL hexane. After continuously agitating for 6 h, the extraction solvent was collected and the amount of DMPBD extracted was analyzed by GC/MS. The entrapment efficiency was calculated using Equation (1):

$$EE\% = (w_t/w_i) \times 100 \quad (1)$$

where $w_t$ is the weight of DMPBD extracted from the fiber blend and $w_i$ is the initial weight of DMPBD loaded into the fiber blend.

### *2.5. In Vitro Release*

The in vitro release of DMPBD from the fiber membrane was conducted in PBS solution (pH 7.4) at ambient temperature. The pre-weighed fiber samples with a diameter of 16 mm were placed onto a regenerated cellulose membrane and mounted between the donor and receptor chambers of Franz diffusion cell. At specific time intervals of 0.17, 0.33, 0.5, 0.67, 0.83, 1, 2, 4, 6, 8, 10, 12, 24, and 48 h, 1 mL of solution was withdrawn and directly replaced with an equal amount of fresh PBS solution. The collected samples were subsequently extracted with 500 μL hexane twice. After solvent extraction, the hexane layer was collected, combined, and evaporated. The obtained residue was redissolved in hexane, and the amount of DMPBD released was quantified using GC/MS.

Release kinetics were assessed using the Ritger–Peppas equation [30], which is often used to describe the release from the polymeric system. An initial 60% of the cumulative release data were fitted to Equation (2):

$$M_t/M_\infty = kt^n \quad (2)$$

where $M_t/M_\infty$ is the fraction of compound released at time t, k is the rate constant, and n is the release exponent, which identifies the release mechanism. When $n \leq 0.5$, the release is governed by a Fickian diffusion mechanism. When $n \geq 1.0$, the release mechanism follows the case II transport. The mechanism lies between the previous means for $0.5 < n < 1$ and is considered an anomalous non-Fickian transport.

### 2.6. In Vitro Antibacterial Test

The in vitro antibacterial activity of the fiber blend was tested using a disk diffusion method. *Staphylococcus aureus* (*S. aureus*) (ATCC 25923) and *Escherichia coli* (*E. coli*) (ATCC 25922) were chosen as representatives for gram-positive and gram-negative bacterial strains frequently involved in wound infections [31]. These bacteria were cultured at 37 °C for 24 h. Subsequently, the bacterial suspension of $1 \times 10^8$ CFU/mL was spread over the Mueller–Hinton agar (MHA) plate. The fiber membranes (Ø 6 mm) were sterilized with UV light for 30 min prior to testing. Then, the fiber membranes were mounted on the agar plates and incubated at 37 °C for 24 h. The antibacterial activity of the fiber blend membranes was thereafter calculated from the diameter of the clear inhibition zone. The experiments were conducted in triplicate, and the results were reported as the mean ± standard deviation.

### 2.7. In Vitro Cytotoxicity Test

Human primary fibroblast cells (Normal, Human, Adult (HDFa), ATCC® PCS-201-012™, Manassas, VA, USA) and immortalized human keratinocytes (HaCaT, human keratinocyte cells, ATCC® Number PCS-200-011™, 300493, CLS, Eppelheim, Germany) were cultured in Dulbecco's modified Eagle's medium (DMEM) with 10% fetal bovine serum (FBS), 1 mM sodium pyruvate, 100 U/mL penicillin, and 100 µg/mL streptomycin. The cells were cultured at 37 °C and 5% $CO_2$ and were serially passaged at 70–80% confluence. Then, the experiments were performed with subconfluent cells at passage three in the proliferation phase.

In this test, cytotoxicity was evaluated by MTS assay. The cultured cells were seeded 1 day prior to the test in 96-well plates (10,000 cells/well for fibroblast and 8000 cells/well for keratinocyte). The fiber membranes were immersed in PBS for 24 h to obtain the extraction media at 0.63, 1.25, 2.5, and 5 mg/mL. The extract was then filtered through the 0.22 µm syringe filter. Then, cells were exposed to the extraction media at the different concentrations. PBS was used as a negative control for the test. The supernatant was removed after 24 h of incubation, and cells were rinsed twice with PBS. Fresh media (100 µL) containing MTS (20 µL) was added in each well plate. Then, cells were incubated at 5% $CO_2$ at 37 °C for another 3 h. For viable cells, the colorless MTS reagent was converted into a soluble-colored 1-(4,5-dimethylthiazol-2-yl)-3,5-diphenylformazan product, which absorbed light at 490 nm. Cell viability was then calculated based on a change in absorbance at 490 nm compared to the negative control.

### 2.8. Statistical Analysis

The experimental data are reported as mean ± standard derivation (SD). The results were analyzed by one-way ANOVA. The $p$-values of less than 0.05 were statistically accepted as significant.

## 3. Results and Discussion

### 3.1. Fabrication and Characterization of Fiber Blend Membranes

The oil-free PLA/PEO fiber blend membranes were successfully prepared through the electrospinning of the PLA/PEO mixture solutions at the polymer concentrations of 8%, 10%, and 12% $w/v$. The weight ratios of PLA and PEO were 9:1, 8:2, and 7:3 $w/w$. The effects of polymer concentration and PLA/PEO weight ratio on fiber size and morphology were studied by SEM. As shown in Figure 1, the fibers prepared at the polymer concentrations of 8% and 10% exhibited a grooved and wrinkled structure at every blend ratio. The formation of this secondary surface morphology is attributed to the fast solvent evaporation (dichloromethane) at the early stage of electrospinning, followed by phase separation and the creation of tiny holes on the fiber surfaces. After the elongation and solidification of the polymer jet, the created voids turned themselves into grooves and wrinkles on the fiber surfaces [32]. On the contrary, at 12% of the polymer blend, uniform fibers with a smooth surface were obtained. This suggests that the low polymer concentration is preferable for the formation of grooves and wrinkles, as previously reported by Liu et al. [33]. Hence,

12% of the polymer blend concentration was chosen to prepare the essential oil-loaded fibrous membranes for further investigation. There are no obvious differences in the SEM images at the same polymer concentration, suggesting that the PLA/PEO weight ratio had no apparent effect on fiber morphology. The *Plai* oil-loaded fiber blend membranes at different PLA/PEO ratios were obtained by electrospinning the polymer blend solutions (12% *w*/*v*) incorporated with *Plai* oil. As seen in Figure 2a–c, the oil-loaded fibers had a round shape with smooth surfaces similar to those without *Plai* oil, revealing that the loading of *Plai* oil did not affect fiber morphology.

The average fiber diameters were determined from the SEM micrographs and are summarized in Table 1. At the same polymer concentration but different PLA/PEO blend ratios, the fiber diameters were not statistically different, suggesting that the blend ratio exerted no apparent effect on the fiber size. When the polymer concentration increased, the mean diameter increased as a result of the increase in molecular entanglement and viscosity of the solution. On the contrary, the mean diameters of the *Plai* oil-loaded fiber blend membranes were relatively smaller than those without *Plai* oil. This is possibly due to the decrease in solution viscosity after incorporating *Plai* oil into the blend solution.

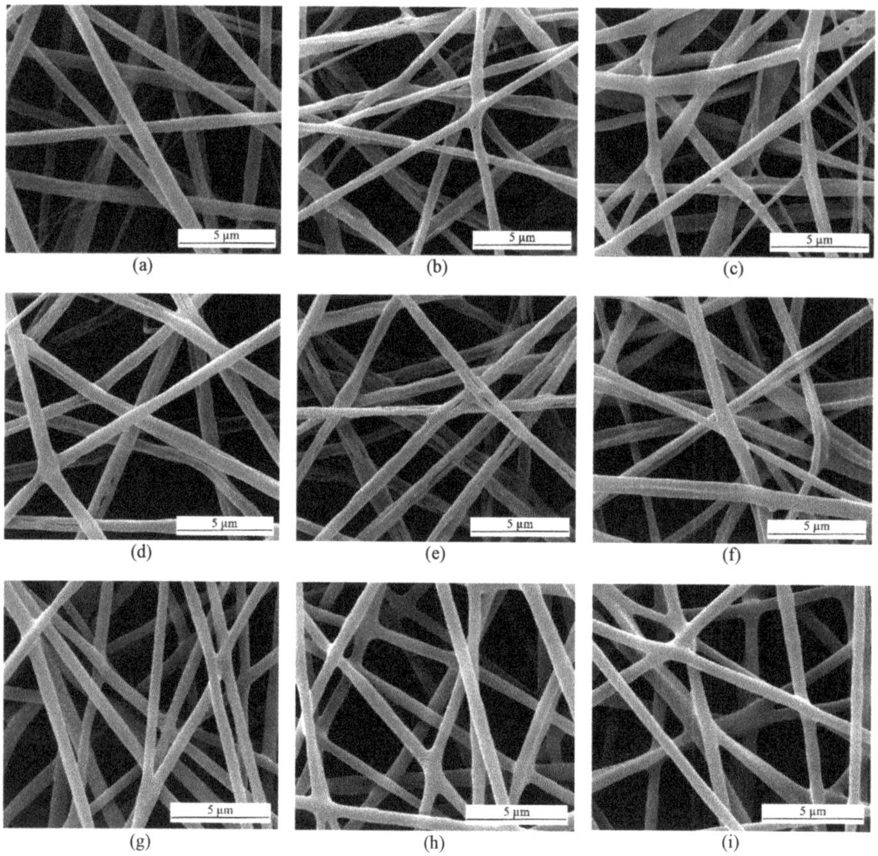


**Figure 1.** SEM images of the PLA/PEO fiber blend prepared at different polymer concentrations and PLA/PEO ratios: (**a**) 8% PLA/PEO (90:10), (**b**) 8% PLA/PEO (80:20), (**c**) 8% PLA/PEO (70:30), (**d**) 10% PLA/PEO (90:10), (**e**) 10% PLA/PEO (80:20), (**f**) 10% PLA/PEO (70:30), (**g**) 12% PLA/PEO (90:10), (**h**) 12% PLA/PEO (80:20), and (**i**) 12% PLA/PEO (70:30).

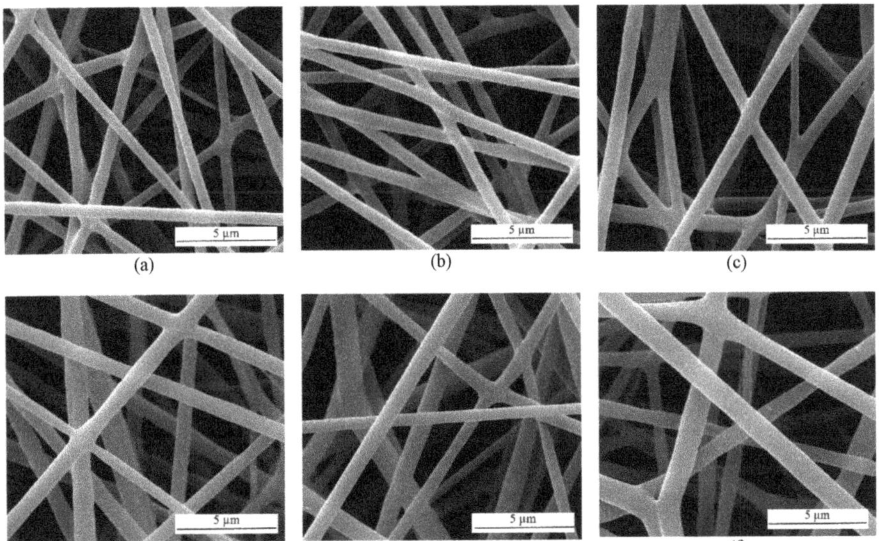


(a) (b) (c) (d) (e) (f)

**Figure 2.** SEM images of the *Plai* oil-loaded fiber blend prepared at 12% polymer concentration with different PLA/PEO ratios: (**a**) PLA/PEO (90:10) + *Plai* oil, (**b**) PLA/PEO (80:20) + *Plai* oil, (**c**) PLA/PEO (70:30) + *Plai* oil, and their corresponding images after three months storage: (**d**) PLA/PEO (90:10) + *Plai* oil, (**e**) PLA/PEO (80:20) + *Plai* oil, and (**f**) PLA/PEO (70:30) + *Plai* oil.

**Table 1.** Mean fiber diameters of electrospun membranes.

| Fiber Membranes | Mean Diameter (μm) |
|---|---|
| 8% PLA/PEO (90:10) | $0.53 \pm 0.12^a$ |
| 8% PLA/PEO (80:20) | $0.54 \pm 0.09^a$ |
| 8% PLA/PEO (70:30) | $0.57 \pm 0.12^a$ |
| 10% PLA/PEO (90:10) | $0.61 \pm 0.14^b$ |
| 10% PLA/PEO (80:20) | $0.62 \pm 0.12^b$ |
| 10% PLA/PEO (70:30) | $0.64 \pm 0.14^b$ |
| 12% PLA/PEO (90:10) | $0.71 \pm 0.13^c$ |
| 12% PLA/PEO (80:20) | $0.72 \pm 0.12^c$ |
| 12% PLA/PEO (70:30) | $0.74 \pm 0.11^c$ |
| 12% PLA/PEO (90:10) + *Plai* oil | $0.67 \pm 0.12^d$ |
| 12% PLA/PEO (80:20) + *Plai* oil | $0.68 \pm 0.13^d$ |
| 12% PLA/PEO (70:30) + *Plai* oil | $0.70 \pm 0.12^d$ |

The statistically significant differences ($p < 0.05$) are indicated with different superscript letters.

The water contact angle measurement was performed to investigate the surface wettability of the prepared fibrous membranes after blending PLA with PEO at different weight ratios. As evidenced by the water contact angle measurement in Figure 3, the average contact angle of the fibers tended to decrease with the increasing amount of PEO. The mean contact angles of the PLA fiber and the blend at the PLA/PEO weight ratios of 90:10, 80:20, and 70:30 were 143.7° ± 3.1°, 112.9° ± 3.3°, 77.2° ± 2.8°, and 41.5° ± 3.6°, respectively. As the weight amount of PEO increased, the contact angle decreased, similar to the previous report by Athanasoulia et al. [34]. The contact angle of the blend at the PLA/PEO ratio of 70:30 was the lowest, indicating the most hydrophilic surface and suggesting the highest water uptake ability. Increasing those properties is particularly important for the dressing application as they enhance the wound exudate absorption, thereby promoting the wound healing process. Hence, the blend at the 70:30 weight ratio was chosen for further study. After loading *Plai* oil into the fiber blend, the contact angle increased to 47.8° ± 4.4°, indicating a slightly more hydrophobic surface due to the lipophilic nature of *Plai* oil. However, the fiber blend membrane still exhibited high surface wettability, even in the presence of *Plai* oil.

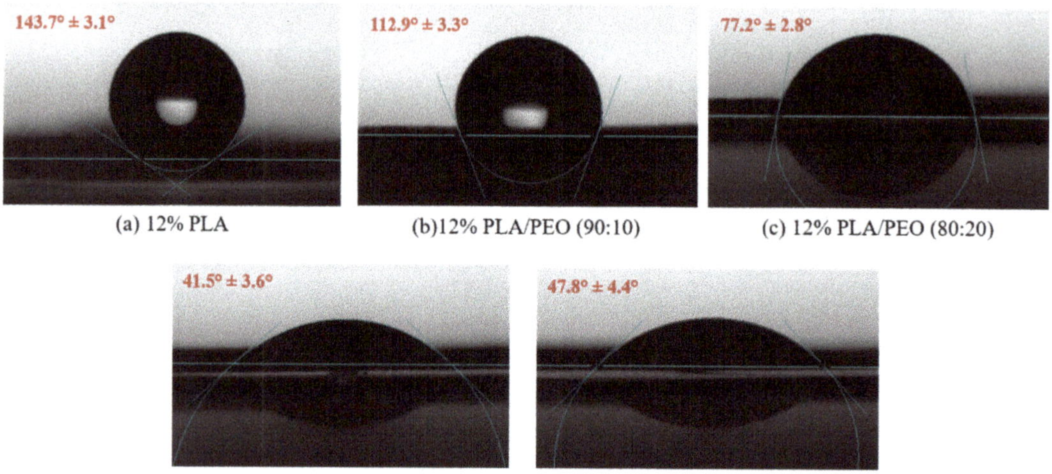


**Figure 3.** Water contact angles of (**a**) 12% PLA, (**b**) 12% PLA/PEO (90:10), (**c**) 12% PLA/PEO (80:20), (**d**) 12% PLA/PEO (70:30), and (**e**) 12% PLA/PEO (70:30) + *Plai* oil fiber membranes.

Although the hydrophilic *Plai* oil-loaded PVP electrospun fiber mat has already been established [28], it has a limited storage life of less than one week due to the hygroscopic property of PVP, which causes the fibers to melt and fuse together. On the contrary, the *Plai* oil-loaded fiber blend membrane prepared within this work was found to be stable for at least three months. No change in fiber morphology was found in the SEM images of the membranes after three months of storage (Figure 2d–f). This indicates that the PLA/PEO blend is an excellent polymeric matrix for fabricating hydrophilic fiber-based wound dressing. Its physical stability against moisture in the air remained high, even after long-term storage. Furthermore, its wettability can also be promptly tailored by altering the blend ratio.

To further confirm the suitability of the PLA/PEO blend as a platform for *Plai* oil loading, the entrapment efficiency of *Plai* oil was evaluated. The entrapment efficiency was determined based on the amount of DMPBD entrapped within the fiber blend. The DMPBD underwent solvent extraction from the *Plai* oil-loaded fiber blend membrane and was then quantified by GC/MS. The entrapment efficiency was found to be as high as 94.6 ± 3.2%, indicating that the electrospinning did not cause any apparent changes to the bioactive compound. This high entrapment efficiency can be primarily attributed to the excellent miscibility of *Plai* oil with PLA [29]. In the previous report on the *Plai* oil-loaded HPβCD/PVP nanofiber, the entrapment efficiency was found to be only 55.5% [28]. This further confirms that the PLA/PEO blend is a good matrix for *Plai* oil loading.

Structural determination and compatibility between *Plai* oil and the blend were studied using Fourier-transform infrared spectroscopy (FTIR). As shown in Figure 4, the PLA fibrous membrane exhibited strong characteristic peaks of C=O, stretching of the carbonyl group at 1755 $cm^{-1}$, and corresponding bending at 1267 $cm^{-1}$. It also showed strong peaks of C–O–C as well as stretching of the ester entity at 1184 $cm^{-1}$ and 1088 $cm^{-1}$. C–H stretching peaks at 2996 $cm^{-1}$ and 2946 $cm^{-1}$ as well as C–H bending at 1455 $cm^{-1}$ are also typically found in the PLA spectrum. The pristine PEO showed typical C–H stretching peaks of the methylene group at 2948 $cm^{-1}$ and 2884 $cm^{-1}$, C–H bending at 1466 $cm^{-1}$,

$CH_2$ wagging at 1340 $cm^{-1}$, and $CH_2$ rocking and twisting at 960 $cm^{-1}$. Its spectrum also exhibited the C–O–C stretching peaks of the ether moiety at 1108 $cm^{-1}$ and 1059 $cm^{-1}$, and bending at 841 $cm^{-1}$. As a blend, the PLA/PEO fibrous membrane showed all of the characteristic peaks of the pristine PLA and PEO at the same positions, suggesting that blending did not change the chemical structures of both polymers. After incorporating *Plai* oil into the fiber blend membrane, all of the characteristic peaks of PLA and PEO were still present in the same position as the blend without oil loading, suggesting that *Plai* oil also did not affect the chemical structures of both polymers. Furthermore, it also means that the interactions between *Plai* oil and the polymer blend were not strong enough to shift the peak positions. As the majority of chemical components in *Plai* oil are monoterpenes, their typical C–H stretching and bending peaks overlapped with those of PLA and PEO. This resulted in a slight increase in peak intensity at those stretching and bending regions after incorporation of *Plai* oil. The additional C–H stretching peak can also be observed at around 2906 $cm^{-1}$ after the oil loading. Furthermore, the oil-loaded fiber blend membrane also showed vibrational peaks of aromatic constituents between 1575 $cm^{-1}$ and 1513 $cm^{-1}$ [29,35]. This confirms the presence of DMPBD in the fiber blend membrane.

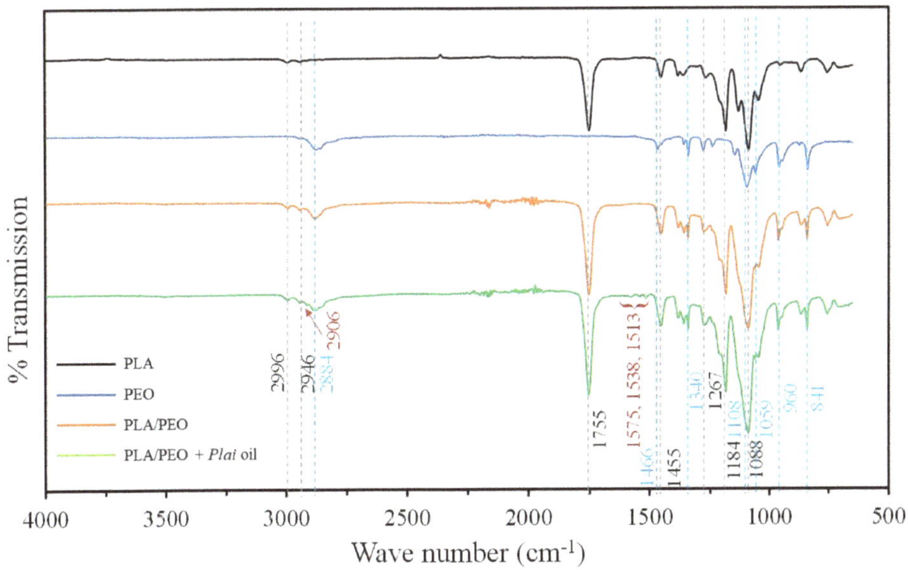


**Figure 4.** FTIR spectrum of pristine PLA, pristine PEO, the PLA/PEO fiber blend, and the oil-loaded PLA/PEO fiber blend membranes.

### 3.2. In Vitro Release Study

The cumulative release of DMPBD from the essential oil-loaded PLA and PLA/PEO (70:30) fiber blend membranes is depicted in Figure 5. The release of DMPBD for both membranes was fast at the initial state, mainly due to the burst release of the loosely bound DMPBD located at or near the essentially large fiber surfaces. After the initial fast desorption, a sustained release of the DMPBD embedded within the fibers was observed. The release at this region gradually decreased until reaching equilibrium.

At the early stage of release, the fiber blend showed a steeper slope than that of the PLA-based membrane. Afterward, the release of the marker compound from the blend increased continuously and nearly reached a plateau at an interval of 12 h. On the contrary, only ca. 80% of DMPBD was discharged from the PLA fibrous membrane at 12 h and almost 100% was released after 24 h. These findings reveal that the fiber blend exhibited a faster discharge of DMPBD. The higher hydrophilicity and water uptake ability of the blend membrane is believed to be a driving force in accelerating the release of non-polar DMPBD from the blend matrix. Contrarily, the release of DMPBD was delayed by the more hydrophobic nature of PLA. Although both membranes showed different release rates, their cumulative release eventually reached 100%. This high release may well arise from the large surface-area-to-volume ratio and the highly accessible interconnected pores of the fibrous membranes. Thus, the liquid medium can thoroughly penetrate the membrane. Therefore, the active compound can be released effectively from the matrix. The faster release of the active compound from the blend membrane will be beneficial for wound dressing applications to exert immediate local anesthetic and analgesic on wounds. On the other hand, slower release from the PLA membrane will facilitate transdermal application to prolong the pain-relieving and anti-inflammatory effects on the muscles [28]. This indicates that the blending of different proper polymers can modulate the properties of the material as well as tailor their applications.

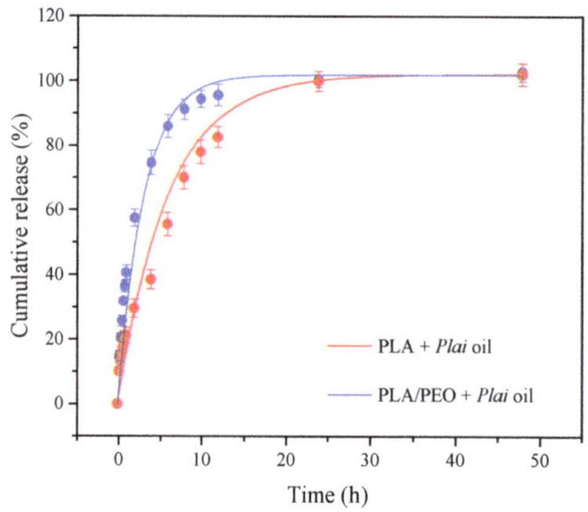


**Figure 5.** Cumulative release of DMPBD from the *Plai* oil-loaded fibrous membranes.

To explain the release mechanism, the Ritger and Peppas model was used to fit the release data. The correlation coefficients ($R^2$) signifying the goodness of the curve fitting were found to be 0.9923 and 0.9905 for the PLA/PEO blend and PLA fibrous membranes, respectively, revealing that the release kinetics of DMPBD were well-fitted to the Ritger and Peppas model. The calculated kinetics exponent ($n$) value was used to categorize the release profile. When $n$ is lower than 0.5, the release is governed by a Fickian diffusion mechanism. When $n$ is equal to 1.0, the release follows a swelling-controlled mechanism (case II transport). When $n$ is between 0.5 and 1.0, the release is defined as an anomalous non-Fickian transport. It combines the swelling-controlled drug release with diffusion. The $n$ value was found to be 0.46 for the PLA-based membrane, suggesting that the release of DMPBD was diffusion-controlled. The release depends on the concentration gradient of the drug between the release media and the polymer matrix [27]. In the case of the PLA/PEO blend membrane, the $n$ value was 0.53, meaning that the blend exhibited an anomalous (non-Fickian) transport mechanism, in which the release was caused by diffusion and matrix swelling [36] due to the presence of hydrophilic PEO. The combination of both phenomena explains the faster release of DMPBD from the fiber blend membrane.

### 3.3. Antibacterial Test

Antibacterial activity of the fiber blend membranes was tested against two representative bacterial strains (i.e., *S. aureus*, *E. coli*) typically found in wound infection by the disk diffusion method. The fibrous membranes were mounted and incubated on the MHA agar plates for 24 h. Then, the clear inhibition zones were measured. In Figure 6, the PLA/PEO fiber blend, as a control, shows no inhibition zones against both bacterial strains, but the oil-loaded membrane exhibits the clear zones with the diameters of 17.33 ± 0.29 and 12.67 ± 0.76 mm against *S. aureus* and *E. coli*, respectively. This indicates that the *Plai* oil is responsible for the inhibition rather than the PLA/PEO blend. Therefore, the prepared *Plai* oil-loaded fiber blend membrane can be considered an alternative antibacterial wound dressing.

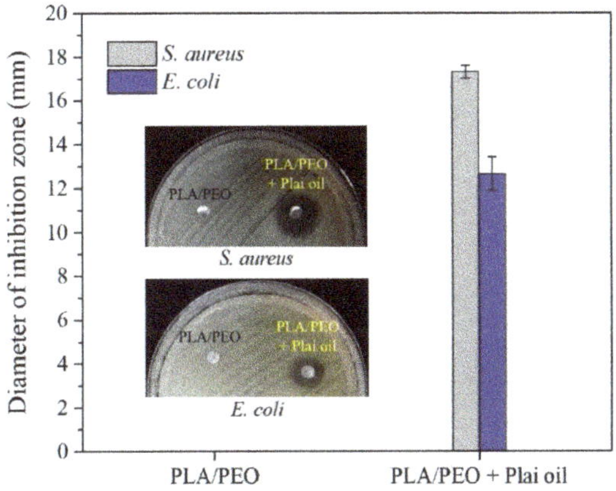


**Figure 6.** Antibacterial activity of PLA/PEO and essential oil-loaded PLA/PEO fiber blend membranes against *S. aureus* and *E. coli*.

### 3.4. In Vitro Cytotoxicity

The indirect in vitro cytotoxicity study of the PLA/PEO blend fiber membrane with and without *Plai* oil was carried out through MTS assay on human dermal fibroblast and keratinocyte cells (HaCat). The membranes were cut into pieces and immersed in PBS buffer for 24 h to obtain the extract media at 0.63, 1.25, 2.5, and 5 mg/mL. After 24 h incubation of the tested cells to the extraction media, the cell viability was evaluated, as shown in Figure 7. A cell viability higher than 80% is considered non-toxic in this test. The viability of both fibroblast and keratinocyte cells after treatment with the extraction media of both membranes at every concentration was in the range of 95–116%, indicating that both materials are non-toxic to the tested cells. This suggests that the present *Plai* oil-loaded fiber blend membrane is safe for use as a wound dressing material.

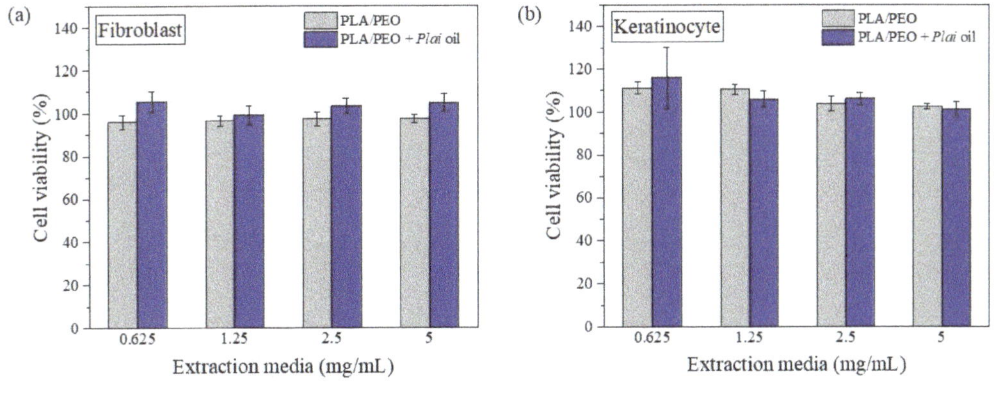


**Figure 7.** Cell viability (%) of (**a**) human dermal fibroblast and (**b**) human keratinocyte after 24 h exposure to the different concentrations of extraction media of the PLA/PEO and *Plai* oil-loaded PLA/PEO fiber blend membranes.

## 4. Conclusions

In this study, *Plai* essential oil with analgesic, anti-inflammatory, and antibacterial activities was successfully loaded into the electrospun PLA/PEO fiber blend. The membrane obtained consisted of smooth fibers with diameter in the nanometer range. The hydrophilicity of the membrane can be simply modulated by blending PLA with PEO at different weight ratios. Although the membrane was hydrophilic, the *Plai* oil entrapment efficiency and physical stability against moisture in the air were still high, implying that the blend is a good platform for the loading of *Plai* oil. The hydrophilic fiber blend membrane loaded with *Plai* oil initially showed a fast release of DMPBD, followed by a sustained release in the following hours. This will be beneficial for wound care to exert immediate local anesthetic and analgesic on wounds and to maintain these effects for several hours. As the fibrous membrane had a large surface to volume ratio and highly interconnected macropores, the liquid medium could thoroughly impregnate the membrane. As such, the complete release of DMPBD was as expected. Further in vitro antibacterial and indirect cytotoxicity tests also showed that the *Plai* oil-loaded fiber blend membrane could inhibit the growth of both *S. aureus* and *E. coli* and exhibited no toxicity to human skin cells. This suggests that the present fiber blend membrane has promising potential for use in a new generation of analgesic and antibacterial wound dressings.

**Author Contributions:** Conceptualization and methodology, D.D., P.Y. and L.C.; validation, D.D., V.T. and L.C.; investigation, P.S., V.T. and A.M.; writing—original draft preparation, D.D. and P.Y.; writing—review and editing, D.D., P.Y., P.P. and S.T.; visualization, D.D., P.Y. and P.P.; supervision, S.T.; project administration, D.D. and L.C.; funding acquisition, P.P. and S.T. All authors have read and agreed to the published version of the manuscript.

**Funding:** This work was financially supported by the Thailand Science Research and Innovation (TSRI), Chulabhorn Research Institute (grant No. 313/2229); the Kasetsart University Research and Development Institute (grant No. FF(KU)25.64); the Graduate School Fellowship Program, Kasetsart University; and the International SciKU Branding (ISB), Faculty of Science, Kasetsart University.

**Institutional Review Board Statement:** Not applicable.

**Informed Consent Statement:** Not applicable.

**Data Availability Statement:** Not applicable.

**Conflicts of Interest:** The authors declare no conflict of interest.

## References

1. Koontongkaew, S.; Poachanukoon, O.; Sireeratawong, S.; Dechatiwongse Na Ayudhya, T.; Khonsung, P.; Jaijoy, K.; Soawakontha, R.; Chanchai, M. Safety evaluation of Zingiber cassumunar Roxb. rhizome extract: Acute and chronic toxicity studies in rats. *Int. Sch. Res. Not.* **2014**, *2014*, 632608.
2. Pongprayoon, U.; Tuchinda, P.; Claeson, P.; Sematong, T.; Reutrakul, V.; Soontornsaratune, P. Topical anti-inflammatory activity of the major lipophilic constituents of the rhizome of Zingiber cassumunar. Part II: Hexane extractives. *Phytomedicine* **1997**, *3*, 323–326. [CrossRef]
3. Chongmelaxme, B.; Sruamsiri, R.; Dilokthornsakul, P.; Dhippayom, T.; Kongkaew, C.; Saokaew, S.; Chuthaputti, A.; Chaiyakunapruk, N. Clinical effects of Zingiber cassumunar (Plai): A systematic review. *Complement. Ther. Med.* **2017**, *35*, 70–77. [CrossRef]
4. Cheechareoan, S.; Pathanawiriyasirikul, T.; Manmee, C.; Janpol, K. Efficacy of Plai cream in adult patients with muscle strain: A randomized, double-blind, placebo-controlled trial. *J. Med. Assoc. Thail.* **2016**, *99*, 147–152.
5. Leelarungrayub, J.; Manorsoi, J.; Manorsoi, A. Anti-inflammatory activity of niosomes entrapped with Plai oil (Zingiber cassumunar Roxb.) by therapeutic ultrasound in a rat model. *Int. J. Nanomed.* **2017**, *12*, 2469–2476. [CrossRef]
6. Panthong, A.; Kanjanapothi, D.; Niwatananant, W.; Tuntiwachwuttikul, P.; Reutrakul, V. Anti-inflammatory activity of compound D {(E)-4-(3′,4′-dimethoxyphenyl)but-3-en-2-ol} isolated from Zingiber cassumunar Roxb. *Phytomedicine* **1997**, *4*, 207–212. [CrossRef]
7. Masuda, T.; Jitoe, A.; Mabry, T.J. Isolation and structure determination of cassumunarins A, B, and C: New anti-inflammatory antioxidants from a tropical ginger, Zingiber cassumunar. *J. Am. Oil Chem. Soc.* **1995**, *72*, 1053–1057. [CrossRef]
8. Ozaki, Y.; Kawahara, N.; Harada, M. Anti-inflammatory effect of Zingiber cassumunar Roxb. and its active principles. *Chem. Pharm. Bull.* **1991**, *39*, 2353–2356. [CrossRef] [PubMed]

9. Jeenapongsa, R.; Yoovathaworn, K.; Sriwatanakul, K.M.; Pongprayoon, U.; Sriwatanakul, K. Anti-inflammatory activity of (E)-1-(3,4-dimethoxyphenyl) butadiene from Zingiber cassumunar Roxb. *J. Ethnopharmacol.* **2003**, *87*, 143–148. [CrossRef]
10. Prakatthagomol, W.; Sirithunyalug, J.; Okonogi, S. Comparison of antibacterial activity against food-borne bacteria of Alpinia galanga, Curcuma longa, and Zingiber cassumunar. *CMU J. Nat. Sci.* **2012**, *11*, 177–186.
11. Boonyanugomol, W.; Kraisriwattana, K.; Rukseree, K.; Boonsam, K.; Narachai, P. In vitro synergistic antibacterial activity of the essential oil from Zingiber cassumunar Roxb against extensively drug-resistant Acinetobacter baumannii strains. *J. Infect. Public Health* **2017**, *10*, 586–592. [CrossRef]
12. Taechowisan, T.; Suttichokthanakorn, S.; Phutdhawong, W.S. Antibacterial and cytotoxicity activities of phenylbutanoids from Zingiber cassumunar Roxb. *J. Appl. Pharm. Sci.* **2018**, *8*, 121–127.
13. Alven, S.; Buyana, B.; Feketshane, Z.; Aderibigbe, B.A. Electrospun nanofibers/nanofibrous scaffolds loaded with silver nanoparticles as effective antibacterial wound dressing materials. *Pharmaceutics* **2021**, *13*, 964. [CrossRef]
14. Stoica, A.E.; Chircov, C.; Grumezescu, A.M. Nanomaterials for wound dressings: An up-to-date overview. *Molecules* **2020**, *25*, 2699. [CrossRef]
15. Keshvardoostchokami, M.; Majidi, S.S.; Huo, P.; Ramachandran, R.; Chen, M.; Liu, B. Electrospun nanofibers of natural and synthetic polymers as artificial extracellular matrix for tissue engineering. *Nanomaterials* **2021**, *11*, 21. [CrossRef]
16. Tanzli, E.; Ehrmann, A. Electrospun nanofibrous membranes for tissue engineering and cell growth. *Appl. Sci.* **2021**, *11*, 6929. [CrossRef]
17. Politi, S.; Carotenuto, F.; Rinaldi, A.; Di Nardo, P.; Manzari, V.; Albertini, M.C.; Araneo, R.; Ramakrishna, S.; Teodori, L. Smart ECM-based electrospun biomaterials for skeletal muscle regeneration. *Nanomaterials* **2020**, *10*, 1781. [CrossRef] [PubMed]
18. Li, G.; Zhao, M.; Xu, F.; Yang, B.; Li, X.; Meng, X.; Teng, L.; Sun, F.; Li, Y. Synthesis and Biological Application of Polylactic Acid. *Molecules* **2020**, *25*, 5023. [CrossRef] [PubMed]
19. Im, S.H.; Im, D.H.; Park, S.J.; Chung, J.J.; Jung, Y.; Kim, S.H. Stereocomplex Polylactide for Drug Delivery and Biomedical Applications: A Review. *Molecules* **2021**, *26*, 2846. [CrossRef]
20. Boey, J.Y.; Mohamad, L.; Khok, Y.S.; Tay, G.S.; Baidurah, S. A Review of the Applications and Biodegradation of Polyhydroxyalkanoates and Poly(lactic acid) and Its Composites. *Polymers* **2021**, *13*, 1544. [CrossRef]
21. Honarbakhsh, S.; Pourdeyhimi, B. Scaffolds for drug delivery, part I: Electrospun porous poly(lactic acid) and poly(lactic acid)/poly(ethylene oxide) hybrid scaffolds. *J. Mater. Sci.* **2011**, *46*, 2874–2881. [CrossRef]
22. Heunis, T.; Bshena, O.; Klumperman, B.; Dicks, L. Release of bacteriocins from nanofibers prepared with combinations of poly(d,l-lactide) (PDLLA) and poly(ethylene oxide) (PEO). *Int. J. Mol. Sci.* **2011**, *12*, 2158–2173. [CrossRef] [PubMed]
23. Abid, S.; Hussain, T.; Nazir, A.; Zahir, A.; Ramakrishna, S.; Hameed, M.; Khenoussi, N. Enhanced antibacterial activity of PEO-chitosan nanofibers with potential application in burn infection management. *Int. J. Biol. Macromol.* **2019**, *135*, 1222–1236. [CrossRef] [PubMed]
24. Eskitoros-Togay, Ş.M.; Bulbul, Y.E.; Tort, S.; Demirtaş Korkmaz, F.; Acartürk, F.; Dilsiz, N. Fabrication of doxycycline-loaded electrospun PCL/PEO membranes for a potential drug delivery system. *Int. J. Pharm.* **2019**, *565*, 83–94. [CrossRef]
25. Wang, B.; Li, H.; Yao, Q.; Zhang, Y.; Zhu, X.; Xia, T.; Wang, J.; Li, G.; Li, X.; Ni, S. Local in vitro delivery of rapamycin from electrospun PEO/PDLLA nanofibers for glioblastoma treatment. *Biomed. Pharmacother.* **2016**, *83*, 1345–1352. [CrossRef]
26. Dai, R.; Lim, L.-T. Release of allyl isothiocyanate from mustard seed meal powder entrapped in electrospun PLA–PEO nonwovens. *Food Res. Int.* **2015**, *77*, 467–475. [CrossRef]
27. Locilento, D.A.; Mercante, L.A.; Andre, R.S.; Mattoso, L.H.C.; Luna, G.L.F.; Brassolatti, P.; Anibal, F.d.F.; Correa, D.S. Biocompatible and biodegradable electrospun nanofibrous membranes loaded with grape seed extract for wound dressing application. *J. Nanomater.* **2019**, *2019*, 2472964. [CrossRef]
28. Tonglairoum, P.; Chuchote, T.; Ngawhirunpat, T.; Rojanarata, T.; Opanasopit, P. Encapsulation of Plai oil/2-hydroxypropyl-β-cyclodextrin inclusion complexes in polyvinylpyrrolidone (PVP) electrospun nanofibers for topical application. *Pharm. Dev. Technol.* **2014**, *19*, 430–437. [CrossRef] [PubMed]
29. Wongkanya, R.; Teeranachaideekul, V.; Makarasen, A.; Chuysinuan, P.; Yingyuad, P.; Nooeaid, P.; Techasakul, S.; Chuenchom, L.; Dechtrirat, D. Electrospun poly(lactic acid) nanofiber mats for controlled transdermal delivery of essential oil from Zingiber cassumunar Roxb. *Mater. Res. Express* **2020**, *7*, 055305. [CrossRef]
30. Ritger, P.L.; Peppas, N.A. A simple equation for description of solute release I. Fickian and non-fickian release from non-swellable devices in the form of slabs, spheres, cylinders or discs. *J. Control Release* **1987**, *5*, 23–36. [CrossRef]
31. Negut, I.; Grumezescu, V.; Grumezescu, A.M. Treatment strategies for infected wounds. *Molecules* **2018**, *23*, 2392. [CrossRef]
32. Zaarour, B.; Zhu, L.; Jin, X. A Review on the secondary surface morphology of electrospun nanofibers: Formation mechanisms, characterizations, and applications. *Chem. Sel.* **2020**, *5*, 1335–1348. [CrossRef]
33. Liu, W.; Huang, C.; Jin, X. Tailoring the grooved texture of electrospun polystyrene nanofibers by controlling the solvent system and relative humidity. *Nanoscale Res. Lett.* **2014**, *9*, 350. [CrossRef] [PubMed]
34. Athanasoulia, I.-G.; Tarantili, P.A. Preparation and characterization of polyethylene glycol/poly (L-lactic acid) blends. *Pure Appl. Chem.* **2017**, *89*, 141–152. [CrossRef]
35. Lim, J.S.; Park, K.; Chung, G.S.; Kim, J.H. Effect of composition ratio on the thermal and physical properties of semicrystalline PLA/PHB HHx composites. *Mater. Sci. Eng. C* **2013**, *33*, 2131–2137. [CrossRef] [PubMed]
36. Fu, Y.; Kao, W.J. Drug release kinetics and transport mechanisms of non-degradable and degradable polymeric delivery systems. *Expert Opin. Drug Deliv.* **2010**, *7*, 429–444. [CrossRef]

membranes

Article

# Bacterial Infection-Mimicking Three-Dimensional Phagocytosis and Chemotaxis in Electrospun Poly(ε-caprolactone) Nanofibrous Membrane

**Seung-Jun Lee [1,2,†], Perry Ayn Mayson A Maza [1,2,†], Gyu-Min Sun [1,2], Petr Slama [3], In-Jeong Lee [4] and Jong-Young Kwak [1,2,4,*]**

1 Department of Pharmacology, School of Medicine, Ajou University, Worldcup-Ro 164, Suwon 16499, Korea; blacksky@naver.com (S.-J.L.); perryayn@gmail.com (P.A.M.A.M.); rbals3922@ajou.ac.kr (G.-M.S.)
2 Department of Biomedical Sciences, The Graduate School, Ajou University, Suwon 16499, Korea
3 Department of Animal Morphology, Physiology and Genetics, Faculty of AgriSciences, Mendel University in Brno, Zemedelska 1, 613 00 Brno, Czech Republic; petr.slama@mendelu.cz
4 3D Immune System Imaging Core Center, Ajou University, Suwon 16499, Korea; injeong@ajou.ac.kr
* Correspondence: jykwak@ajou.ac.kr; Tel.: +82-31-219-5064; Fax: +82-31-219-5069
† Authors contributed equally.

**Abstract:** In this study, we developed a three-dimensional (3D) in vitro infection model to investigate the crosstalk between phagocytes and microbes in inflammation using a nanofibrous membrane (NM). Poly(ε-caprolactone) (PCL)-NMs (PCL-NMs) were generated via electrospinning of PCL in chloroform. *Staphylococcus aureus* and phagocytes were able to adhere to the nanofibers and phagocytes engulfed *S. aureus* in the PCL-NM. The migration of phagocytes to *S. aureus* was evaluated in a two-layer co-culture system using PCL-NM. Neutrophils, macrophages and dendritic cells (DCs) cultured in the upper PCL-NM layer migrated to the lower PCL-NM layer containing bacteria. DCs migrated to neutrophils that cultured with bacteria and then engulfed neutrophils in two-layer system. In addition, phagocytes in the upper PCL-NM layer migrated to bacteria-infected MLE-12 lung epithelial cells in the lower PCL-NM layer. *S. aureus*-infected MLE-12 cells stimulated the secretion of tumor necrosis factor-α and IL-1α in 3D culture conditions, but not in 2D culture conditions. Therefore, the PCL-NM-based 3D culture system with phagocytes and bacteria mimics the inflammatory response to microbes in vivo and is applicable to the biomimetic study of various microbe infections.

**Keywords:** nanofiber; poly(caprolactone); 3D culture; neutrophil; dendritic cell; inflammation; *Staphylococcus aureus*

**Citation:** Lee, S.-J.; Maza, P.A.M.A.; Sun, G.-M.; Slama, P.; Lee, I.-J.; Kwak, J.-Y. Bacterial Infection-Mimicking Three-Dimensional Phagocytosis and Chemotaxis in Electrospun Poly(ε-caprolactone) Nanofibrous Membrane. *Membranes* **2021**, *11*, 569. https://doi.org/10.3390/membranes11080569

Academic Editor: Andrea Ehrmann

Received: 30 June 2021
Accepted: 26 July 2021
Published: 28 July 2021

**Publisher's Note:** MDPI stays neutral with regard to jurisdictional claims in published maps and institutional affiliations.



## 1. Introduction

The recognition of pathogenic microorganisms is the first step of host defense. Components derived from bacteria, such as N-formylmethionine-leucyl-phenylalanine (fMLP), trigger chemotaxis and activation of phagocytes such as neutrophils and macrophages in infectious tissues [1]. Early-recruited neutrophils phagocytose the bacteria and subsequent chemoattractant signals from the activated and necrotic neutrophils can directly and indirectly recruit more neutrophils and other phagocytes, including macrophages and dendritic cells (DCs) [2]. In addition, a variety of infected host cells, such as fibroblasts, endothelial cells and epithelial cells, secrete phagocyte chemoattractants [2,3].

Observing the migration of phagocytes in infected tissue is technically challenging. Numerous in vitro studies of host immune responses to bacteria have been carried out under two-dimensional (2D) co-culture of phagocytes and bacteria, although in tissues, these exist and interact under three-dimensional (3D) conditions. Investigation of cell functions such as phagocytosis and cell migration requires accurate mimicry of the in vivo microenvironment. The migration rate significantly relies on the chemoattractant gradient in tissue. The transwell-based (Boyden chamber) assay is widely used to assess cell migration; however, it provides a bulk end-point measurement and does not measure directionality of cell migration [4]. The assay is not appropriate for assessing migration of cells to a variety of bacteria because non-adherent bacteria grow as a suspension culture in the bottom chamber and can float, adhere, or migrate freely to the upper chamber through the permeable filter of transwell, thus inhibiting chemoattractant gradient formation. Microfluidic devices have been used to study the migration of phagocytes in stable chemokine gradients [5]. A previously developed DC migration assay in fibronectin-coated micro-channels is complementary to more complex 3D migration systems [6,7]. In addition, it has been reported that DCs and T cells readily migrate through the pores in composite macroporous poly(ethylene glycol) hydrogel scaffolds infused with collagen [8]. Recently, Huh et al. developed a microfluidic lung-on-a-chip system consisting of a human alveolar cell layer and a human pulmonary microvascular endothelial cell layer separated by a porous membrane made of poly(dimethylsiloxane) (PDMS) and demonstrated that this system mimics the innate cellular response of bacteria to pulmonary infection [9].

Phagocytes continue to migrate through the extracellular matrix (ECM), a 3D fiber mesh, until they reach an infected or inflamed tissue. Nanofibrous scaffold mimics the configuration of native ECM of tissues [10]. For optimal cellular infiltration, we developed a hybrid electrospun poly(ε-caprolactone) (PCL)-nanofibrous membrane (NM) (PCL-NM) consisting of nano- as well as submicron-scale fibers and we used the PCL-NM to mimic the naturally occurring crosstalk between immune and cancer cells [11]. Additionally, we demonstrated that neutrophils migrate against interleukin (IL)-8 in an electrospun PCL-NM [12]. PCL is a Food and Drug Administration (FDA)-approved biocompatible and biodegradable aliphatic polyester [13]. Furthermore, DCs cultured on PCL nanofibers show a baseline inactive form [11], indicating that PCL is a suitable material for immune cell culture. Thus, we used electrospun PCL-NM to mimic the 3D fibrillar environment in studying the inflammatory response to bacterial infection.

*S. aureus* is a leading cause of human bacterial diseases, including skin and soft tissue infections, food-borne illness, toxic shock syndrome and sepsis. A number of 3D in vitro culture systems have been developed to study the interactions between bacteria and cells [14]. *S. aureus* infection in skin tissue and bone has been studied using 3D co-culture models [15,16]. Ding et al. demonstrated that macrophages can transmigrate across a polyethylene terephthalate membrane to reach the epithelial cell layer cultured with *S. aureus* [17]. In the present study, we used the in vitro infection system to analyze the early events associated with *S. aureus* infection of phagocytes. Cells in 3D cell culture form multilayers of cells, whereas cells grown in 2D form a monolayer. In this study, we designed a chemotaxis assay system using PCL-NM-based two layers, attempting to mimic inflammatory microenvironment, to evaluate early recruitment of phagocytes induced by *S. aureus*. The PCL-NM-based 3D co-culture system can be used to evaluate in vivo-mimicking bacteria-host interactions and immune responses.

## 2. Materials and Methods

### 2.1. Materials

PCL ($M_n$ = 700,000–900,000), chloroform, PKH26, PKH67, 4-(2-hydroxyethyl)-1-piperazineethanesulfonic acid (HEPES), lipopolysaccharide (LPS), thioglycollate, dimethyl sulfoxide (DMSO) and 4′,6-diamidino-2-phenylindole (DAPI) were purchased from Sigma-Aldrich (St. Louis, MO, USA). Dulbecco's modified Eagle's medium (DMEM), RPMI-1640 medium, fetal bovine serum (FBS), penicillin/streptomycin, glutamine and 0.05% trypsin-ethylenediaminetetraacetic acid (EDTA) were purchased from Gibco (Rockville, MD, USA). Granulocyte macrophage-colony stimulating factor (GM-CSF), M-CSF, IL-4 and antibodies against CD11, Ly6G and F4/80 were obtained from R&D Systems Inc. (Minneapolis, MN, USA). Boc-Met-Leu-Phe (Boc-MLF) and Trp-Arg-Trp-Trp-Trp-Trp-$NH_2$ (WRW4) were obtained from Tocris bioscience (Bristol, UK). CellTracker Red CMTPX dye was purchased from Molecular Probes, Inc. (Eugene, OR, USA). PDMS was purchased from Dow-Corning Korea, Inc. (Seoul, Korea).

### 2.2. Electrospinning and Fabrication of the PCL Nanofibers

Porous PCL nanofibers were prepared according to the procedure published earlier [11,18]. Briefly, the polymer for electrospinning was dissolved in 99.5% pure chloroform at a concentration of 8.8 wt% and stirred for 5 h to obtain a homogeneous solution. PCL-NMs were fabricated by electrospinning (NanoNC, Seoul, Korea). Two-nozzle spinnerets were used with an average flow rate of approximately 8 μL/min, using a syringe pump. Each nozzle had an inner diameter of approximately 210 μm (27G). Nanofibers were collected onto a rotating metallic mandrel at 100 rpm at ambient temperature for 4 h. The nozzle tip-to-collector distance was set at 20 cm, with an electrical potential of 17.5 kV from the grounded collector plate. An aluminum plate was used for fabricating conventionally electrospun PCL-NM. Membrane thickness was measured using a high-precision caliper (Mitsutoyo Co., Kawasaki, Japan). The morphology of electrospun fibers was observed using scanning electron microscopy (SEM) with a model SEM4500 (Sec, Suwon, Korea). The electrospun PCL-NMs were sterilized by soaking in a solution of 70% ethanol for 12 h and dried under UV exposure for 12 h.

### 2.3. Preparation of Phagocytes from Bone Marrow

BALB/c mice were maintained in accordance with institutional animal care and use guidelines. All experimental procedures involving mice were performed according to protocols approved by the Committee on the Ethics of Animal Experiments of the Ajou University Medical Center (Permit Number: 2015–0028). Bone marrow (BM) cells were harvested from the femurs and tibias of the mice and BM-derived neutrophils were purified using anti-Ly6G antibody-bound microbeads [19]. BM-derived DCs were obtained from BM precursors, as described previously [20]. Briefly, BM cells plated in complete RPMI-1640 containing recombinant murine GM-CSF (20 ng/mL) plus recombinant murine IL-4 (20 ng/mL). On the third day of incubation, the same concentration of IL-4 and GM-CSF was added and half of the DC culture medium was refreshed. All BM cells were harvested on day 7 of incubation. By immunomagnetic selection of $CD11c^+$ cells, BMDCs were purified to >90% using CD11c Microbeads Ultrapure (Miltenyi Biotec, Bergisch-Gladbach, Germany). BM-derived macrophages were obtained by culturing of mouse BM cells with recombinant murine M-CSF (20 ng/mL) for 7 days [21]. BM-derived DCs and macrophages were purified using anti-CD11c and anti-F4/80 antibody-bound microbeads, respectively. The purity of neutrophils, DCs and macrophages was verified to be >90% by flow cytometry using surface markers, Ly6G, CD11c and F4/80, respectively.

### 2.4. Preparation of Peritoneal Neutrophils and Macrophages

BALB/c mice were treated with 3% thioglycollate in distilled water (intraperitoneal injection) for 6 h and 3 days to purify peritoneal neutrophils and macrophages, respectively, which were isolated by lavage with phosphate-buffered saline (PBS) and purified with anti-Ly6G- and anti-F4/80 antibody-bound microbeads, respectively.

### 2.5. MLE-12 Cell Culture

MLE-12 mouse lung epithelial type II cells were purchased from American Type Culture Collection (ATCC) (Manassas, VA, USA) and cultured in ATCC-formulated DMEM supplemented with 10% FBS, 1% penicillin/streptomycin at 37 °C in 5% $CO_2$ incubator. MLE-12 cells were subcultured every 3–4 days and cells at passages 6–10 were used herein.

### 2.6. Bacterial Culture and Adhesion Assay

*S. aureus* strain RN4220 was grown to mid-log phase in Luria-Bertani (LB) broth, washed and resuspended in PBS [22]. The bacterial concentration was estimated by determining colony forming units (CFUs) and the optical density at 600 nm ($OD_{600}$) using a spectrophotometer. The bacterial suspension was diluted to an $OD_{600}$ of 0.1. Bacteria were also stained with PKH67 Green and PKH26 Red Fluorescent Cell Linker Kits (Sigma) and washed with PBS three times. PDMS (100 μL) was poured in an 8-well plate (SPL, Seoul, Korea) and kept on a slide warmer at 100 °C for 3 min. The PCL-NM (1.1 cm × 1.3 cm) was attached to the surface of gel-state PDMS on an 8-well plate and immersed in DMEM (700 μL) for 6 h to increase cell adhesion at 37 °C. For bacterial culture in the PCL-NM, a bacteria suspension ($1 \times 10^7$ CFU/200 μL) was seeded on the surface of a wet PCL-NM. After incubation at 37 °C and under 5% $CO_2$ for the indicated times to allow the bacteria to settle and attach to the nanofibers, the PCL-NMs were gently rinsed thrice with PBS to remove not-adherent bacteria.

### 2.7. Laser Confocal Microscopy

Fluorescence-labeled phagocytes and *S. aureus* on a culture dish and in the PCL-NM were observed using a K1 confocal microscope (Nanoscope, Daejeon, Korea) at the 3D immune system imaging core facility of Ajou University. Images was analyzed using the ImageJ software.

### 2.8. SEM

Phagocytes and/or bacteria cultured in the PCL-NM were affixed to aluminum mounts with double-sided carbon tape (SPI Supplies Inc., Seoul, Korea) and coated with gold-sputter. Cell structure and nanofiber morphology were observed using SEM.

### 2.9. FACS Analysis of Phagocytosis

Neutrophils, macrophages and DCs were labeled with the PKH67 fluorescent dye and *S. aureus* were labeled with the PKH26 fluorescent dye. The cells were washed thrice with PBS. The labeled phagocytes ($1 \times 10^5$) were allowed to adhere to the culture dish for 1 h and cultured with labeled *S. aureus* ($1 \times 10^6$ CFU) in RPMI 1640 medium (200 μL) in the absence or presence of 5% FBS at 37 °C in a humidified atmosphere containing 5% $CO_2$. For the phagocytic assay in the PCL-NM, labeled phagocytes ($1 \times 10^5$) were top-seeded on the PCL-NM surface and incubated for 1 h; then, the bacterial suspension ($1 \times 10^6$ CFU/ 200 μL) was added. For the assay with formyl peptide receptor (FPR) inhibitors, Boc-MLF and WRW4 were dissolved in DMSO and PBS, respectively, and 0.1% DMSO solution was used as a control. Phagocytes and bacteria were detached from the PCL-NM by soaking in trypsin-EDTA solution for 5 min. The percentage of PKH26-stained cells among total PKH67-stained phagocytes was analyzed using flow cytometry.

### 2.10. Live Imaging of Phagocytosis

The dynamic interactions between *S. aureus* and phagocytes were visualized using a K1 confocal microscopy. Neutrophils, macrophages and DCs ($1 \times 10^5$) were labeled with CellTracker and *S. aureus* cells ($1 \times 10^7$ CFU) were labeled with PKH67 fluorescent dye. Labeled phagocytes were top-seeded on PCL-NMs in 3D culture and on culture dishes in 2D culture. After a 4 h incubation of phagocytes ($1 \times 10^5$/100 μL) for cell attachment to the culture dish or PCL-NM, *S. aureus* suspension ($1 \times 10^6$ CFU/100 μL) in RPMI-1640 culture media was added to the labeled phagocytes. Time-lapse videos of phagocytes engulfing bacteria were generated by taking pictures every 60 s for 1 h.

### 2.11. Migration Assay in PCL-NM-Based Two-Layer System

We assembled a two-layer system of two PCL-NMs for cell migration, as shown in supplementary material Figure S1, in a modified Transwell chamber [18]. To construct the upper PCL-NM in the apical chamber, the polycarbonate filter of a Transwell insert was removed and replaced with a 5-mm-diameter circular section of PCL-NM, which were attached with gel-state PDMS. Neutrophils, macrophages and DCs were labeled with PKH26 dye and 10 μL of cell suspension ($1 \times 10^5$) was seeded onto the PCL-NM in the apical chamber. After 4 h, the scaffold was washed with PBS. For the lower PCL-NM layer,

PDMS (220 μL) was poured in the lower cultivation well to maintain constant distance from upper surface of the lower membrane to the lower surface of the upper membrane and a PCL-NM was attached on top of the surface of the gel-state PDMS. *S. aureus* suspension ($1 \times 10^7$ CFU) was added to the surface of the PCL-NM in the basal chamber and bacteria were allowed to adhere for 2 h. The phagocyte- and *S. aureus*-seeded membranes were then assembled layer-by-layer by placing the insert in the well. The chamber contained 600 μL of medium and phagocytes were allowed to migrate at 37 °C in a humidified $CO_2$ atmosphere. Migrated PKH26-labeled cells in the lower PCL-NM layer were counted under a confocal microscope in three random fields at a magnification of 100×. A lower PCL-NM layer without bacteria served as a control to evaluate migration of the phagocytes.

### 2.12. *Cytokine Assay*

Secretion of cytokines and chemokines was determined using a Proteome Profiler Mouse Cytokine Array Kit (R&D Systems). Neutrophils and lung epithelial cells ($1 \times 10^5$) in the presence or absence of *S. aureus* ($1 \times 10^7$ CFU) were cultured on a culture dish and in the PCL-NM. After culture for the indicated times, supernatants were taken and assayed as described in the user manual. Chemiluminescence signal density was quantified with ImageJ. IL-8 levels were measured by ELISA (R&D Systems) according to manufacturer's instruction.

### 2.13. *Statistical Analysis*

The results are presented as means ± standard deviations (SDs). Student's *t*-test was used to compare the means of paired or unpaired samples. A *p*-value of <0.05 was considered significant.

## 3. Results

### 3.1. *Culture of S. aureus and Phagocytes in Electrospun PCL-NM*

The PCL-NM had an average thickness of 57.0 ± 5.5 μm ($n = 10$). The ultrastructure of PCL-NM was analyzed by SEM to assess the fiber size, fiber uniformity and surface topography [11]. The electrospun samples were homogeneous and the majority of fiber diameters in the membranes was between 400 nm and 2 μm (data not shown). The most frequent fiber diameter was between 200 nm and 600 nm, but substantial microfibers with a diameter of 1–10 μm were deposited in the PCL-NM (data not shown) [18]. Thus, microfibers in the PCL-NM introduced larger pores than nanofibers. The mercury intrusion data for electrospun PCL-NM showed pores with a diameter of approximately 10–100 μm (data not shown) [11]. The large pores in the nano-submicron PCL-NM were expected to allow phagocytes to infiltrate deep into the membrane [11].

First, we evaluated the attachment of the bacteria to the nanofibers by SEM in an experiment using bacteria alone. *S. aureus* were top-seeded on a PCL-NM and allowed to infiltrate. *S. aureus* are Gram-positive round cells of 0.5–1 μm diameter. As shown in Figure 1A, the bacteria were well distributed throughout the membrane and attached along the nanofiber surfaces. Small bacterial agglomerates were seen between two or more fibers in areas of fiber cross over. Bacteria cultured for 12 h formed aggregates inside the PCL-NM. Bacterial viability and the proliferation of PKH-labeled and unlabeled cells are identical, indicating that PKH fluorescent dyes are nontoxic to bacteria [23]. In addition, the fluorescence intensity of PKH-labeled bacteria remains constant during the lag phase, whereas replication results in a 50% decrease in fluorescence with each cell division [23]. As shown in Figure 1B, the density of PKH-labeled bacteria exhibiting fluorescence in the PCL-NM increased up to 12 h, but then decreased at 24 h. Next, we quantified bacteria after detachment from nanofibers. The numbers of *S. aureus* adhering to the nanofibers increased in a time-dependent manner up to 12 h, but then decreased at 24 h (Figure 1C).

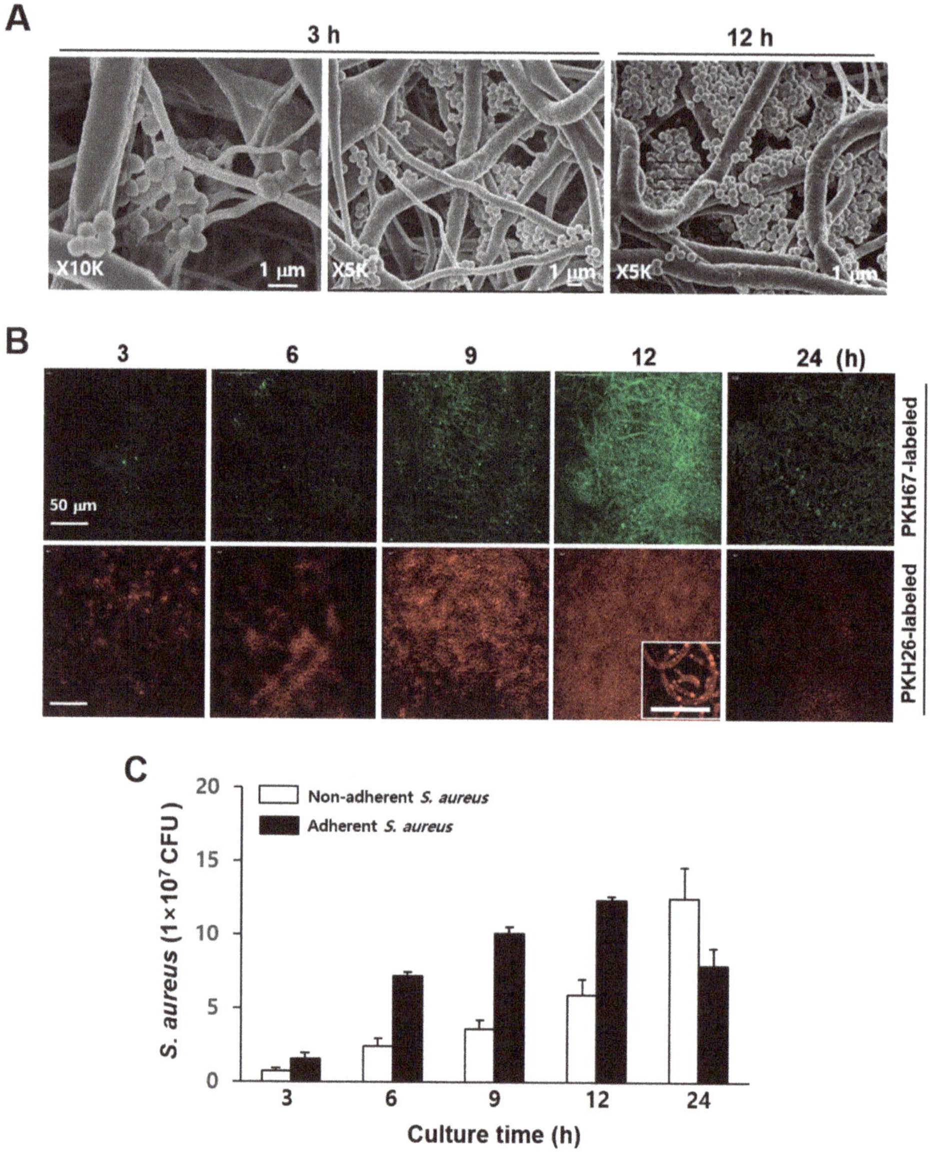


**Figure 1.** Adhesion of *S. aureus* to the PCL-NM. (**A**) *S. aureus* ($1 \times 10^7$ CFU) were top-seeded on the PCL-NM for the indicated times and adhesion of *S. aureus* to the nanofibers was evaluated by SEM. Images are representative of three independent experiments. (**B**) PKH67- and PKH26-labeled *S. aureus* ($1 \times 10^7$ CFU) were top-seeded on the PCL-NM and cultured for the indicated times ($n = 3$). The adhesion and proliferation of fluorescence-labeled *S. aureus* in the PCL-NM were evaluated by confocal microscopy. High magnification image (×1000) in the box shows numerous spots along the nanofibers. (**C**) *S. aureus* ($1 \times 10^7$ CFU/200 μL) were cultured in the PCL-NM for the indicated times. Non-adherent bacteria were gently recovered from culture media (50 μL) and adherent bacteria on nanofibers were harvested by vigorous pipetting after the addition of culture media (100 μL) to the PCL-NM. The numbers of non-adherent and adherent bacteria in the PCL-NM were estimated by determining CFUs. Data are presented as means ± SDs ($n = 3$).

In comparison, the numbers of non-adherent *S. aureus* increased up to 24 h. These results indicated that the bacteria could proliferate on the fibers, producing agglomerates, but are not able to maintain large colonies in the PCL-NM after prolonged culture, possibly because of the small diameter (400 nm–2 μm) of the nanofibers. To rule out the toxic effect of PCL-NM, PCL-NM alone was soaked in culture media for 24 h and bacteria were cultured in the media for 24 h. We confirmed that the growth of *S. aureus* was not affected (data not shown). The spreading and infiltration of bacteria and neutrophils in the PCL-NM were evaluated 1 h after cells were seeded on the membranes. A focus-stacking analysis of confocal microscopy images captured at 3 μm intervals showed that fluorescent-labeled bacteria and neutrophils infiltrated the membrane from the upper surface to a depth 15–30 μm (Figure 2A). Bacteria and neutrophils were homogeneously distributed at a similar level inside the PCL-NM. Moreover, the green fluorescence of bacteria co-localized with red fluorescence of neutrophils, indicating internalization of the bacteria by the neutrophils (Figure 2B). In addition, co-localization of bacteria with neutrophils, macrophages and DCs cultured in the PCL-NM was confirmed by SEM (Figure 2C). PCL nanofibers have various fiber diameters and pore sizes, which depend on polymer concentration, flow rate and others. Fabrication of PCL nanofibers through electrospinning facilitated PCL-NM to allow the infiltration of bacteria and phagocytes. Therefore, the PCL-NM can be used for the 3D culture of bacteria and phagocytes.

### 3.2. Engulfment of *S. aureus* by Phagocytes in the PCL-NM

In live cell imaging analysis, PKH67-labeled *S. aureus* were caught and ingested by CellTracker red CMTPX-labeled neutrophils and DCs in 2D as well as 3D culture. In monolayer co-culture, neutrophils adhered to the surface of culture dish but did not show chemotactic migration toward bacteria because the latter were floating around in the media (Supplementary Video S1). On the other hand, the DCs adherent on the plastic surface showed dynamic changes in intracellular red dye, which we first hypothesized to represent cell spreading to pick up bacteria; however, careful observation learned that the changes in dye levels were located inside the cytoplasm and did not represent dynamic cell extensions (Supplementary Video S2). In contrast, single-cell tracking in the PCL-NM revealed changes in neutrophil morphology and migration, with directional movement toward the bacteria (Supplementary Video S3). Neutrophils extended long protrusions of cell body along the nanofibers and then retracted the protrusion, indicating transition from the spreading to the migratory state. When assuming a cell size of 10 μm, the leading edge elongated several folds. Time-lapse microscopy clearly showed CellTracker-labeled DCs crawling along the nanofibers to nearby green-fluorescent PKH67-labeled *S. aureus*, with high motility to pick up the bacteria (Supplementary Video S4). These results suggested that co-culture of bacteria and phagocytes in the PCL-NM allows assessing phagocytosis of bacteria in 3D condition.

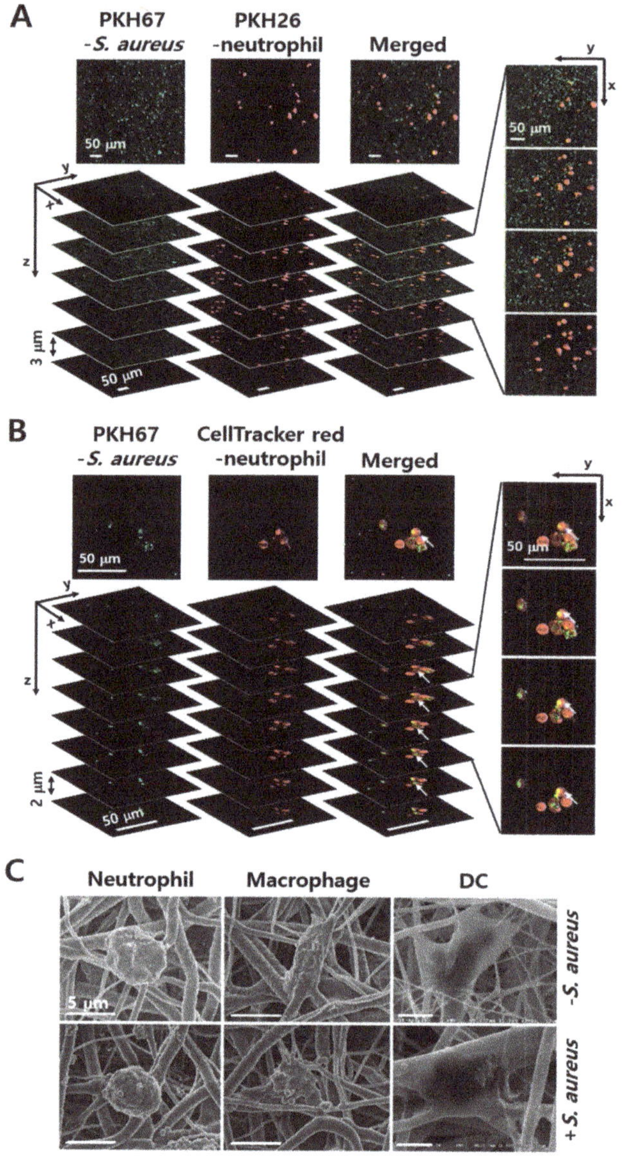


**Figure 2.** Co-culture of phagocytes and *S. aureus* in the PCL-NM. (**A**,**B**) Peritoneal neutrophils were labeled with PKH26 (**A**) and CellTracker red (**B**). The labeled neutrophils ($1 \times 10^5$) and PKH67-labeled *S. aureus* ($1 \times 10^6$ CFU) were top-seeded on the PCL-NM and cultured for 1 h in the membranes ($n = 3$). Co-localization of neutrophils and bacteria was evaluated by focus-stacking images of the PCL-NM. Images of 3 μm (**A**) and 2 μm (**B**) in thickness from the top to bottom are shown. Arrows indicate green-fluorescent spots inside of red-fluorescent areas. (**C**) Neutrophils, macrophages and DCs ($1 \times 10^5$) were top-seeded on the PCL-NM and cultured with (+) or without (−) *S. aureus* ($1 \times 10^6$ CFU) for 1 h in the PCL-NM. The localization of *S. aureus* and phagocytes in the PCL-NM was evaluated by SEM. Data are representative of three independent experiments.

### 3.3. Differential Rate of Phagocytosis in 2D and 3D Culture Conditions

It has been demonstrated that there is a significant difference in phagocytosis of *S. aureus* between adherent and suspended neutrophils; the phagocytosis by adherent neutrophils is largely unaffected by opsonization with serum [24]. We compared the rate of phagocytosis of live bacteria between phagocytes cultured in 2D and 3D conditions. Flow cytometry can be used for the fluorescence quantitation of the phagocytosis of labeled bacteria [23]. PKH26-labeled neutrophils were incubated for 1 h on a culture dish and in the PCL-NM and cultured with PKH67-labeled *S. aureus* for 2 h. Flow cytometry assays showed that double positive-stained cells correspond to neutrophils that internalized bacteria and the fluorescence intensity of PKH67-labeled *S. aureus* remained constant during 2 h of culture in the presence of neutrophils on a culture dish and in the PCL-NM (Figure 3A). Thus, we can rule out the possibility that fluorescence was lost during bacterial culture and that the rate of phagocytosis was underestimated in 2D and 3D conditions. The uptake of bacteria by BM-derived neutrophils occurred in a time-dependent manner on culture dishes and in the PCL-NMs containing 5% FBS (Figure 3B). Next, we measured the phagocytosis rate on culture dishes and in the PCL-NMs containing serum-free media. The phagocytic rates of neutrophils on the culture dish and in the PCL-NM were significantly lower in serum-free media than in FBS-containing media. The reduction in the rate of phagocytosis in serum-free media was greater in the PCL-NMs than on culture dishes. The levels of phagocytosis 4 h after the co-culture of neutrophils and bacteria increased to 80–90% on culture dishes containing serum-free media and FBS-containing media. In contrast, the uptake of bacteria by neutrophils after 4 h of bacterial and neutrophil co-culture was significantly lower in serum-free condition than in FBS-containing media in the PCL-NM (32.0 ± 5.6% vs. 62.3 ± 8.0%, $p < 0.05$). The uptake of bacteria by BM-derived and peritoneal phagocytes in the presence of FBS occurred in a time-dependent manner (Figure 3C). Half of total peritoneal neutrophils (50.3 ± 5.1%) and the majority of peritoneal neutrophils (79.5 ± 6.5%) were double positively stained after 1 h and 2 h of 2D culture, respectively. In contrast, in the PCL-NM, the number of BM-derived and peritoneal neutrophils that had phagocytosed bacteria was lower than that in 2D culture. BM-derived macrophages and DCs rapidly phagocytosed bacteria, but the phagocytic rate and ability in the PCL-NM were lower than those on the culture dish.

### 3.4. Effects of FPR Inhibitors on the Phagocytosis of S. aureus in 2D and 3D Culture Conditions

Phagocytes sense chemotactic gradients produced by bacteria, triggering phagocytosis. The bacterial peptide, fMLP is a highly potent leukocyte chemoattractant and chemotactic activity of phagocytes is mediated by FPR [25]. We next compared the effect of Boc-MLF and WRW4, selective FPR1 and FPR2 inhibitors in murine neutrophils [26], on phagocytosis by neutrophils and macrophages in 2D and 3D culture conditions. Boc-MLF and WRW4 decreased the phagocytic rate of neutrophils in the PCL-NM when compared to that of neutrophils grown in control condition (44 ± 5% and 45 ± 6% vs. 60 ± 8%, respectively, $p < 0.05$) (Figure 3D). Similar results were obtained for macrophages treated with Boc-MLF and WRW4 when compared to macrophages without inhibitors (35 ± 4% and 30 ± 3% vs. 50 ± 7%, respectively, $p < 0.05$). In contrast, no inhibition of phagocytosis was observed when FPR inhibitors were added to co-culture of *S. aureus* and neutrophils or macrophages in 2D culture. These results suggested that phagocyte migration plays a role in phagocytosis in 3D, but not in 2D culture.

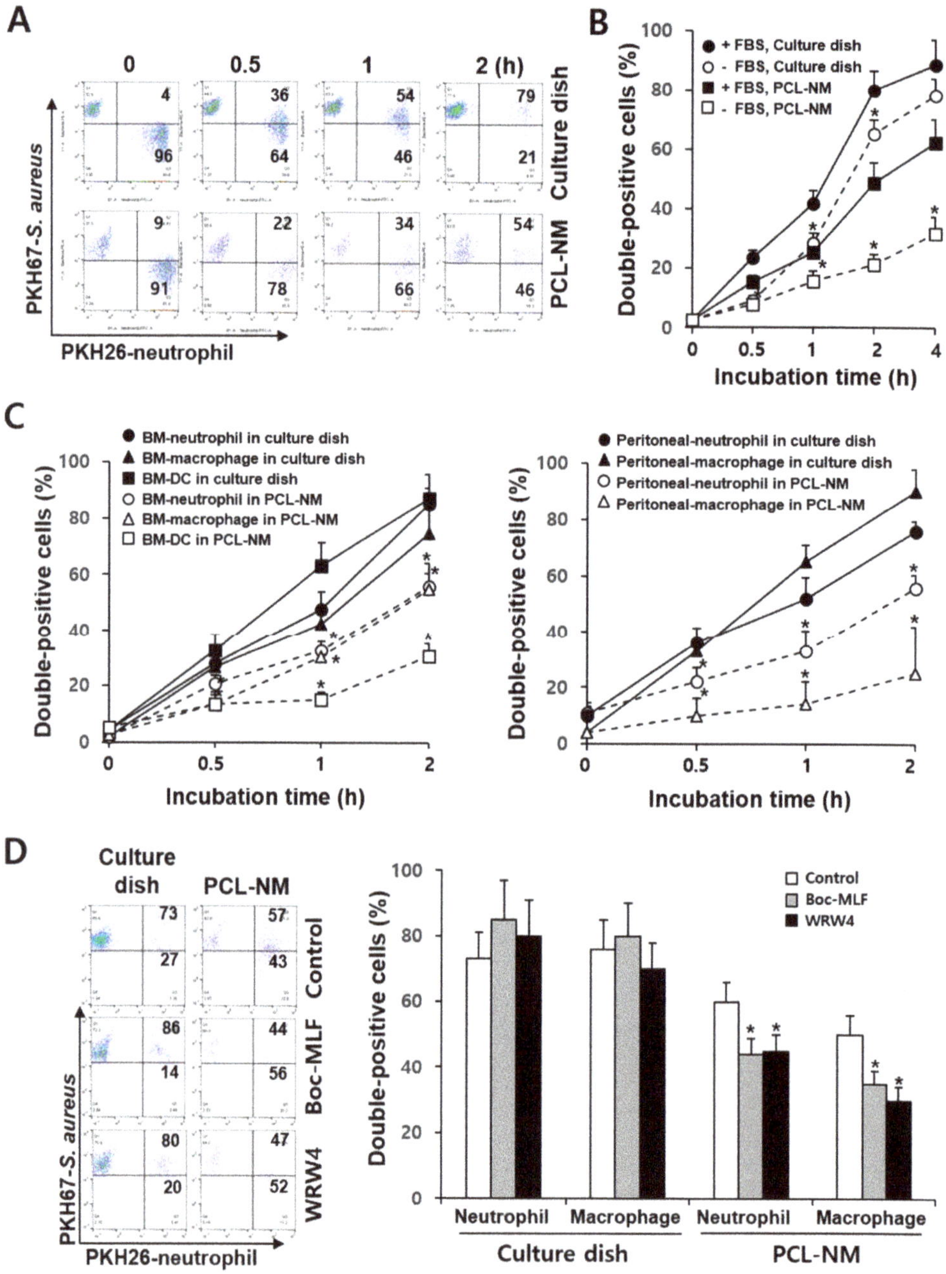


**Figure 3.** Phagocytosis of *S. aureus* by neutrophils, macrophages and DCs in the PCL-NM. (**A**) PKH67-labeled *S. aureus* ($1 \times 10^6$ CFU) was cultured with PKH26-labeled BM-derived neutrophils ($1 \times 10^5$) for the indicated times on a culture dish and in the PCL-NM and double-positive cells among PKH26-positive cells were detected by flow cytometry ($n = 4$). Numbers represent the percentages of cells within the indicated gates of PKH26-labeled neutrophils. (**B**) BM-derived

neutrophils were stained with PKH26 and cultured with PKH67-labeled *S. aureus* (1 × $10^6$ CFU) in the presence (+FBS) or absence (−FBS) of 5% FBS for the indicated times on a culture dish and in the PCL-NM. Rates of phagocytosis were evaluated by counting double-positive cells among PKH26-labeled phagocytes. Data are presented as means ± SDs ($n$ = 3). Asterisks (*) denote significant differences compared to culture with FBS ($p < 0.05$). (**C**) BM- and peritoneal-derived phagocytes were prepared as described in the Materials and methods. BM- and peritoneal-derived neutrophils, macrophages and DCs (1 × $10^5$) were stained with PKH26 and cultured with PKH67-stained *S. aureus* (1 × $10^6$ CFU) for the indicated times in the presence of 5% FBS on a culture dish and in the PCL-NM. Rates of phagocytosis were evaluated by counting double-positive cells among PKH26-labeled neutrophils, macrophages and DCs. Data are presented as means ± SDs ($n$ = 4). Asterisks (*) denote significant differences compared to culture on a culture dish ($p < 0.05$). (**D**) Phagocytosis of *S. aureus* (1 × $10^6$ CFU) by BM-derived neutrophils and macrophages (1 × $10^5$) in the PCL-NM containing 5% FBS was measured in the presence of 0.1% DMSO (Control), Boc-MLF (20 μM) and WRW4 (4 μM) ($n$ = 4). Numbers in the left panel represent the percentage of cells within the indicated gates of PKH26-labeled neutrophils. Data are presented as means ± SDs. Asterisks (*) denote significant differences compared to the control ($p < 0.05$).

### *3.5. 3D Migration of Phagocytes to S. aureus in PCL-NM-Based Two-Layer Culture System*

As shown in supplementary Figure S1, 3D migration of phagocytes to bacteria was assessed quantitatively using PCL-NM-based two-layer system in a Transwell chamber. PKH26-labeled phagocytes and PKH67-labeled bacteria were cultured in the upper and lower PCL-NM layers, respectively. After indicated times of incubation, the numbers of phagocytes that had migrated from the upper to the lower PCL-NM were counted. As shown in Figure 4A, increasing numbers of neutrophils and DCs were detected in the lower PCL-NM over time, while the numbers of phagocytes in the upper PCL-NM gradually decreased. Neutrophils migrated rapidly in response to *S. aureus*, when compared to DCs. Next, we performed a similar assay, in which equal numbers of PKH26-labeled neutrophils and PKH67-labeled macrophages were top-seeded on the upper PCL-NM and bacteria on the lower PCL-NM. While both macrophages and neutrophils migrated to the lower PCL-NM, the numbers and rates of migration of macrophages were lower than those of neutrophils (Figure 4B).

### *3.6. Neutrophil-Induced Recruitment of More Neutrophils to S. aureus in PCL-NM-Based Two-Layer Culture System*

Neutrophil migration is mediated by chemokines derived from bacteria and phagocytes. When PKH67-labeled neutrophils were pretreated with FPR inhibitors and then cultured with PKH26-stained bacteria in the presence of the inhibitors in the PCL-NM-based two-layer system, the numbers of PKH67-positive neutrophils in the lower PCL-NM were significantly decreased when compared to those of untreated neutrophils (Figure 5A). The effects of FPR inhibitors on migration were only obvious in the first hour after co-culture of neutrophils in the upper PCL-NM and bacteria in the lower PCL-NM, indicating that the early phase of migration is mediated by bacteria-derived chemoattractants, such as fMLP. It is possible that the secretion of chemokines by neutrophils migrated from the upper to the lower PCL-NM may trigger a later, second phase of neutrophil migration. It has been shown that DAPI staining does not impact the growth of *Listeria monocytogenes* in culture media and macrophages [27]. Thus, we measured the migration rate of PKH26-labeled neutrophils in the upper PCL-NM using cultures of PKH67-labeled neutrophils and/or DAPI-stained bacteria in the lower PCL-NM. More PKH26-labeled neutrophils were observed in the lower PCL-NM containing both *S. aureus* and neutrophils than in the lower PCL-NM containing *S. aureus* or neutrophils alone (Figure 5B). Neutrophils cultured with bacteria in PCL-NM as well as in culture dish secreted IL-8 into the medium (Figure 5C). The concentrations of IL-8 secreted by neutrophils cultured with bacteria in the PCL-NM were lower than those in 2D culture, but increased in a time-dependent manner (Figure 5D). These results indicated that neutrophil-derived chemokines in the lower compartment, such as IL-8, enhance the migration of neutrophils from the upper to the lower PCL-NM.

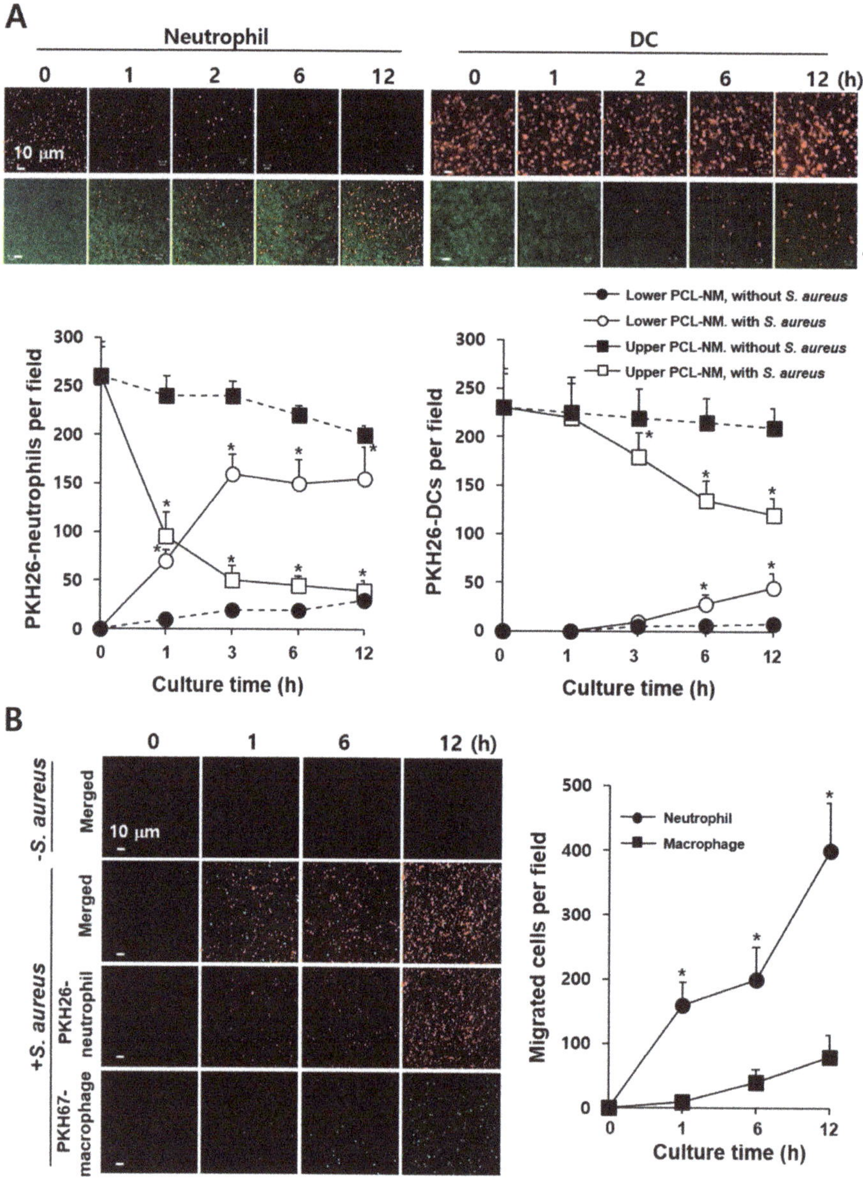


**Figure 4.** Migration of phagocytes to *S. aureus* in the PCL-NM-based two-layer culture system. (**A**) PKH26-labeled BM-derived neutrophils and DCs ($1 \times 10^5$) and PKH67-labeled *S. aureus* ($1 \times 10^7$ CFU) were seeded and cultured in the upper and lower PCL-NMs, respectively, for the indicated times ($n = 4$). The graphs depict the numbers of red-fluorescent spots detected in the upper (■, □) and lower (•, ○) PCL-NMs when bacteria were absent (•, ■) or present (○, □) in the lower PCL-NM. Data are presented as means ± SDs ($n = 4$). Asterisks (*) denote significant differences compared to culture without bacteria ($p < 0.05$). (**B**) PKH26-labeled peritoneal neutrophils ($1 \times 10^5$) and PKH67-labeled peritoneal macrophages ($1 \times 10^5$) were seeded in the upper PCL-NM and *S. aureus* ($1 \times 10^7$ CFU) was seeded in the lower PCL-NM. Phagocytes in the upper PCL-NM were cultured in the presence (+ *S. aureus*) or absence (−*S. aureus*) of bacteria in the lower PCL-NM for

the indicated times ($n = 3$). The graphs depict the numbers of red- (Neutrophil) and green- (Macrophage) fluorescent spots detected in the lower PCL-NM containing the bacteria. Data are represented as means ± SDs ($n = 3$). Asterisks (*) denote significant differences compared to culture with macrophages ($p < 0.05$).

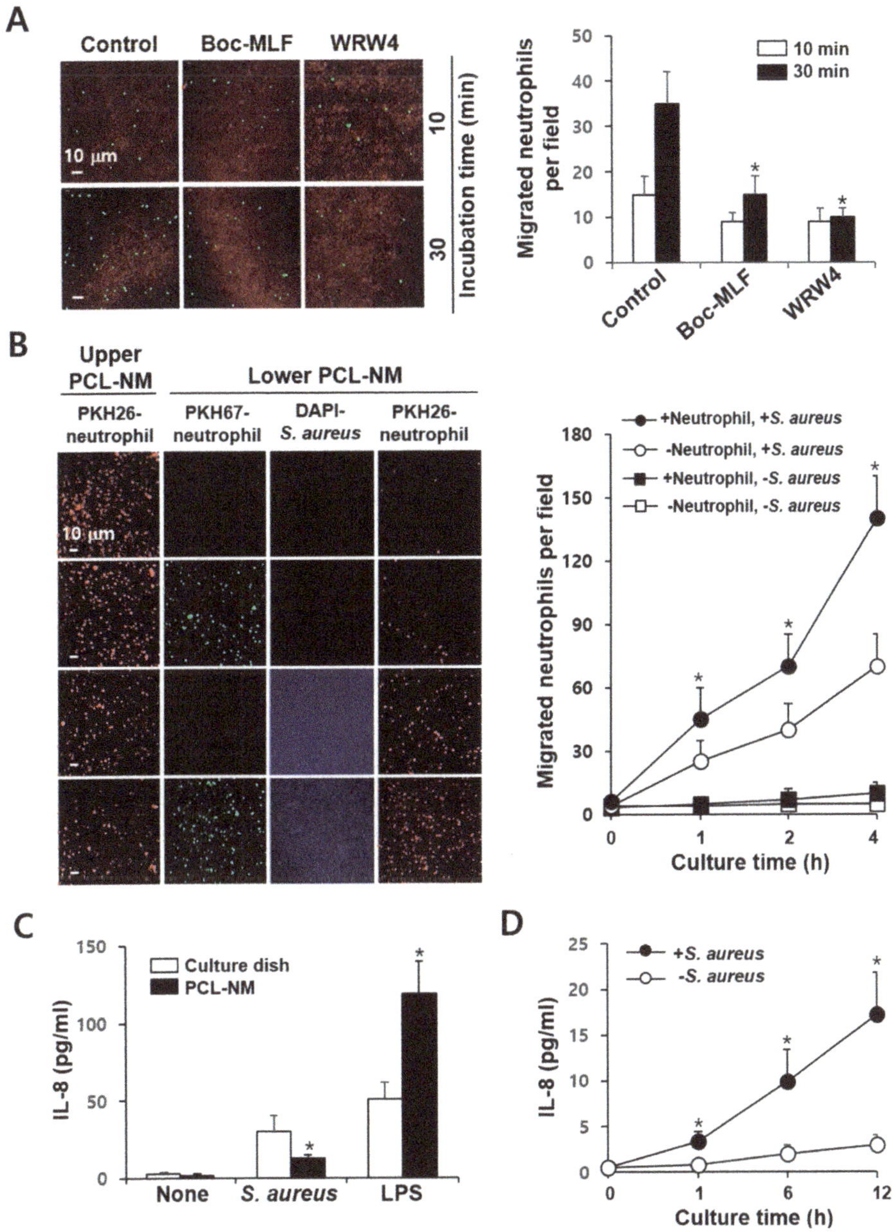


**Figure 5.** More neutrophils are recruited when *S. aureus* is cultured with neutrophils than when it is cultured alone. (**A**) PKH67-labeled BM derived neutrophils ($1 \times 10^5$) were top-seeded and pretreated with 0.1% DMSO (Control), Boc-MLF

(20 μM) and WRW4 (4 μM) for 1 h in the upper PCL-NM. The migration rates of control and FRP inhibitor-pretreated neutrophils to the lower PCL-NM containing PKH26-labeled *S. aureus* ($1 \times 10^7$) were measured for the indicated times in the presence of 0.1% DMSO (control), Boc-MLF (20 μM) and WRW4 (4 μM) ($n = 4$). Green-fluorescent spots in the lower PCL-NM were counted. Asterisks (*) denote significant differences compared to the control ($p < 0.05$). (**B**) PKH26-labeled neutrophils ($1 \times 10^5$) in the upper PCL-NM were cultured with or without PKH67-labeled neutrophils ($1 \times 10^5$) and/or DAPI-stained *S. aureus* ($1 \times 10^7$ CFU) in the lower PCL-NM ($n = 4$). Red-fluorescent spots were detected in the lower PCL-NM. The graphs depict the numbers of red-fluorescent spots detected in the lower PCL-NM in the absence (○, □) or presence (•, ■) of neutrophils with (○, •) or without (□, ■) *S. aureus*. Data are presented as means ± SDs ($n = 4$). Asterisks (*) denote significant differences compared to the culture condition lacking neutrophil with *S. aureus* in the lower PCL-NM ($p < 0.05$). (**C**) Neutrophils ($1 \times 10^5$) were cultured in media alone (None), with *S. aureus* ($1 \times 10^7$ CFU) and with LPS (1 μg/mL) for 12 h on a culture dish and in the PCL-NM. Data are presented as means ± SDs ($n = 3$). Asterisks (*) denote significant differences compared to untreated control ($p < 0.05$). (**D**) Neutrophils ($1 \times 10^5$) were cultured with (+) or without (−) *S. aureus* ($1 \times 10^7$ CFU) for the indicated times in the PCL-NM. IL-8 concentrations in culture media were measured by ELISA. Data are presented as means ± SDs ($n = 4$). Asterisks (*) denote significant differences compared to culture without *S. aureus* ($p < 0.05$).

### *3.7. Engulfment of Neutrophils by DCs in the Presence of S. aureus in PCL-NM-Based Co-Culture Condition*

Co-culture of DCs and neutrophils showed internalization of live neutrophils by DCs [28]. This study confirmed that DCs engulfed live neutrophils in the presence or absence of *S. aureus* when neutrophils and DCs were co-cultured at a ratio of 10:1 on a culture dish (Figure 6A). However, it is unknown whether the engulfment of viable neutrophils by DCs occurs in bacteria-infected tissues in vivo. Moreover, we cannot rule out the possibility of experimental artifact during co-culture of different types of phagocytes in 2D culture. Thus, we investigated whether DCs in the upper PCL-NM migrate to and are able to uptake bacteria-engulfing neutrophils in the lower PCL-NM using a two-layer culture system. *S. aureus* was labeled with DAPI and cultured with PKH26-labeled neutrophils for 2 h in the lower PCL-NM. Neutrophils and bacteria in the lower PCL-NM were further cultured for 4 h with PKH67-labeled DCs in the upper PCL-NM. As shown in Figure 6B, DCs migrated to the lower PCL-NM containing both neutrophils and bacteria, but did not migrate to the lower PCL-NM harboring neutrophils alone. Focus-stacking images of the lower PCL-NM showed that PKH67-labeled DCs extended dendrites to PKH26-labeled neutrophils. In z-stack analysis of confocal microscopic images, the red fluorescence of CellTracker red-labeled neutrophils co-localized with green fluorescence of CellTracker green-labeled DCs, indicating internalization of the neutrophils by the DCs (Figure 6C). This result indicated that DCs migrate to neutrophils phagocytosing bacteria and then engulf neutrophils in 3D in vitro culture.

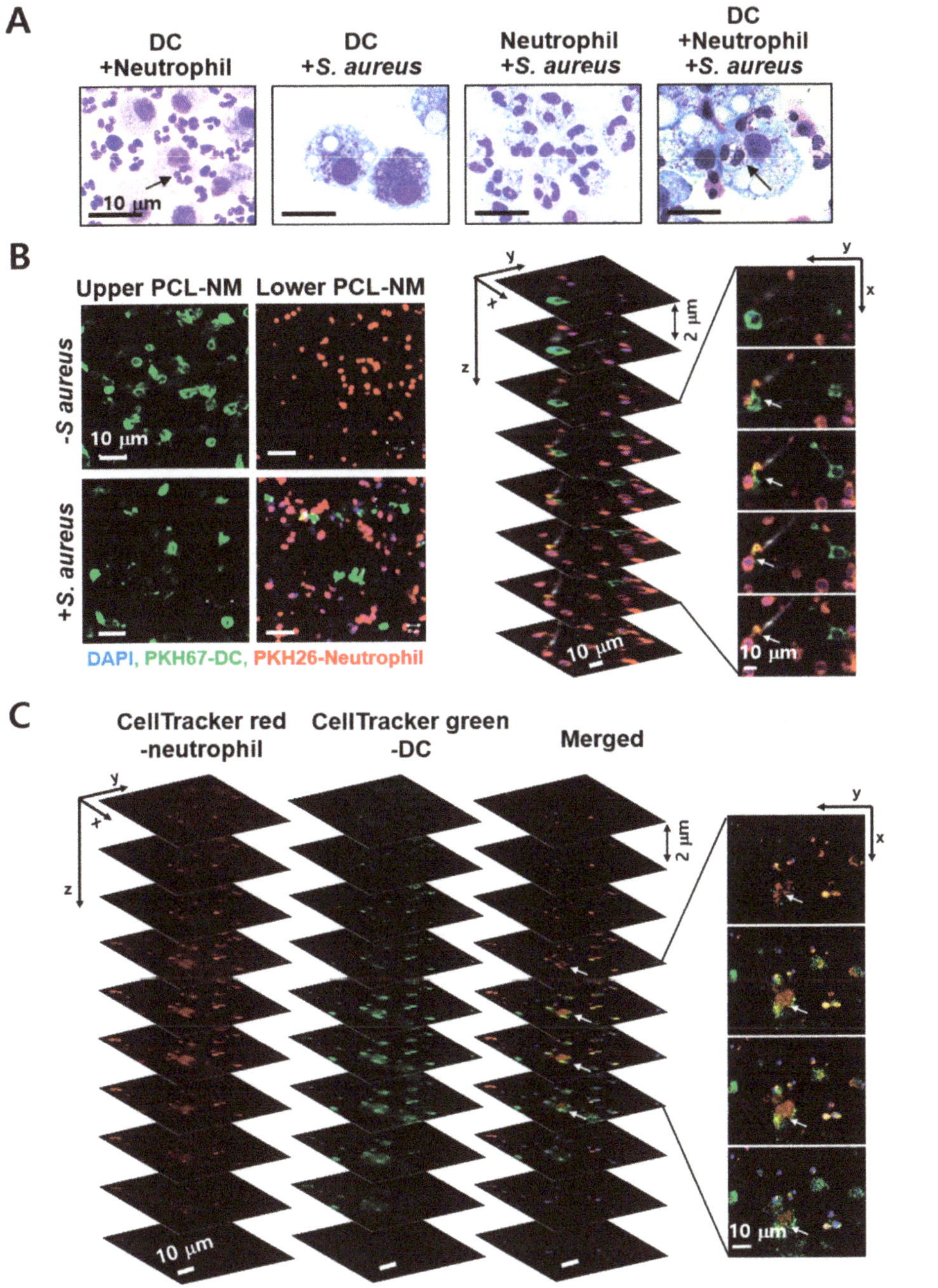


**Figure 6.** Engulfment of neutrophils by DCs in 2D and 3D culture conditions. (**A**) DCs ($2 \times 10^5$) and/or neutrophils ($1 \times 10^6$) were cultured with or without *S. aureus* ($5 \times 10^6$ CFU) for 6 h in culture dishes and stained with Giemsa solution ($n = 3$). Arrows indicate live neutrophils inside of DCs. (**B**) PKH26-labeled neutrophils ($2 \times 10^6$) were cultured for 2 h with

or without DAPI-stained *S. aureus* ($2 \times 10^7$ CFU) in the lower PCL-NM. PKH67-labeled DCs ($2 \times 10^5$) in the upper PCL-NM were cultured for 4 h with bacteria and neutrophils in the lower PCL-NM ($n = 3$). Green-fluorescent spots were detected in the lower PCL-NM (in the left panel). The right panel depicts focus-stacking images of the lower PCL-NM. (**C**) CellTracker green-labeled DCs in the upper PCL-NM were cultured with CellTracker red-labeled neutrophils and DAPI-stained bacteria as described in figure B ($n = 3$). Arrows indicate red-fluorescent spots inside of green-fluorescent areas in the lower PCL-NM. Images show stacks of 2 µm in thickness from the top to bottom.

### *3.8. Migration of Phagocytes to S. aureus–Infected Epithelial Cells in PCL-NM-Based Two-Layer Culture System*

Fibroblasts, epithelial cells and endothelial cells are potentially important targets for *S. aureus* infection [29]. PKH67-labeled MLE-12 cells and DAPI-stained *S. aureus* were co-cultured for 12 h in the lower PCL-NM layer and then, PKH26-labeled phagocytes were top-seeded on the upper PCL-NM layer. The migration of phagocytes to infected epithelial cells was measured. As shown in Figure 7A, the neutrophils migrated toward the lung epithelial cells when these were infected with *S. aureus*. Immunofluorescence analysis of the upper PCL-NM showed that the numbers of red-fluorescent neutrophils decreased but those of green-fluorescent epithelial cells from the lower PCL-NM increased in a time-dependent manner (Figure 7B). In contrast, *S. aureus*-infected epithelial cells were not detected in the upper PCL-NM when it had not been seeded with neutrophils (data not shown). These results indicated that neutrophils in the upper PCL-NM migrated to the lower PCL-NM and damaged epithelial cells in the lower PCL-NM floated up to the upper PCL-NM. On the basis of these observations, we tested the production of chemokines by lung epithelial cells treated with *S. aureus* in 2D and 3D culture conditions. IL-8 secretion by MLE-12 cells was increased in cells cultured with *S. aureus* in both 2D and 3D culture conditions (Figure 7C). MLE-12 cells treated with *S. aureus* in the PCL-NM, but not on culture dishes, stimulated the secretion of tumor necrosis factor (TNF)-$\alpha$ and IL-1$\alpha$ (Figure 7D). This result indicated that *S. aureus* infection of epithelial cells in 3D culture condition induces proinflammatory response.

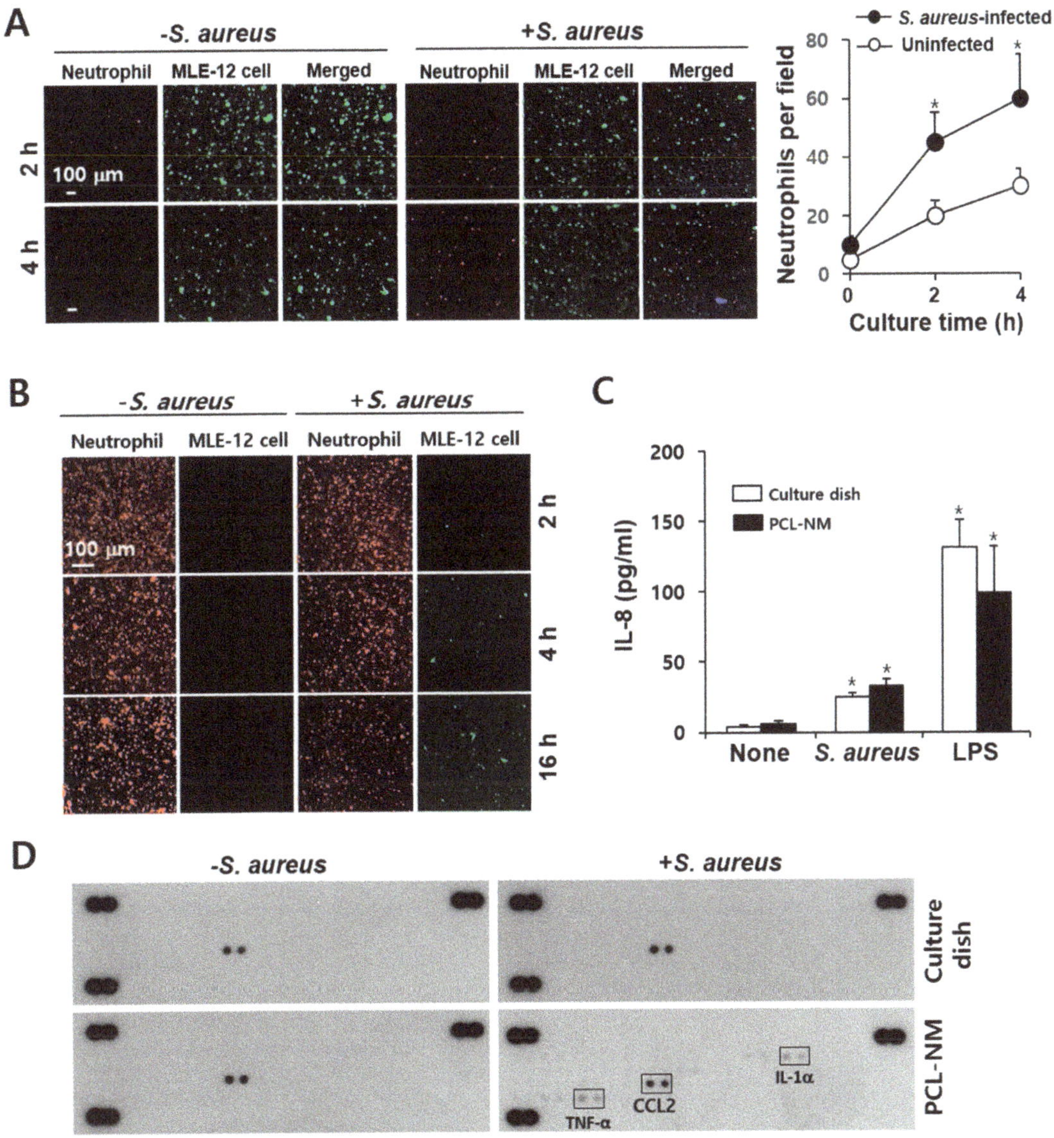


**Figure 7.** Migration of neutrophils to *S. aureus*-infected epithelial cells in the PCL-NM-based two-layer culture system. (**A**) PKH67-labeled MLE-12 cells ($1 \times 10^5$) were incubated with or without DAPI-stained *S. aureus* ($1 \times 10^7$ CFU) for 12 h in the lower PCL-NM and washed with culture media. PKH26-labeled peritoneal neutrophils ($1 \times 10^5$) were seeded on the upper PCL-NM and cultured for the indicated times with untreated and bacteria-infected MLE-12 in the lower PCL-NM. Red-fluorescent spots were detected in the lower PCL-NM containing PKH67-labeled uninfected and *S. aureus*-infected MLE-12 cells (Epithelial cell) ($n = 3$). The graphs depict the numbers of red-fluorescent spots detected in the lower PCL-NM. Data are presented as means ± SDs ($n = 3$). Asterisks (*) denote significant differences compared to uninfected MLE-12 cells ($p < 0.05$). (**B**) Green-fluorescent spots of MLE-12 cells (Epithelial cell) were detected on the upper PCL-NM containing

PKH26-labeled neutrophils (Neutrophil) ($n = 3$). (**C**) MLE-12 cells ($1 \times 10^5$) were cultured with media alone (None), *S. aureus* ($1 \times 10^7$ CFU) and LPS (1 μg/mL) for 12 h on a culture dish and in the PCL-NM. The secretion of IL-8 was measured by ELISA. Data are presented as means ± SDs ($n = 3$). Asterisks (*) denote significant differences compared to None ($p < 0.05$). (**D**) MLE-12 cells ($1 \times 10^5$) were cultured with (+) or without (−) *S. aureus* ($1 \times 10^7$ CFU) for 12 h on a culture dish and in the PCL-NM. Cytokines and chemokines secreted in the culture media were detected using cytokine and chemokine arrays ($n = 3$).

## 4. Discussion

The ability of phagocytes to uptake bacteria in vitro may depend on culture conditions and methodology. To date, no method for evaluating phagocytosis of bacteria in 3D co-culture system, which may mimic bacterial infection in vivo, has been reported. Phagocytosis of *S. aureus* by human neutrophils in suspension has been shown to depend on opsonization, while adherent neutrophils do readily internalize bacteria, independent of opsonization [24]. In addition, persistence of free extracellular bacteria in suspension culture has been shown to the poor killing [24]. Adherence to ECM or plasma proteins of host cells is mediated by bacterial surface proteins. In addition, fibronectin-binding proteins of *S. aureus* bind to fibronectin on the surface of epithelial cells [29]. Bacterial adhesion to nanofibers occurs by a number of mechanisms. The adhesion of bacteria to nanofibers is influenced by the surface chemistry, fiber morphology, fiber diameter and physical characteristics of the fiber surface [30]. In addition, specific interactions between bacteria and nanofibers involve macromolecules on the bacterial surface, such as fimbriae and pili [31]. Nanofibers with an average diameter close to bacterial sizes were found to offer support for bacterial adhesion [31], although it is unknown which protein mediates the binding. In this study, *S. aureus* and phagocytes adhered onto the surface of electrospun PCL nanofibers and infiltrated the PCL-NM, allowing 3D evaluation of phagocytosis in PCL-NM-based in vitro culture system. The cell number and ratio of phagocytes to bacteria were the same in 2D and 3D culture conditions, but phagocytosis was lower in 3D than in 2D conditions. In addition, the effect of FBS on the phagocytosis of *S. aureus* by neutrophils in the PCL-NM was more evident than that in culture dishes. Lu et al. demonstrated that more *S. aureus* are phagocytosed by adherent neutrophils compared with those in suspension and that serum opsonization of *S. aureus* enhances phagocytosis by neutrophils in suspension, but has no apparent impact on the phagocytosis of bacteria by adherent neutrophils [24]. They suggested that adherent neutrophils are primed for enhanced phagocytosis [24]. Thus, phagocytes adhered to nanofibers may be less primed for phagocytosis than adherent cells on the culture dish. It is also possible that both phagocytes and bacteria adhere to nanofibers and distribute in 3D space, rather than on the 2D surface. Thus, phagocytes need to sense chemotactic signals from bacteria in 3D culture.

Inflammation is characterized by an influx of phagocytes to the site of infection. While neutrophils are very rapidly recruited into infection site, macrophages constitute the majority of immune cells residing in the tissue at steady state. In a mouse infection model, DCs were actively recruited to infected tissue and phagocytosed *S. aureus*, but they did not actively kill the bacteria [32]. Phagocytes reach the site of infection through directed migration along an increasing gradient of chemoattractants, which are derived from bacteria as well as secreted by activated cells including leukocytes and epithelial cells. In vitro studies have been crucial in understanding cell migration, but the direct exposure of phagocytes to bacteria is in most cases poorly justified. Methods such as Transwell and microfluidic-based assays have been used to evaluate the migration of immune cells in response to chemokines and the migration of phagocytes in vitro is usually measured by chemotaxis through a filter in Transwell chambers. Neutrophils transmigrate through the pulmonary alveolar-endothelial barrier by passing through pores in the PDMS membrane and engulf bacteria in a lung-on-a-chip microfluidic system [9]. When compared to the Transwell assay and lung-on-a-chip microfluidic system, the 3D PCL-NM system has some advantages. First, it is easy to observe the movement of phagocytes over time in the PCL-NM. Second, 3D migratory behavior of neutrophils, macrophages and DCs can be evaluated. We found that neutrophils and DCs on nanofibers become elongated and motile in the vicinity of bacteria and travel along the nanofiber to reach the nearest bacteria. Finally, it is easy to quantitatively evaluate the rates of migration of phagocytes and the phagocytosis of bacteria.

Cell components released from *S. aureus* may act as a phagocyte chemoattractant and bacteria-activated phagocytes secrete cytokines and chemokines, thereby sustaining and prolonging inflammation [33,34]. It has been demonstrated that FPR is required for neutrophil migration to and killing of bacteria, but not for bacterial phagocytosis [35]. This study showed that FPR inhibitors partially inhibited phagocytosis of *S. aureus* by neutrophils and macrophages in 3D, but not in 2D culture, suggesting that FPR agonist released from bacteria may be involved in the detection of bacteria by phagocytes only in 3D culture. Host-derived chemotactic factors, such as IL-8 have been shown to recruit additional neutrophils to areas of infection [36]. This chemokine may be crucial during the early phase of bacterial infection. Thus, the first wave of neutrophil migration is dependent on bacterial chemokines and the second wave of migration of neutrophils, macrophages and DCs is dependent on chemokines secreted from already migrated neutrophils in the infected regions. In this study, the migration of phagocytes through the microstructure of the PCL-NM should be correlated to chemotactic gradients. In this aspect, there was increased secretion of chemokines in the PCL-NM, indicating that phagocytes were further recruited to chemokine-secreting cells via a chemoattractant gradient in the PCL-NM-based two-layer system. Therefore, the compartmentalization and directional orientation of the two-layer system can be applied for the analysis of migration of phagocytes in response to a chemotactic gradient.

Neutrophils and macrophages migrated to *S. aureus*-infected lung epithelial cells in the PCL-NM-based two-layer culture system. By using this system, we can compare the capacity of various bacteria in addition to *S. aureus* to cause phagocyte influx into an infected area. The secretion of TNF-$\alpha$ and IL-1$\alpha$ was increased in lung epithelial cells cultured with *S. aureus* in 3D, but not in 2D culture. TNF-$\alpha$ functions as an initiator of the inflammatory response. Thus, an in vivo inflammation-mimicking response occurred in 3D culture condition. In addition to neutrophil migration to the lower PCL-NM containing epithelial cells and bacteria, epithelial cells were also detected in the upper PCL-NM. The detachment of *S. aureus*-infected epithelial cells was observed only in the presence of migrating neutrophils from the upper to the lower PCL-NM. It is possible that activated neutrophils produce reactive oxygen species and release cytoplasmic granules, which results in significant host cell damage or death as well as the killing of bacteria. This finding suggests that epithelial integrity is lost in 3D culture condition, as observed in infected lung tissue.

In conclusion, the PCL-NM-based culture system provides a 3D space for cell attachment and migration. In the PCL-NM, cell-derived products can diffuse easily and tight cell-to-cell contact can be avoided. Technically, the PCL-NM can be easily assembled into the two-layer system, which mimics multilayers of cells in tissue. Moreover, PCL nanofibers are inert and biocompatible in immune cell culture because the adhesion of DCs to PCL nanofibers does not affect their activation status [12,13]. Finally, 3D migration to surrounding bacteria and phagocytosis processes in the PCL-NM can be visualized though a live cell-imaging setup. Thus, our PCL-NM-based 3D culture system is widely applicable to the biomimetic study of various microbe infections.

**Supplementary Materials:** The following are available online at https://www.mdpi.com/article/10.3390/membranes11080569/s1, Figure S1: Workflow for setup of a migration assay with two layers of PCL-NMs. Video S1: Time-lapse images of neutrophils cultured with *S. aureus* on a culture dish depicting phagocytosis of PKH67-labeled *S. aureus* (green) by CellTracker red-labeled neutrophils (red). Video S2: Time-lapse images of DCs cultured with *S. aureus* on a culture dish depicting phagocytosis of PKH67-labeled *S. aureus* (green) by CellTracker red-labeled DCs (red). Video S3: Time-lapse images of neutrophils cultured with *S. aureus* in PCL-NM depicting phagocytosis of PKH67-labeled *S. aureus* (green) by CellTracker red-labeled neutrophils (red). Video S4: Time-lapse images of DCs cultured with *S. aureus* in PCL-NM depicting phagocytosis of PKH67-labeled *S. aureus* (green) by CellTracker red-labeled DCs (red).

**Author Contributions:** S.-J.L. and P.A.M.A.M. performed experiments; G.-M.S. produced nanofibers; J.-Y.K. supervised research and obtained funding; J.-Y.K., I.-J.L. and P.S. designed and analyzed experiments, interpreted the data. J.-Y.K. and I.-J.L. wrote the paper. All authors have read and agreed to the published version of the manuscript.

**Funding:** This work was supported by a grant of the Korea Health Technology R&D Project through the Korea Health Industry Development Institute (KHIDI), funded by the Ministry of Health & Welfare, Republic of Korea (HI16C0992 and HR16C0001) and by Basic Science Research Capacity Enhancement Project through Korea Basic Science Institute (National Research Facilities and Equipment Center) grant funded by the Ministry of Education (2019R1A6C1010003).

**Institutional Review Board Statement:** All experiments were approved by the Institutional Animal Care and Use Committee of Ajou University (IACUC approval number 2015-0028, approved on 21 August 2018).

**Informed Consent Statement:** Not Applicable.

**Data Availability Statement:** Not Applicable.

**Conflicts of Interest:** The authors declare no other relevant affiliations or financial involvement with any organization or entity with a financial interest or any financial conflict on the subject matter discussed in the manuscript. The authors report no known conflicts of interest associated with this publication.

## References

1. Schiffmann, E.; Corcoran, B.A.; Wahl, S.M. N-formylmethionyl peptides as chemoattractants for leucocytes. *Proc. Natl. Acad. Sci. USA* **1975**, *72*, 1059–1062. [CrossRef] [PubMed]
2. Gouwy, M.; Struyf, S.; Proost, P.; Van Damme, J. Synergy in cytokine and chemokine networks amplifies the inflammatory response. *Cytokine Growth Factor Rev.* **2005**, *16*, 561–580. [CrossRef]
3. Kim, N.D.; Luster, A.D. The role of tissue resident cells in neutrophil recruitment. *Trends Immunol.* **2015**, *36*, 547–555. [CrossRef] [PubMed]
4. Boyden, S. The chemotactic effect of mixtures of antibody and antigen on polymorphonuclear leucocytes. *J. Exp. Med.* **1962**, *115*, 453–466. [CrossRef] [PubMed]
5. Sackmann, E.K.; Berthier, E.; Young, E.W.; Shelef, M.A.; Wernimont, S.A.; Huttenlocher, A.; Beebe, D.J. Microfluidic kit-on-a-lid: A versatile platform for neutrophil chemotaxis assays. *Blood* **2012**, *120*, e45–e53. [CrossRef]
6. Faure-André, G.; Vargas, P.; Yuseff, M.I.; Heuzé, M.; Diaz, J.; Lankar, D.; Steri, V.; Manry, J.; Hugues, S.; Vascotto, F.; et al. Regulation of dendritic cell migration by CD74, the MHC class II-associated invariant chain. *Science* **2008**, *322*, 1705–1710. [CrossRef] [PubMed]
7. Heuzé, M.L.; Collin, O.; Terriac, E.; Lennon-Duménil, A.M.; Piel, M. Cell migration in confinement: A micro-channel-based assay. *Methods Mol. Biol.* **2011**, *769*, 415–434.
8. Stachowiak, A.N.; Irvine, D.J. Inverse opal hydrogel-collagen composite scaffolds as a supportive microenvironment for immune cell migration. *J. Biomed. Mater. Res. A* **2008**, *85*, 815–828. [CrossRef] [PubMed]
9. Huh, D.; Matthews, B.D.; Mammoto, A.; Montoya-Zavala, M.; Hsin, H.Y.; Ingber, D.E. Reconstituting organ-level lung functions on a chip. *Science* **2010**, *328*, 1662–1668. [CrossRef] [PubMed]
10. Pham, Q.P.; Sharma, U.; Mikos, A.G. Electrospinning of polymeric nanofibers for tissue engineering applications: A review. *Tissue Eng.* **2006**, *12*, 1197–1211. [CrossRef]
11. Kim, T.E.; Kim, C.G.; Kim, J.S.; Jin, S.; Yoon, S.; Bae, H.R.; Kim, J.H.; Jeong, Y.H.; Kwak, J.Y. Three-dimensional culture and interaction of cancer cells and dendritic cells in an electrospun nano-submicron hybrid fibrous scaffold. *Int. J. Nanomed.* **2016**, *11*, 823–835.
12. Jin, S.; Park, T.M.; Kim, C.H.; Kim, J.S.; Le, B.D.; Jeong, Y.H.; Kwak, J.Y.; Yoon, S. Three-dimensional migration of neutrophils through an electrospun nanofibrous membrane. *Biotechniques* **2015**, *58*, 285–292. [CrossRef]
13. Woodruff, M.A.; Hutmacher, D.W. The Return of a Forgotten Polymer-Polycaprolactone in the 21st Century. *Prog. Polym. Sci.* **2010**, *35*, 1217–1256. [CrossRef]
14. Mountcastle, S.E.; Cox, S.C.; Sammons, R.L.; Jabbari, S.; Shelton, R.M.; Kuehne, S.A. A review of co-culture models to study the oral microenvironment and disease. *J. Oral Microbiol.* **2020**, *12*, 1773122. [CrossRef] [PubMed]
15. Popov, L.; Kovalski, J.; Grandi, G.; Bagnoli, F.; Amieva, M.R. Three-dimensional human skin models to understand *Staphylococcus aureus* skin colonization and infection. *Front. Immunol.* **2014**, *5*, 41. [CrossRef]

16. Parente, R.; Possetti, V.; Schiavone, M.L.; Campodoni, E.; Menale, C.; Loppini, M.; Doni, A.; Bottazzi, B.; Mantovani, A.; Sandri, M. 3D cocultures of osteoblasts and *Staphylococcus aureus* on biomimetic bone scaffolds as a tool to investigate the host–pathogen interface in osteomyelitis. *Pathogens* **2021**, *10*, 837. [CrossRef]
17. Ding, P.; Wu, H.; Fang, L.; Wu, M.; Liu, R. Transmigration and phagocytosis of macrophages in an airway infection model using four-dimensional techniques. *Am. J. Respir. Cell Mol. Biol.* **2014**, *51*, 1–10. [CrossRef]
18. Oh, Y.S.; Choi, M.H.; Shin, J.I.; Maza, P.A.M.A.; Kwak, J.Y. Co-culturing of endothelial and cancer cells in a nanofibrous scaffold-based two-layer system. *Int. J. Mol. Sci.* **2020**, *11*, 4128. [CrossRef]
19. Swamydas, M.; Luo, Y.; Dorf, M.E.; Lionakis, M.S. Isolation of mouse neutrophils. *Curr. Protoc. Immunol.* **2015**, *110*, 3.20.1–3.20.15. [CrossRef]
20. Maza, P.A.M.A.; Lee, J.H.; Kim, Y.S.; Sun, G.M.; Sung, Y.J.; Ponomarenko, L.P.; Stonik, V.A.; Ryu, M.; Kwak, J.Y. Inotodiol from *Inonotus obliquus* chaga mushroom induces atypical maturation in dendritic cells. *Front. Immunol.* **2021**, *12*, 650841. [CrossRef]
21. Pineda-Torra, I.; Gage, M.; de Juan, A.; Pello, O.M. Isolation, culture, and polarization of murine bone marrow-derived and peritoneal macrophages. *Methods Mol. Biol.* **2015**, *1339*, 101–109. [PubMed]
22. Park, K.H.; Kurokawa, K.; Zheng, L.; Jung, D.J.; Tateishi, K.; Jin, J.O.; Ha, N.C.; Kang, H.J.; Matsushita, M.; Kwak, J.Y.; et al. Human serum mannose-binding lectin senses wall teichoic acid glycopolymer of *Staphylococcus aureus*, which is restricted in infancy. *J. Biol. Chem.* **2010**, *285*, 27167–27175. [CrossRef] [PubMed]
23. Raybourne, R.B.; Bunning, V.K. Bacterium-host cell interactions at the cellular level: Fluorescent labeling of bacteria and analysis of short-term bacterium-phagocyte interaction by flow cytometry. *Infect. Immun.* **1994**, *62*, 665–672. [CrossRef] [PubMed]
24. Lu, T.; Porter, A.R.; Kennedy, A.D.; Kobayashi, S.D.; DeLeo, F.R. Phagocytosis and killing of *Staphylococcus aureus* by human neutrophils. *J. Innate Immun.* **2014**, *6*, 639–649. [CrossRef]
25. Ye, R.D.; Boulay, F.; Wang, J.M.; Dahlgren, C.; Gerard, C.; Parmentier, M.; Serhan, C.N.; Murphy, P.M. International union of basic and clinical pharmacology. LXXIII. Nomenclature for the formyl peptide receptor (FPR) family. *Pharmacol. Rev.* **2009**, *61*, 119–161.
26. Skovbakke, S.L.; Winther, M.; Gabl, M.; Holdfeldt, A.; Linden, S.; Wang, J.M.; Dahlgren, C.; Franzyk, H.; Forsman, H. The peptidomimetic Lau-(Lys-βNSpe)$_6$-NH$_2$ antagonizes formyl peptide receptor 2 expressed in mouse neutrophils. *Biochem. Pharmacol.* **2016**, *119*, 56–65. [CrossRef]
27. Warren, S.E.; Armstrong, A.; Hamilton, M.K.; Mao, D.P.; Leaf, I.A.; Miao, E.A.; Aderem, A. Cutting Edge: Cytosolic bacterial DNA activates the inflammasome via Aim2. *J. Immunol.* **2010**, *185*, 818–821. [CrossRef]
28. Alfaro, C.; Suarez, N.; Oñate, C.; Perez-Gracia, J.L.; Martinez-Forero, I.; Hervas-Stubbs, S.; Rodriguez, I.; Perez, G.; Bolaños, E.; Palazon, A.; et al. Dendritic cells take up and present antigens from viable and apoptotic polymorphonuclear leukocytes. *PLoS ONE* **2011**, *6*, e29300. [CrossRef]
29. Sinha, B.; François, P.P.; Nüsse, O.; Foti, M.; Hartford, O.M.; Vaudaux, P.; Foster, T.J.; Lew, D.P.; Herrmann, M.; Krause, K.H. Fibronectin-binding protein acts as *Staphylococcus aureus* invasin via fibronectin bridging to integrin α5β1. *Cell. Microbiol.* **1999**, *1*, 101–117. [CrossRef] [PubMed]
30. Mortimer, C.J.; Burke, L.; Wright, C.J. Microbial interactions with nanostructures and their importance for the development of electrospun nanofibrous materials used in regenerative medicine and filtration. *J. Microb. Biochem. Technol.* **2016**, *8*, 195–201. [CrossRef]
31. Abrigo, M.; Kingshott, P.; McArthur, S.L. Electrospun polystyrene fiber diameter influencing bacterial attachment, proliferation, and growth. *ACS Appl. Mater. Interfaces* **2015**, *7*, 7644–7652. [CrossRef] [PubMed]
32. Schindler, D.; Gutierrez, M.G.; Beineke, A.; Rauter, Y.; Rohde, M.; Foster, S.; Goldmann, O.; Medina, E. Dendritic cells are central coordinators of the host immune response to *Staphylococcus aureus* bloodstream infection. *Am. J. Pathol.* **2012**, *181*, 1327–1337. [CrossRef]
33. Shimada, T.; Park, B.G.; Wolf, A.J.; Brikos, C.; Goodridge, H.S.; Becker, C.A.; Reyes, C.N.; Miao, E.A.; Aderem, A.; Götz, F.; et al. *Staphylococcus aureus* evades lysozyme-based peptidoglycan digestion that links phagocytosis, inflammasome activation, and IL-1β secretion. *Cell. Host Microbe* **2010**, *7*, 38–49.
34. Spaan, A.N.; Surewaard, B.G.; Nijland, R.; van Strijp, J.A. Neutrophils versus *Staphylococcus aureus*: A biological tug of war. *Annu. Rev. Microbiol.* **2013**, *67*, 629–650. [CrossRef]
35. Liu, M.; Chen, K.; Yoshimura, T.; Liu, Y.; Gong, W.; Wang, A.; Gao, J.; Murphy, P.M.; Wang, J.M. Formylpeptide receptors are critical for rapid neutrophil mobilization in host defense against *Listeria monocytogenes*. *Sci. Rep.* **2012**, *2*, 786. [CrossRef]
36. Sadik, C.D.; Kim, N.D.; Luster, A.D. Neutrophils cascading their way to inflammation. *Trends Immunol.* **2011**, *32*, 452–460. [CrossRef]

MDPI
St. Alban-Anlage 66
4052 Basel
Switzerland
Tel. +41 61 683 77 34
Fax +41 61 302 89 18
www.mdpi.com

*Membranes* Editorial Office
E-mail: membranes@mdpi.com
www.mdpi.com/journal/membranes

www.ingramcontent.com/pod-product-compliance
Lightning Source LLC
Chambersburg PA
CBHW061504210926
51850CB00008B/1813
*9783036577876*